Biostatistical Methods
in Epidemiology

Biostatistical Methods in Epidemiology

STEPHEN C. NEWMAN

A Wiley-Interscience Publication

JOHN WILEY & SONS, INC.

New York • Chichester • Weinheim • Brisbane • Singapore • Toronto

This book is printed on acid-free paper. ∞

Copyright © 2001 by John Wiley & Sons, Inc. All rights reserved.

Published simultaneously in Canada.

For ordering and customer service, call 1-800-CALL-WILEY.

Library of Congress Cataloging-in-Publication Data:

Newman, Stephen C., 1952–
    Biostatistical methods in epidemiology / Stephen C. Newman.
      p. cm.—(Wiley series in probability and statistics. Biostatistics section)
    Includes bibliographical references and index.
    ISBN 0-471-36914-4 (cloth : alk. paper)
    1. Epidemiology—Statistical methods. 2. Cohort analysis. I. Title. II. Series.
RA652.2.M3 N49 2001
    614.4'07'27—dc21   2001028222

Printed in the United States of America

10 9 8 7 6 5 4 3 2 1

To Sandra

Contents

Preface

The aim of this book is to provide an overview of statistical methods that are important in the analysis of epidemiologic data, the emphasis being on nonregression techniques. The book is intended as a classroom text for students enrolled in an epidemiology or biostatistics program, and as a reference for established researchers. The choice and organization of material is based on my experience teaching biostatistics to epidemiology graduate students at the University of Alberta. In that setting I emphasize the importance of exploring data using nonregression methods prior to undertaking a more elaborate regression analysis. It is my conviction that most of what there is to learn from epidemiologic data can usually be uncovered using nonregression techniques.

I assume that readers have a background in introductory statistics, at least to the stage of simple linear regression. Except for the Appendices, the level of mathematics used in the book is restricted to basic algebra, although admittedly some of the formulas are rather complicated expressions. The concept of confounding, which is central to epidemiology, is discussed at length early in the book. To the extent permitted by the scope of the book, derivations of formulas are provided and relationships among statistical methods are identified. In particular, the correspondence between odds ratio methods based on the binomial model, and hazard ratio methods based on the Poisson model are emphasized (Breslow and Day, 1980, 1987). Historically, odds ratio methods were developed primarily for the analysis of case-control data. Students often find the case-control design unintuitive, and this can adversely affect their understanding of the odds ratio methods. Here, I adopt the somewhat unconventional approach of introducing odds ratio methods in the setting of closed cohort studies. Later in the book, it is shown how these same techniques can be adapted to the case-control design, as well as to the analysis of censored survival data. One of the attractive features of statistics is that different theoretical approaches often lead to nearly identical numerical results. I have attempted to demonstrate this phenomenon empirically by analyzing the same data sets using a variety of statistical techniques.

I wish to express my indebtedness to Allan Donner, Sander Greenland, John Hsieh, David Streiner, and Stephen Walter, who generously provided comments on a draft manuscript. I am especially grateful to Sander Greenland for his advice on the topic of confounding, and to John Hsieh who introduced me to life table theory when I was

a student. The reviewers did not have the opportunity to read the final manuscript and so I alone am responsible for whatever shortcomings there may be in the book. I also wish to acknowledge the professionalism and commitment demonstrated by Steve Quigley and Lisa Van Horn of John Wiley & Sons. I am most interested in receiving your comments, which can be sent by e-mail using a link at the website www.stephennewman.com.

Prior to entering medicine and then epidemiology, I was deeply interested in a particularly elegant branch of theoretical mathematics called Galois theory. While studying the historical roots of the topic, I encountered a monograph having a preface that begins with the sentence "I wrote this book for myself." (Hadlock, 1978). After this remarkable admission, the author goes on to explain that he wanted to construct his own path through Galois theory, approaching the subject as an enquirer rather than an expert. Not being formally trained as a mathematical statistician, I embarked upon the writing of this book with a similar sense of discovery. The learning process was sometimes arduous, but it was always deeply rewarding. Even though I wrote this book partly "for myself," it is my hope that others will find it useful.

STEPHEN C. NEWMAN

Edmonton, Alberta, Canada
May 2001

Biostatistical Methods
in Epidemiology

CHAPTER 1

Introduction

In this chapter some background material from the theory of probability and statistics is presented that will be useful throughout the book. Such fundamental concepts as probability function, random variable, mean, and variance are defined, and several of the distributions that are important in the analysis of epidemiologic data are described. The Central Limit Theorem and normal approximations are discussed, and the maximum likelihood and weighted least squares methods of parameter estimation are outlined. The chapter concludes with a discussion of different types of random sampling. The presentation of material in this chapter is informal, the aim being to give an overview of some key ideas rather than provide a rigorous mathematical treatment. Readers interested in more complete expositions of the theoretical aspects of probability and statistics are referred to Cox and Hinkley (1974), Silvey (1975), Casella and Berger (1990), and Hogg and Craig (1994). References for the theory of probability and statistics in a health-related context are Armitage and Berry (1994), Rosner (1995), and Lachin (2000). For the theory of sampling, the reader is referred to Kish (1965) and Cochran (1977).

1.1 PROBABILITY

1.1.1 Probability Functions and Random Variables

Probability theory is concerned with mathematical models that describe phenomena having an element of uncertainty. Problems amenable to the methods of probability theory range from the elementary, such as the chance of randomly selecting an ace from a well-shuffled deck of cards, to the exceedingly complex, such as predicting the weather. Epidemiologic studies typically involve the collection, analysis, and interpretation of health-related data where uncertainty plays a role. For example, consider a survey in which blood sugar is measured in a random sample of the population. The aims of the survey might be to estimate the average blood sugar in the population and to estimate the proportion of the population with diabetes (elevated blood sugar). Uncertainty arises because there is no guarantee that the resulting esti-

1

mates will equal the true population values (unless the entire population is enrolled in the survey).

Associated with each probability model is a random variable, which we denote by a capital letter such as X. We can think of X as representing a potential data point for a proposed study. Once the study has been conducted, we have actual data points that will be referred to as realizations (outcomes) of X. An arbitrary realization of X will be denoted by a small letter such as x. In what follows we assume that realizations are in the form of numbers so that, in the above survey, diabetes status would have to be coded numerically—for example, 1 for present and 0 for absent. The set of all possible realizations of X will be referred to as the sample space of X. For blood sugar the sample space is the set of all nonnegative numbers, and for diabetes status (with the above coding scheme) the sample space is $\{0, 1\}$. In this book we assume that all sample spaces are either continuous, as in the case of blood sugar, or discrete, as in the case of diabetes status. We say that X is continuous or discrete in accordance with the sample space of the probability model.

There are several mathematically equivalent ways of characterizing a probability model. In the discrete case, interest is mainly in the probability mass function, denoted by $P(X = x)$, whereas in the continuous case the focus is usually on the probability density function, denoted by $f(x)$. There are important differences between the probability mass function and the probability density function, but for present purposes it is sufficient to view them simply as formulas that can be used to calculate probabilities. In order to simplify the exposition we use the term probability function to refer to both these constructs, allowing the context to make the distinction clear. Examples of probability functions are given in Section 1.1.2. The notation $P(X = x)$ has the potential to be confusing because both X and x are "variables." We read $P(X = x)$ as the probability that the discrete random variable X has the realization x. For simplicity it is often convenient to ignore the distinction between X and x. In particular, we will frequently use x in formulas where, strictly speaking, X should be used instead.

The correspondence between a random variable and its associated probability function is an important concept in probability theory, but it needs to be emphasized that it is the probability function which is the more fundamental notion. In a sense, the random variable represents little more than a convenient notation for referring to the probability function. However, random variable notation is extremely powerful, making it possible to express in a succinct manner probability statements that would be cumbersome otherwise. A further advantage is that it may be possible to specify a random variable of interest even when the corresponding probability function is too difficult to describe explicitly. In what follows we will use several expressions synonymously when describing random variables. For example, when referring to the random variable associated with a binomial probability function we will variously say that the random variable "has a binomial distribution," "is binomially distributed," or simply "is binomial."

We now outline a few of the key definitions and results from introductory probability theory. For simplicity we focus on discrete random variables, keeping in mind that equivalent statements can be made for the continuous case. One of the defining properties of a probability function is the identity

$$\sum_x P(X = x) = 1 \qquad (1.1)$$

where here, and in what follows, the summation is over all elements in the sample space of X. Next we define two fundamental quantities that will be referred to repeatedly throughout the book. The mean of X, sometimes called the expected value of X, is defined to be

$$E(X) = \sum_x x \, P(X = x) \qquad (1.2)$$

and the variance of X is defined to be

$$\text{var}(X) = \sum_x [x - E(X)]^2 P(X = x). \qquad (1.3)$$

It is important to note that when the mean and variance exist, they are constants, not random variables. In most applications the mean and variance are unknown and must be estimated from study data. In what follows, whenever we refer to the mean or variance of a random variable it is being assumed that these quantities exist—that is, are finite constants.

Example 1.1   Consider the probability function given in Table 1.1. Evidently (1.1) is satisfied. The sample space of X is $\{0, 1, 2\}$, and the mean and variance of X are

$$E(X) = (0 \times .20) + (1 \times .50) + (2 \times .30) = 1.1$$

and

$$\text{var}(X) = [(0 - 1.1)^2 .20] + [(1 - 1.1)^2 .50] + [(2 - 1.1)^2 .30] = .49.$$

Transformations can be used to derive new random variables from an existing random variable. Again we emphasize that what is meant by such a statement is that we can derive new probability functions from an existing probability function. When the probability function at hand has a known formula it is possible, in theory, to write down an explicit formula for the transformed probability function. In practice, this

TABLE 1.1   Probability Function of X

x	$P(X = x)$
0	.20
1	.50
2	.30

TABLE 1.2   Probability Function of Y

y	$P(Y = y)$
5	.20
7	.50
9	.30

may lead to a very complicated expression, which is one of the reasons for relying on random variable notation.

Example 1.2   With X as in Example 1.1, consider the random variable $Y = 2X + 5$. The sample space of Y is obtained by applying the transformation to the sample space of X, which gives $\{5, 7, 9\}$. The values of $P(Y = x)$ are derived as follows: $P(Y = 7) = P(2X + 5 = 7) = P(X = 1) = .50$. The probability function of Y is given in Table 1.2.

The mean and variance of Y are

$$E(Y) = (5 \times .20) + (7 \times .50) + (9 \times .30) = 7.2$$

and

$$\mathrm{var}(Y) = [(5 - 7.2)^2.20] + [(7 - 7.2)^2.50] + [(9 - 7.2)^2.30] = 1.96.$$

Comparing Examples 1.1 and 1.2 we note that X and Y have the same probability values but different sample spaces.

Consider a random variable which has as its only outcome the constant β, that is, the sample space is $\{\beta\}$. It is immediate from (1.2) and (1.3) that the mean and variance of the random variable are β and 0, respectively. Identifying the random variable with the constant β, and allowing a slight abuse of notation, we can write $E(\beta) = \beta$ and $\mathrm{var}(\beta) = 0$. Let X be a random variable, let α and β be arbitrary constants, and consider the random variable $\alpha X + \beta$. Using (1.2) and (1.3) it can be shown that

$$E(\alpha X + \beta) = \alpha E(X) + \beta \tag{1.4}$$

and

$$\mathrm{var}(\alpha X + \beta) = \alpha^2 \, \mathrm{var}(X). \tag{1.5}$$

Applying these results to Examples 1.1 and 1.2 we find, as before, that $E(Y) = 2(1.1) + 5 = 7.2$ and $\mathrm{var}(Y) = 4(.49) = 1.96$.

Example 1.3   Let X be an arbitrary random variable with mean μ and variance σ^2, where $\sigma > 0$, and consider the random variable $(X - \mu)/\sigma$. With $\alpha = 1/\sigma$ and

$\beta = -\mu/\sigma$ in (1.4) and (1.5), it follows that

$$E\left(\frac{X - \mu}{\sigma}\right) = 0$$

and

$$\text{var}\left(\frac{X - \mu}{\sigma}\right) = 1.$$

In many applications it is necessary to consider several related random variables. For example, in a health survey we might be interested in age, weight, and blood pressure. A probability function characterizing two or more random variables simultaneously is referred to as their joint probability function. For simplicity we discuss the case of two discrete random variables, X and Y. The joint probability function of the pair of random variables (X, Y) is denoted by $P(X = x, Y = y)$. For the present discussion we assume that the sample space of the joint probability function is the set of pairs $\{(x, y)\}$, where x is in the sample space of X and y is in the sample space of Y. Analogous to (1.1), the identity

$$\sum_x \sum_y P(X = x, Y = y) = 1 \tag{1.6}$$

must be satisfied. In the joint distribution of X and Y, the two random variables are considered as a unit. In order to isolate the distribution of X, we "sum over" Y to obtain what is referred to as the marginal probability function of X,

$$P(X = x) = \sum_y P(X = x, Y = y).$$

Similarly, the marginal probability function of Y is

$$P(Y = y) = \sum_x P(X = x, Y = y).$$

From a joint probability function we are to able obtain marginal probability functions, but the process does not necessarily work in reverse. We say that X and Y are independent random variables if $P(X = x, Y = y) = P(X = x) P(Y = y)$, that is, if the joint probability function is the product of the marginal probability functions. Other than the case of independence, it is not generally possible to reconstruct a joint probability function in this way.

Example 1.4 Table 1.3 is an example of a joint probability function and its associated marginal probability functions. For example, $P(X = 1, Y = 3) = .30$. The marginal probability function of X is obtained by summing over Y, for example,

$$P(X = 1) = P(X = 1, Y = 1) + P(X = 1, Y = 2) + P(X = 1, Y = 3) = .50.$$

TABLE 1.3 Joint Probability Function of X and Y

		$P(X = x, Y = y)$		
		y		
x	1	2	3	$P(X = x)$
0	.02	.06	.12	.20
1	.05	.15	.30	.50
2	.03	.09	.18	.30
$P(Y = y)$	.10	.30	.60	1

It is readily verified that X and Y are independent, for example, $P(X = 1, Y = 2) = .15 = P(X = 1) P(Y = 2)$.

Now consider Table 1.4, where the marginal probability functions of X and Y are the same as in Table 1.3 but where, as is easily verified, X and Y are not independent.

We now present generalizations of (1.4) and (1.5). Let $X_1, X_2, \ldots, X_n$ be arbitrary random variables, let $\alpha_1, \alpha_2, \ldots, \alpha_n, \beta$ be arbitrary constants, and consider the random variable $\sum_{i=1}^{n} \alpha_i X_i + \beta$. It can be shown that

$$E\left(\sum_{i=1}^{n} \alpha_i X_i + \beta \right) = \sum_{i=1}^{n} \alpha_i E(X_i) + \beta \tag{1.7}$$

and, if the X_i are independent, that

$$\text{var}\left(\sum_{i=1}^{n} \alpha_i X_i + \beta \right) = \sum_{i=1}^{n} \alpha_i^2 \, \text{var}(X_i). \tag{1.8}$$

In the case of two independent random variables X_1 and X_2,

$$E(X_1 + X_2) = E(X_1) + E(X_2)$$
$$E(X_1 - X_2) = E(X_1) - E(X_2)$$

TABLE 1.4 Joint Probability Function of X and Y

		$P(X = x, Y = y)$		
		y		
x	1	2	3	$P(X = x)$
0	.01	.05	.14	.20
1	.06	.18	.26	.50
2	.03	.07	.20	.30
$P(Y = y)$	.10	.30	.60	1

and

$$\text{var}(X_1 + X_2) = \text{var}(X_1 - X_2) = \text{var}(X_1) + \text{var}(X_2). \qquad (1.9)$$

If $X_1, X_2, \ldots, X_n$ are independent and all have the same distribution, we say the X_i are a sample from that distribution and that the sample size is n. Unless stated otherwise, it will be assumed that all samples are simple random samples (Section 1.3). With the distribution left unspecified, denote the mean and variance of X_i by μ and σ^2, respectively. The sample mean is defined to be

$$\overline{X} = \frac{1}{n} \sum_{i=1}^{n} X_i.$$

Setting $\alpha_i = 1/n$ and $\beta = 0$ in (1.7) and (1.8), we have

$$E(\overline{X}) = \mu \qquad (1.10)$$

and

$$\text{var}(\overline{X}) = \frac{\sigma^2}{n}. \qquad (1.11)$$

1.1.2   Some Probability Functions

We now consider some of the key probability functions that will be of importance in this book.

Normal (Gaussian)
For reasons that will become clear after we have discussed the Central Limit Theorem, the most important distribution is undoubtedly the normal distribution. The normal probability function is

$$f(z|\mu, \sigma) = \frac{1}{\sigma\sqrt{2\pi}} \exp\left[\frac{-(z - \mu)^2}{2\sigma^2}\right]$$

where the sample space is all numbers and exp stands for exponentiation to the base e. We denote the corresponding normal random variable by Z. A normal distribution is completely characterized by the parameters μ and $\sigma > 0$. It can be shown that the mean and variance of Z are μ and σ^2, respectively.

When $\mu = 0$ and $\sigma = 1$ we say that Z has the standard normal distribution. For $0 < \gamma < 1$, let z_γ denote that point which cuts off the upper γ-tail probability of the standard normal distribution; that is, $P(Z \geq z_\gamma) = \gamma$. For example, $z_{.025} = 1.96$. In some statistics books the notation z_γ is used to denote the lower γ-tail. An important property of the normal distribution is that, for arbitrary constants α and $\beta > 0$, $(Z - \alpha)/\beta$ is also normally distributed. In particular this is true for $(Z - \mu)/\sigma$ which, in view of Example 1.3, is therefore standard normal. This explains why statistics books only need to provide values of z_γ for the standard normal distribution rather than a series of tables for different values of μ and σ.

Another important property of the normal distribution is that it is additive. Let $Z_1, Z_2, \ldots, Z_n$ be independent normal random variables and suppose that Z_i has mean μ_i and variance σ_i^2 ($i = 1, 2, \ldots, n$). Then the random variable $\sum_{i=1}^n Z_i$ is also normally distributed and, from (1.7) and (1.8), it has mean $\sum_{i=1}^n \mu_i$ and variance $\sum_{i=1}^n \sigma_i^2$.

Chi-Square

The formula for the chi-square probability function is complicated and will not be presented here. The sample space of the distribution is all nonnegative numbers. A chi-square distribution is characterized completely by a single positive integer r, which is referred to as the degrees of freedom. For brevity we write $\chi_{(r)}^2$ to indicate that a random variable has a chi-square distribution with r degrees of freedom. The mean and variance of the chi-square distribution with r degrees of freedom are r and $2r$, respectively.

The importance of the chi-square distribution stems from its connection with the normal distribution. Specifically, if Z is standard normal, then Z^2, the transformation of Z obtained by squaring, is $\chi_{(1)}^2$. More generally, if Z is normal with mean μ and variance σ^2 then, as remarked above, $(Z - \mu)/\sigma$ is standard normal and so $[(Z - \mu)/\sigma]^2 = (Z - \mu)^2/\sigma^2$ is $\chi_{(1)}^2$. In practice, most chi-square distributions with 1 degree of freedom originate as the square of a standard normal distribution. This explains why the usual notation for a chi-square random variable is X^2, or sometimes χ^2.

Like the normal distribution, the chi-square distribution has an additive property. Let $X_1^2, X_2^2, \ldots, X_n^2$ be independent chi-square random variables and suppose that X_i^2 has r_i degrees of freedom ($i = 1, 2, \ldots, n$). Then $\sum_{i=1}^n X_i^2$ is chi-square with $\sum_{i=1}^n r_i$ degrees of freedom. As a special case of this result, let $Z_1, Z_2, \ldots, Z_n$ be independent normal random variables, where Z_i has mean μ_i and variance σ_i^2 ($i = 1, 2, \ldots, n$). Then $(Z_i - \mu_i)^2/\sigma_i^2$ is $\chi_{(1)}^2$ for all i, and so

$$X^2 = \sum_{i=1}^n \frac{(Z_i - \mu_i)^2}{\sigma_i^2} \tag{1.12}$$

is $\chi_{(n)}^2$.

Binomial

The binomial probability function is

$$P(A = a|\pi) = \binom{r}{a} \pi^a (1 - \pi)^{r-a}$$

where the sample space is the (finite) set of integers $\{0, 1, 2, \ldots, r\}$. A binomial distribution is completely characterized by the parameters π and r which, for convenience, we usually write as (π, r). Recall that, for $0 \le a \le r$, the binomial coefficient is defined to be

$$\binom{r}{a} = \frac{r!}{a!\,(r-a)!}$$

where $r! = r\,(r-1) \cdots 2 \cdot 1$. We adopt the usual convention that $0! = 1$. The binomial coefficient $\binom{r}{a}$ equals the number of ways of choosing a items out of r without regard to order of selection. For example, the number of possible bridge hands is $\binom{52}{13} = 6.35 \times 10^{11}$. It can be shown that

$$\sum_{a=0}^{r} \binom{r}{a} \pi^a (1-\pi)^{r-a} = [\pi + (1-\pi)]^r = 1$$

and so (1.1) is satisfied. The mean and variance of A are πr and $\pi(1-\pi)r$, respectively; that is,

$$E(A) = \sum_{a=0}^{r} a \binom{r}{a} \pi^a (1-\pi)^{r-a} = \pi r$$

and

$$\mathrm{var}(A) = \sum_{a=0}^{r} (a - \pi r)^2 \binom{r}{a} \pi^a (1-\pi)^{r-a} = \pi(1-\pi)r.$$

Like the normal and chi-square distributions, the binomial distribution is additive. Let $A_1, A_2, \ldots, A_n$ be independent binomial random variables and suppose that A_i has parameters $\pi_i = \pi$ and r_i ($i = 1, 2, \ldots, n$). Then $\sum_{i=1}^{n} A_i$ is binomial with parameters π and $\sum_{i=1}^{n} r_i$. A similar result does not hold when the π_i are not all equal.

The binomial distribution is important in epidemiology because many epidemiologic studies are concerned with counted (discrete) outcomes. For instance, the binomial distribution can be used to analyze data from a study in which a group of r individuals is followed over a defined period of time and the number of outcomes of interest, denoted by a, is counted. In this context the outcome of interest could be, for example, recovery from an illness, survival to the end of follow-up, or death from some cause. For the binomial distribution to be applicable, two conditions need to be satisfied: The probability of an outcome must be the same for each subject, and subjects must behave independently; that is, the outcome for each subject must be unrelated to the outcome for any other subject. In an epidemiologic study the first condition is unlikely to be satisfied across the entire group of subjects. In this case, one strategy is to form subgroups of subjects having similar characteristics so that, to a greater or lesser extent, there is uniformity of risk within each subgroup. Then the binomial distribution can be applied to each subgroup separately. As an example where the second condition would not be satisfied, consider a study of influenza in a classroom of students. Since influenza is contagious, the risk of illness in one student is not independent of the risk in others. In studies of noninfectious diseases, such as cancer, stroke, and so on, the independence assumption is usually satisfied.

Poisson

The Poisson probability function is

$$P(D = d|v) = \frac{e^{-v}v^d}{d!} \tag{1.13}$$

where the sample space is the (infinite) set of nonnegative integers $\{0, 1, 2, \ldots\}$. A Poisson distribution is completely characterized by the parameter v, which is equal to both the mean and variance of the distribution, that is,

$$E(D) = \sum_{d=0}^{\infty} d \left(\frac{e^{-v}v^d}{d!} \right) = v$$

and

$$\text{var}(D) = \sum_{d=0}^{\infty} (d - v)^2 \left(\frac{e^{-v}v^d}{d!} \right) = v.$$

Similar to the other distributions considered above, the Poisson distribution has an additive property. Let $D_1, D_2, \ldots, D_n$ be independent Poisson random variables, where D_i has the parameter v_i $(i = 1, 2, \ldots, n)$. Then $\sum_{i=1}^{n} D_i$ is Poisson with parameter $\sum_{i=1}^{n} v_i$.

Like the binomial distribution, the Poisson distribution can be used to analyze data from a study in which a group of individuals is followed over a defined period of time and the number of outcomes of interest, denoted by d, is counted. In epidemiologic studies where the Poisson distribution is applicable, it is not the number of subjects that is important but rather the collective observation time experienced by the group as a whole. For the Poisson distribution to be valid, the probability that an outcome will occur at any time point must be "small." Expressed another way, the outcome must be a "rare" event.

As might be guessed from the above remarks, there is a connection between the binomial and Poisson distributions. In fact the Poisson distribution can be derived as a limiting case of the binomial distribution. Let D be Poisson with mean v, and let $A_1, A_2, \ldots, A_i, \ldots$ be an infinite sequence of binomial random variables, where A_i has parameters (π_i, r_i). Suppose that the sequence satisfies the following conditions: $\pi_i r_i = v$ for all i, and the limiting value of π_i equals 0. Under these circumstances the sequence of binomial random variables "converges" to D; that is, as i gets larger the distribution of A_i gets closer to that of D. This theoretical result explains why the Poisson distribution is often used to model rare events. It also suggests that the Poisson distribution with parameter v can be used to approximate the binomial distribution with parameters (π, r), provided $v = \pi r$ and π is "small."

TABLE 1.5 Binomial and Poisson Probability Functions (%)

	Binomial			Poisson
x	$\pi = .2$ $r = 10$	$\pi = .1$ $r = 20$	$\pi = .01$ $r = 200$	$\nu = 2$
0	10.74	12.16	13.40	13.53
1	26.84	27.02	27.07	27.07
2	30.20	28.52	27.20	27.07
3	20.13	19.01	18.14	18.04
4	8.81	8.98	9.02	9.02
5	2.64	3.19	3.57	3.61
6	.55	.89	1.17	1.20
7	.08	.20	.33	.34
8	.01	.04	.08	.09
9	< .01	.01	.02	.02
10	< .01	< .01	< .01	< .01
$\vdots$	—	$\vdots$	$\vdots$	$\vdots$

Example 1.5 Table 1.5 gives three binomial distributions with parameters $(.2, 10)$, $(.1, 20)$, and $(.01, 200)$, so that in each case the mean is 2. Also shown is the Poisson distribution with a mean of 2. The sample spaces have been truncated at 10. As can be seen, as π becomes smaller the Poisson distribution provides a progressively better approximation to the binomial distribution.

1.1.3 Central Limit Theorem and Normal Approximations

Let $X_1, X_2, \ldots, X_n$ be a sample from an arbitrary distribution and denote the common mean and variance by μ and σ^2. It was shown in (1.10) and (1.11) that $\overline{X}$ has mean $E(\overline{X}) = \mu$ and variance $\text{var}(\overline{X}) = \sigma^2/n$. So, from Example 1.3, the random variable $\sqrt{n}(\overline{X} - \mu)/\sigma$ has mean 0 and variance 1. If the X_i are normal then, from the properties of the normal distribution, $\sqrt{n}(\overline{X} - \mu)/\sigma$ is standard normal. The Central Limit Theorem is a remarkable result from probability theory which states that, even when the X_i are not normal, $\sqrt{n}(\overline{X} - \mu)/\sigma$ is "approximately" standard normal, provided n is sufficiently "large." We note that the X_i are not required to be continuous random variables. Probability statements such as this, which become more accurate as n increases, are said to hold asymptotically. Accordingly, the Central Limit Theorem states that $\sqrt{n}(\overline{X} - \mu)/\sigma$ is asymptotically standard normal.

Let A be binomial with parameters (π, n) and let $A_1, A_2, \ldots, A_n$ be a sample from the binomial distribution with parameters $(\pi, 1)$. Similarly, let D be Poisson with parameter ν, where we assume that $\nu = n$, an integer, and let $D_1, D_2, \ldots, D_n$ be a sample from the Poisson distribution with parameter 1. From the additive properties of binomial and Poisson distributions, A has the same distribution as $\sum_{i=1}^{n} A_i$, and D has the same distribution as $\sum_{i=1}^{n} D_i$. It follows from the Central Limit Theorem that, provided n is large, A and D will be asymptotically normal. We illustrate this phenomenon below with a series of graphs.

Let $D_1, D_2, \ldots, D_n$ be independent Poisson random variables, where D_i has the parameter v_i $(i = 1, 2, \ldots, n)$. From the arguments leading to (1.12) and the Central Limit Theorem, it follows that

$$X^2 = \sum_{i=1}^{n} \frac{(D_i - v_i)^2}{v_i} \tag{1.14}$$

is approximately $\chi^2_{(n)}$. More generally, let $X_1, X_2, \ldots, X_n$ be independent random variables where X_i has mean μ_i and variance σ_i^2 $(i = 1, 2, \ldots, n)$. If each X_i is approximately normal then

$$X^2 = \sum_{i=1}^{n} \frac{(X_i - \mu_i)^2}{\sigma_i^2} \tag{1.15}$$

is approximately $\chi^2_{(n)}$.

Example 1.6 Table 1.6(a) gives the exact and approximate values of the lower and upper tail probabilities of the binomial distribution with parameters (.3, 10). In statistics the term "exact" means that an actual probability function is being used to perform calculations, as opposed to a normal approximation. The mean and variance of the binomial distribution are .3(10) $= 3$ and .3(.7)(10) $= 2.1$. The approximate values were calculated using the following approach. The normal approximation to $P(A \le 2 \,|.3)$, for example, equals the area under the standard normal curve to the left of $[(2+.5)-3]/\sqrt{2.1}$, and the normal approximation to $P(A \ge 2 \,|.3)$ equals the area under the standard normal curve to the right of $[(2 - .5) - 3]/\sqrt{2.1}$. The continuity correction factors $\pm.5$ have been included because the normal distribution, which is continuous, is being used to approximate a binomial distribution, which is discrete (Breslow and Day, 1980, §4.3). As can be seen from Table 1.6(a), the exact and approximate values show quite good agreement. Table 1.6(b) gives the results for the

TABLE 1.6(a)   Exact and Approximate Tail Probabilities (%) for the Binomial Distribution with Parameters (.3,10)

	$P(A \le a \,|.3)$		$P(A \ge a \,|.3)$	
a	Exact	Approximate	Exact	Approximate
2	38.28	36.50	85.07	84.97
4	84.97	84.97	35.04	36.50
6	98.94	99.21	4.73	4.22
8	99.99	99.99	.16	.10

TABLE 1.6(b)   Exact and Approximate Tail Probabilities (%) for the Binomial Distribution with Parameters (.3,100)

	$P(A \leq a \,|.3)$		$P(A \geq a \,|.3)$	
a	Exact	Approximate	Exact	Approximate
20	1.65	1.91	99.11	98.90
25	16.31	16.31	88.64	88.50
30	54.91	54.34	53.77	54.34
35	88.39	88.50	16.29	16.31
40	98.75	98.90	2.10	1.91

binomial distribution with parameters (.3,100), which shows even better agreement due to the larger sample size.

Arguments were presented above which show that binomial and Poisson distributions are approximately normal when the sample size is large. The obvious question is, How large is "large"? We approach this matter empirically and present a sample size criterion that is useful in practice. The following remarks refer to Figures 1.1(a)–1.8(a), which show graphs of selected binomial and Poisson distributions. The points in the sample space have been plotted on the horizontal axis, with the corresponding probabilities plotted on the vertical axis. Magnitudes have not been indicated on the axes since, for the moment, we are concerned only with the shapes of distributions. The horizontal axes are labeled with the term "count," which stands for the number of binomial or Poisson outcomes. Distributions with the symmetric, bell-shaped appearance of the normal distribution have a satisfactory normal approximation.

The binomial and Poisson distributions have sample spaces consisting of consecutive integers, and so the distance between neighboring points is always 1. Consequently the graphs could have been presented in the form of histograms (bar charts). Instead are shown as step functions so as to facilitate later comparisons with the remaining graphs in the same figures. Since the base of each step has a length of 1, the area of the rectangle corresponding to that step equals the probability associated with that point in the sample space. Consequently, summing across the entire sample space, the area under each step function equals 1, as required by (1.1). Some of the distributions considered here have tails with little associated probability (area). This is obviously true for the Poisson distributions, where the sample space is infinite and extreme tail probabilities are small. The graphs have been truncated at the extremes of the distributions corresponding to tail probabilities of 1%.

The binomial parameters used to create Figures 1.1(a)–1.5(a) are (.3,10), (.5,10), (.03,100), (.05,100), and (.1,100), respectively, and so the means are 3, 5, and 10. The Poisson parameters used to create Figures 1.6(a)–1.8(a) are 3, 5, and 10, which are also the means of the distributions. As can be seen, for both the binomial and Poisson distributions, a rough guideline is that the normal approximation should be satisfactory provided the mean of the distribution is greater than or equal to 5.

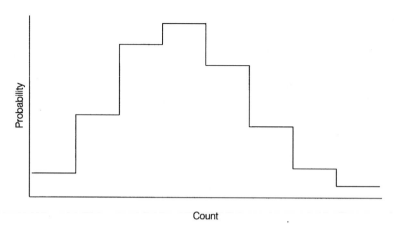

FIGURE 1.1(a)   Binomial distribution with parameters (.3, 10)

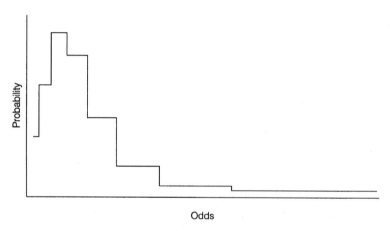

FIGURE 1.1(b)   Odds transformation of binomial distribution with parameters (.3, 10)

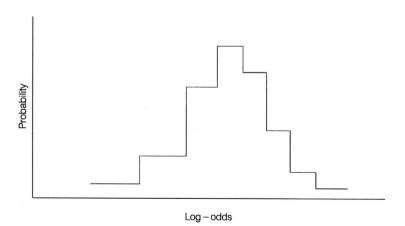

FIGURE 1.1(c)   Log-odds transformation of binomial distribution with parameters (.3, 10)

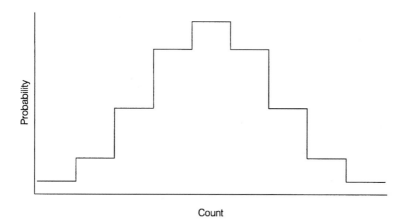

Count

FIGURE 1.2(a)   Binomial distribution with parameters (.5, 10)

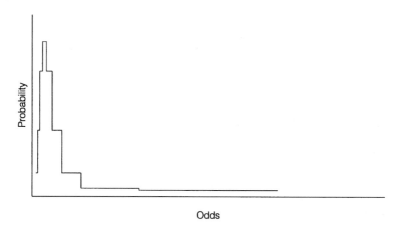

Odds

FIGURE 1.2(b)   Odds transformation of binomial distribution with parameters (.5, 10)

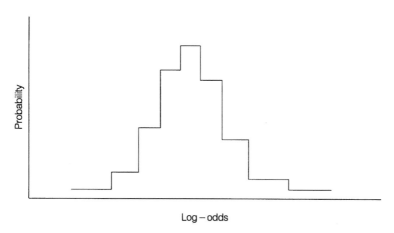

Log – odds

FIGURE 1.2(c)   Log-odds transformation of binomial distribution with parameters (.5, 10)

15

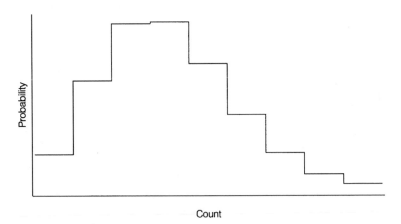

FIGURE 1.3(a) Binomial distribution with parameters (.03, 100)

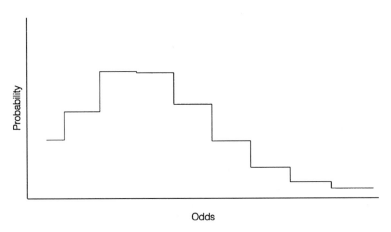

FIGURE 1.3(b) Odds transformation of binomial distribution with parameters (.03, 100)

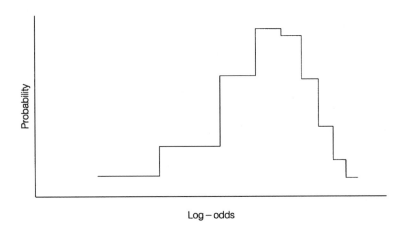

FIGURE 1.3(c) Log-odds transformation of binomial distribution with parameters (.03, 100)

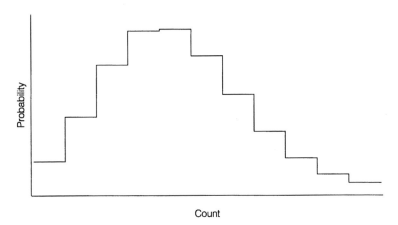

FIGURE 1.4(a) Binomial distribution with parameters (.05, 100)

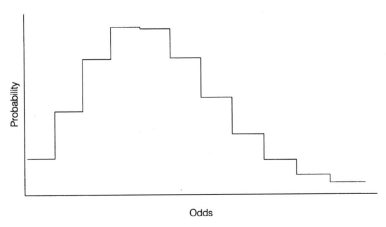

FIGURE 1.4(b) Odds transformation of binomial distribution with parameters (.05, 100)

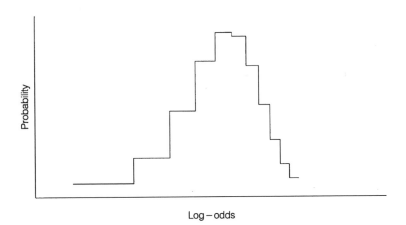

FIGURE 1.4(c) Log-odds transformation of binomial distribution with parameters (.05, 100)

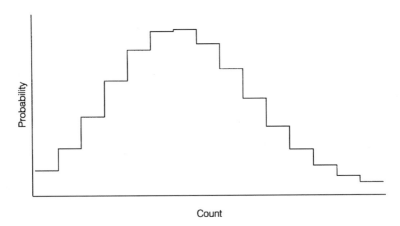

FIGURE 1.5(a) Binomial distribution with parameters (.1, 100)

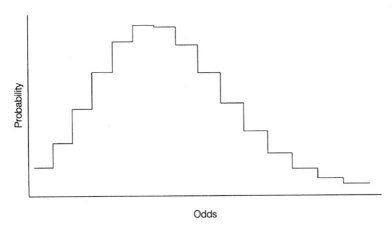

FIGURE 1.5(b) Odds transformation of binomial distribution with parameters (.1, 100)

FIGURE 1.5(c) Log-odds transformation of binomial distribution with parameters (.1, 100)

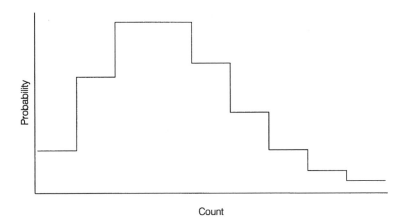

FIGURE 1.6(a)   Poisson distribution with parameter 3

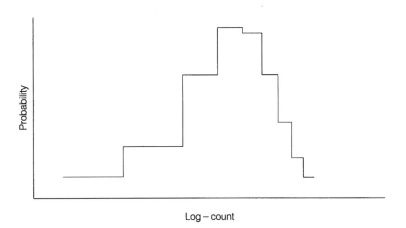

FIGURE 1.6(b)   Log transformation of Poisson distribution with parameter 3

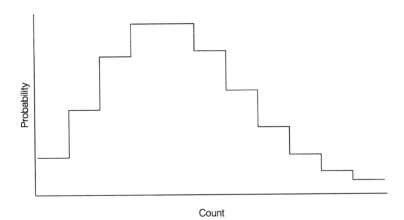

FIGURE 1.7(a)   Poisson distribution with parameter 5

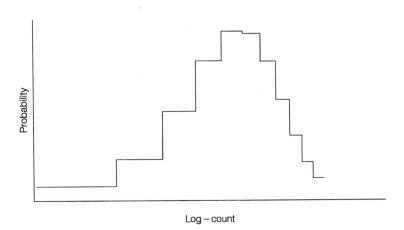

FIGURE 1.7(b)   Log transformation of Poisson distribution with parameter 5

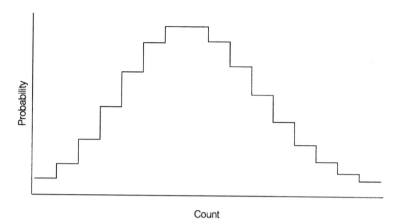

FIGURE 1.8(a)   Poisson distribution with parameter 10

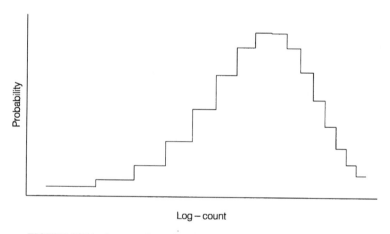

FIGURE 1.8(b)   Log transformation of Poisson distribution with parameter 10

1.2   PARAMETER ESTIMATION

In the preceding section we discussed the properties of distributions in general, and those of the normal, chi-square, binomial, and Poisson distributions in particular. These distributions and others are characterized by parameters that, in practice, are usually unknown. This raises the question of how to estimate such parameters from study data.

   In certain applications the method of estimation seems intuitively clear. For example, suppose we are interested in estimating the probability that a coin will land heads. A "study" to investigate this question is straightforward and involves tossing the coin r times and counting the number of heads, a quantity that will be denoted by a. The question of how large r should be is answered in Chapter 14. The proportion of tosses landing heads a/r tells us something about the coin, but in order to probe more deeply we require a probability model, the obvious choice being the binomial distribution. Accordingly, let A be a binomial random variable with parameters (π, r), where π denotes the unknown probability that the coin will land heads. Even though the parameter π can never be known with certainty, it can be estimated from study data. From the binomial model, an estimate is given by the random variable A/r which, in the present study, has the realization a/r. We denote A/r by $\hat{\pi}$ and refer to $\hat{\pi}$ as a (point) estimate of π. In some of the statistics literature, $\hat{\pi}$ is called an estimator of π, the term estimate being reserved for the realization a/r. In keeping with our convention of intentionally ignoring the distinction between random variables and realizations, we use estimate to refer to both quantities.

The theory of binomial distributions provides insight into the properties of $\hat{\pi}$ as an estimate of π. Since A has mean $E(A) = \pi r$ and variance $\mathrm{var}(A) = \pi(1-\pi)r$, it follows that $\hat{\pi}$ has mean $E(\hat{\pi}) = E(A)/r = \pi$ and variance $\mathrm{var}(\hat{\pi}) = \mathrm{var}(A)/r^2 = \pi(1-\pi)/r$. In the context of the coin-tossing study, these properties of $\hat{\pi}$ have the following interpretations: Over the course of many replications of the study, each based on r tosses, the realizations of $\hat{\pi}$ will be tend to be near π; and when r is large there will be little dispersion of the realizations on either side of π. The latter interpretation is consistent with our intuition that π will be estimated more accurately when there are many tosses of the coin.

With the above example as motivation, we now consider the general problem of parameter estimation. For simplicity we frame the discussion in terms of a discrete random variable, but the same ideas apply to the continuous case. Suppose that we wish to study a feature of a population which is governed by a probability function $P(X = x|\theta)$, where the parameter θ embodies the characteristic of interest. For example, in a population health survey, X could be the serum cholesterol of a randomly chosen individual and θ might be the average serum cholesterol in the population. Let $X_1, X_2, \ldots, X_n$ be a sample of size n from the probability function $P(X = x|\theta)$. A (point) estimate of θ, denoted by $\hat{\theta}$, is a random variable that is expressed in terms of the X_i and that satisfies certain properties, as discussed below. In the preceding example, the survey could be conducted by sampling n individuals at random from the population and measuring their serum cholesterol. For $\hat{\theta}$ we might consider using $\overline{X} = (\sum_{i=1}^{n} X_i)/n$, the average serum cholesterol in the sample.

There is considerable latitude when specifying the properties that $\hat{\theta}$ should be required to satisfy, but in order for a theory of estimation to be meaningful the properties must be chosen so that $\hat{\theta}$ is, in some sense, informative about θ. The first property we would like $\hat{\theta}$ to have is that it should result in realizations that are "near" θ. This is impossible to guarantee in any given study, but over the course of many replications of the study we would like this property to hold "on average." Accordingly, we require the mean of $\hat{\theta}$ to be θ, that is, $E(\hat{\theta}) = \theta$. When this property is satisfied we say that $\hat{\theta}$ is an unbiased estimate of θ, otherwise $\hat{\theta}$ is said to be biased. The second property we would like $\hat{\theta}$ to have is that it should make as efficient use of the data as possible. In statistics, notions related to efficiency are generally expressed in terms of the variance. That is, all other things being equal, the smaller the variance the greater the efficiency. Accordingly, for a given sample size, we require $\text{var}(\hat{\theta})$ to be as small as possible.

In the coin-tossing study the parameter was $\theta = \pi$. We can reformulate the earlier probability model by letting $A_1, A_2, \ldots, A_n$ be independent binomial random variables, each having parameters $(\pi, 1)$. Setting $\overline{A} = (\sum_{i=1}^{n} A_i)/n$ we have $\hat{\pi} = \overline{A}$, and so $E(\overline{A}) = \pi$ and $\text{var}(\overline{A}) = \pi(1-\pi)/n$. Suppose that instead of $\overline{A}$ we decide to use A_1 as an estimate of π; that is, we ignore all but the first toss of the coin. Since $E(A_1) = \pi$, both $\overline{A}$ and A_1 are unbiased estimates of π. However, $\text{var}(A_1) = \pi(1-\pi)$ and so, provided $n > 1$, $\text{var}(A_1) > \text{var}(\overline{A})$. This means that $\overline{A}$ is more efficient than A_1. Based on the above criteria we would choose $\overline{A}$ over A_1 as an estimate of π.

The decision to choose $\overline{A}$ in preference to A_1 was based on a comparison of variances. This raises the question of whether there is another unbiased estimate of π with a variance that is even smaller than $\pi(1-\pi)/n$. We return now to the general case of an arbitrary probability function $P(X = x|\theta)$. For many of the probability functions encountered in epidemiology it can be shown that there is a number $b(\theta)$ such that, for any unbiased estimate $\hat{\theta}$, the inequality $\text{var}(\hat{\theta}) \geq b(\theta)$ is satisfied. Consequently, $b(\theta)$ is at least as small as the variance of any unbiased estimate of θ. There is no guarantee that for given θ and $P(X = x|\theta)$ there actually is an unbiased estimate with a variance this small; but, if we can find one, we clearly will have satisfied the requirement that the estimate has the smallest variance possible.

For the binomial distribution, it turns out that $b(\pi) = \pi(1-\pi)/n$, and so $b(\pi) = \text{var}(\hat{\pi})$. Consequently $\hat{\pi}$ is an unbiased estimate of π with the smallest variance possible (among unbiased estimates). For the binomial distribution, intuition suggests that $\hat{\pi}$ ought to provide a reasonable estimate of π, and it turns out that $\hat{\pi}$ has precisely the properties we require. However, such *ad hoc* methods of defining an estimate cannot always be relied upon, especially when the probability model is complex. We now consider two widely used methods of estimation which ensure that the estimate has desirable properties, provided asymptotic conditions are satisfied.

1.2.1 Maximum Likelihood

The maximum likelihood method is based on a concept that is intuitively appealing and, at first glance, deceptively straightforward. Like many profound ideas, its apparent simplicity belies a remarkable depth. Let $X_1, X_2, \ldots, X_n$ be a sample from the probability function $P(X = x|\theta)$ and consider the observations (realizations) $x_1, x_2, \ldots, x_n$. Since the X_i are independent, the (joint) probability of these observations is the product of the individual probability elements, that is,

$$\prod_{i=1}^{n} P(X_i = x_i|\theta) = P(X_1 = x_1|\theta)\, P(X_2 = x_2|\theta) \cdots P(X_n = x_n|\theta). \quad (1.16)$$

Ordinarily we are inclined to think of (1.16) as a function of the x_i. From this perspective, (1.16) can be used to calculate the probability of the observations provided the value of θ is known. The maximum likelihood method turns this argument around and views (1.16) as a function of θ. Once the data have been collected, values of the x_i can be substituted into (1.16), making it a function of θ alone. When viewed this way we denote (1.16) by $L(\theta)$ and refer to it as the likelihood. For any value of θ, $L(\theta)$ equals the probability of the observations $x_1, x_2, \ldots, x_n$. We can graph $L(\theta)$ as a function of θ to get a visual image of this relationship. The value of θ which is most in accord with the observations, that is, makes them most "likely," is the one which maximizes $L(\theta)$ as a function of θ. We refer to this value of θ as the maximum likelihood estimate and denote it by $\hat{\theta}$.

Example 1.7   Let A_1, A_2, A_3, A_4, A_5 be a sample from the binomial distribution with parameters $(\pi, 1)$, and consider the observations $a_1 = 0$, $a_2 = 1$, $a_3 = 0$, $a_4 = 0$, and $a_5 = 0$. The likelihood is

$$L(\pi) = \prod_{i=1}^{5} \pi^{a_i} (1 - \pi)^{1-a_i} = \pi(1 - \pi)^4.$$

From the graph of $L(\pi)$, shown in Figure 1.9, it appears that $\hat{\pi}$ is somewhere in the neighborhood of .2. Trial and error with larger and smaller values of π confirms that in fact $\hat{\pi} = .2$.

The above graphical method of finding a maximum likelihood estimate is feasible only in the simplest of cases. In more complex situations, in particular when there are several parameters to estimate simultaneously, numerical methods are required, such as those described in Appendix B. When there is a single parameter, the maximum likelihood estimate $\hat{\theta}$ can usually be found by solving the maximum likelihood equation,

$$L'(\hat{\theta}) = 0 \qquad\qquad (1.17)$$

where $L'(\theta)$ is the derivative of $L(\theta)$ with respect to θ.

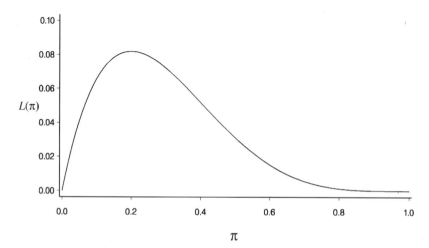

FIGURE 1.9   Likelihood for Example 1.7

Example 1.8 We now generalize Example 1.7. Let $A_1, A_2, \ldots, A_r$ be a sample from the binomial distribution with parameters $(\pi, 1)$, and denote the observations by $a_1, a_2, \ldots, a_r$. The likelihood is

$$L(\pi) = \prod_{i=1}^{r} \pi^{a_i} (1 - \pi)^{1-a_i} = \pi^a (1 - \pi)^{r-a} \tag{1.18}$$

where $a = \sum_{i=1}^{r} a_i$. From the form of the likelihood we see that is not the individual a_i which are important but rather their sum a. Accordingly we might just as well have based the likelihood on $\sum_{i=1}^{r} A_i$, which is binomial with parameters (π, r). In this case the likelihood is

$$L(\pi) = \binom{r}{a} \pi^a (1 - \pi)^{r-a}. \tag{1.19}$$

As far as maximizing (1.19) with respect to π is concerned, the binomial coefficient is irrelevant and so (1.18) and (1.19) are equivalent from the likelihood perspective. It is straightforward to show that the maximum likelihood equation (1.17) simplifies to $a - \hat{\pi}r = 0$ and so the maximum likelihood estimate of π is $\hat{\pi} = a/r$.

Maximum likelihood estimates have very attractive asymptotic properties. Specifically, if $\hat{\theta}$ is the maximum likelihood estimate of θ then $\hat{\theta}$ is asymptotically normal with mean θ and variance $b(\theta)$, where the latter is the lower bound described earlier. As a result, θ satisfies, in an asymptotic sense, the two properties that were proposed above as being desirable features of an estimate—unbiasedness and minimum variance. In addition to parameter estimates, the maximum likelihood approach also provides methods of confidence interval estimation and hypothesis testing. As discussed in Appendix B, included among the latter are the Wald, score, and likelihood ratio tests.

It seems that the maximum likelihood method has much to offer; however, there are two potential problems. First, the maximum likelihood equation may be very complicated and this can make calculating $\hat{\theta}$ difficult in practice. This is especially true when several parameters must be estimated simultaneously. Fortunately, statistical packages are available for many standard analyses and modern computers are capable of handling the computational burden. The second problem is that the desirable properties of maximum likelihood estimates are guaranteed to hold only when the sample size is "large."

1.2.2 Weighted Least Squares

In the coin-tossing study discussed above, we considered a sample $A_1, A_2, \ldots, A_n$ from a binomial distribution with parameters $(\pi, 1)$. Since $E(A_i) = \pi$ we can denote A_i by $\hat{\pi}_i$, and in place of $\overline{A} = \left(\sum_{i=1}^{n} A_i \right)/n$ write $\hat{\pi} = \left(\sum_{i=1}^{n} \hat{\pi}_i \right)/n$. In this way we can express the estimate of π as an average of estimates, one for each i. More generally, suppose that $\hat{\theta}_1, \hat{\theta}_2, \ldots, \hat{\theta}_n$ are independent unbiased estimates of a parameter θ, that is, $E(\hat{\theta}_i) = \theta$ for all i. We do not assume that the $\hat{\theta}_i$ necessarily have the same distribution; in particular, we do not require that the variances $\mathrm{var}(\hat{\theta}_i) = \sigma_i^2$ be equal. We seek a method of combining the individual estimates $\hat{\theta}_i$ of θ into an overall estimate $\hat{\theta}$ which has the desirable properties outlined earlier. (Using the symbol $\hat{\theta}$ for both the weighted least squares and maximum likelihood estimates is a matter of convenience and is not meant to imply any connection between the two estimates.) For constants $w_i > 0$, consider the sum

$$\frac{1}{W} \sum_{i=1}^{n} w_i (\hat{\theta}_i - \hat{\theta})^2 \tag{1.20}$$

where $W = \sum_{i=1}^{n} w_i$. We refer to the w_i as weights and to an expression such (1.20) as a weighted average. It is the relative, not the absolute, magnitude of each w_i that is important in a weighted average. In particular, we can replace w_i with $w_i' = w_i / W$ and obtain a weighted average in which the weights sum to 1. In this way, means (1.2) and variances (1.3) can be viewed as weighted averages.

Expression (1.20) is a measure of the overall weighted "distance" between the $\hat{\theta}_i$ and $\hat{\theta}$. The weighted least squares method defines $\hat{\theta}$ to be that quantity which minimizes (1.20). It can be shown that the weighted least squares estimate of θ is

$$\hat{\theta} = \frac{1}{W} \sum_{i=1}^{n} w_i \hat{\theta}_i \tag{1.21}$$

which is seen to be a weighted average of the $\hat{\theta}_i$. Since each $\hat{\theta}_i$ is an unbiased estimate of θ, it follows from (1.7) that

$$E(\hat{\theta}) = \frac{1}{W} \sum_{i=1}^{n} w_i E(\hat{\theta}_i) = \theta.$$

So $\hat{\theta}$ is also an unbiased estimate of θ, and this is true regardless of the choice of weights. Not all weighting schemes are equally efficient in the sense of keeping the variance var($\hat{\theta}$) to a minimum. The variance σ_i^2 is a measure of the amount of information contained in the estimate $\hat{\theta}_i$. It seems reasonable that relatively greater weight should be given to those $\hat{\theta}_i$ for which σ_i^2 is correspondingly small. It turns out that the weights $w_i = 1/\sigma_i^2$ are optimal in the following sense: The corresponding weighted least squares estimate has minimum variance among all weighted averages of the $\hat{\theta}_i$ (although not necessarily among estimates in general). Setting $w_i = 1/\sigma_i^2$, it follows from (1.8) that

$$\text{var}(\hat{\theta}) = \frac{1}{W^2} \sum_{i=1}^{n} w_i^2 \, \text{var}(\hat{\theta}_i) = \frac{1}{W}. \tag{1.22}$$

Note that up to this point the entire discussion has been based on means and variances. In particular, nothing has been assumed about distributions or sample size. It seems that the weighted least squares method has much to recommend it. Unlike the maximum likelihood approach, the calculations are straightforward, and sample size does not seem to be an issue. However, a major consideration is that we need to know the variances σ_i^2 prior to using the weighted least squares approach, and in practice this information is almost never available. Therefore it is usually necessary to estimate the σ_i^2 from study data, in which case the weights are random variables rather than constants. So instead of (1.21) and (1.22) we have instead

$$\hat{\theta} = \frac{1}{\hat{W}} \sum_{i=1}^{n} \hat{w}_i \hat{\theta}_i \tag{1.23}$$

and

$$\widehat{\mathrm{var}}(\hat{\theta}) = \frac{1}{\hat{W}} \tag{1.24}$$

where $\hat{w}_i = 1/\hat{\sigma}_i^2$ and $\hat{W} = \sum_{i=1}^{n} \hat{w}_i$. When the σ_i^2 are estimated from large samples the desirable properties of (1.21) and (1.22) described above carry over to (1.23) and (1.24), that is, $\hat{\theta}$ is asymptotically unbiased with minimum variance.

1.3 RANDOM SAMPLING

The methods of parameter (point) estimation described in the preceding section, as well as the methods of confidence interval estimation and hypothesis testing to be discussed in subsequent chapters, are based on the assumption that study subjects are selected using random sampling. If subjects are a nonrandom sample, the above methods do not apply. For example, if patients are enrolled in a study of mortality by preferentially selecting those with a better prognosis, the mortality estimates that result will not reflect the experience of the typical patient in the general population. In this section we discuss two types of random sampling that are important in epidemiologic studies: simple random sampling and stratified random sampling. For illustrative purposes we consider a prevalence study (survey) designed to estimate the proportion of the population who have a given disease at a particular time point. This proportion is referred to as the (point) prevalence rate (of the disease), and an individual who has the disease is referred to as a case (of the disease). The binomial distribution can be used to analyze data from a prevalence study. Accordingly, we denote the prevalence rate by π.

1.3.1 Simple Random Sampling

Simple random sampling, the least complicated type of random sampling, is widely used in epidemiologic studies. The cardinal feature of a simple random sample is that all individuals in the population have an equal probability of being selected. For example, a simple random sample would be obtained by randomly selecting names from a census list, making sure that each individual has the same chance of being chosen. Suppose that r individuals are sampled for the prevalence study and that

a of them are cases. The simple random sample estimate of the prevalence rate is $\hat{\pi}_{\text{srs}} = a/r$, which has the variance $\text{var}(\hat{\pi}_{\text{srs}}) = \pi(1 - \pi)/r$.

1.3.2 Stratified Random Sampling

Suppose that the prevalence rate increases with age. Simple random sampling ensures that, on average, the sample will have the same age distribution as the population. However, in a given prevalence study it is possible for a particular age group to be underrepresented or even absent from a simple random sample. Stratified random sampling avoids this difficulty by permitting the investigator to specify the proportion of the total sample that will come from each age group (stratum). For stratified random sampling to be possible it is necessary to know in advance the number of individuals in the population in each stratum. For example, stratification by age could be based on a census list, provided information on age is available. Once the strata have been created, a simple random sample is drawn from each stratum, resulting in a stratified random sample.

Suppose there are n strata. For the ith stratum we make the following definitions: N_i is the number of individuals in the population, π_i is the prevalence rate, r_i is the number of subjects in the simple random sample, and a_i is the number of cases among the r_i subjects ($i = 1, 2, \ldots, n$). Let $N = \sum_{i=1}^{n} N_i$, $a = \sum_{i=1}^{n} a_i$ and

$$r = \sum_{i=1}^{n} r_i. \tag{1.25}$$

For a stratified random sample, along with the N_i, the r_i must also be known prior to data collection. We return shortly to the issue of how to determine the r_i, given an overall sample size of r. For the moment we require only that the r_i satisfy the constraint (1.25). Since a simple random sample is chosen in each stratum, an estimate of π_i is $\hat{\pi}_i = a_i/r_i$, which has the variance $\text{var}(\hat{\pi}_i) = \pi_i(1 - \pi_i)/r_i$. The stratified random sample estimate of the prevalence rate is

$$\hat{\pi}_{\text{str}} = \sum_{i=1}^{n} \left(\frac{N_i}{N}\right)\hat{\pi}_i \tag{1.26}$$

which is seen to be a weighted average of the $\hat{\pi}_i$. Since $E(\hat{\pi}_i) = \pi_i$, it follows from (1.7) that

$$E(\hat{\pi}_{\text{str}}) = \sum_{i=1}^{n} \left(\frac{N_i}{N}\right)\pi_i = \pi$$

and so $\hat{\pi}_{\text{str}}$ is unbiased. Applying (1.8) to (1.26) gives

$$\text{var}(\hat{\pi}_{\text{str}}) = \sum_{i=1}^{n} \left(\frac{N_i}{N}\right)^2 \left[\frac{\pi_i(1 - \pi_i)}{r_i}\right]. \tag{1.27}$$

We now consider the issue of determining the r_i. There are a number of approaches that can be followed, each of which places particular conditions on the r_i. For example, according to the method of optimal allocation, the r_i are chosen so that $\text{var}(\hat{\pi}_{\text{str}})$ is minimized. It can be shown that, based on this criterion,

$$r_i = \left(\frac{N_i \sqrt{\pi_i(1 - \pi_i)}}{\sum_{i=1}^{n} N_i \sqrt{\pi_i(1 - \pi_i)}} \right) r. \tag{1.28}$$

As can be seen from (1.28), in order to determine the r_i it is necessary to know, or at least have reasonable estimates of, the π_i. Since this is one of the purposes of the prevalence study, it is therefore necessary to rely on findings from earlier prevalence studies or, when such studies are not available, have access to informed opinion.

Stratified random sampling should be considered only if it is known, or at least strongly suspected, that the π_i vary across strata. Suppose that, unknown to the investigator, the π_i are all equal, so that $\pi_i = \pi$ for all i. It follows from (1.28) that $r_i = (N_i/N)r$ and hence, from (1.27), that $\text{var}(\hat{\pi}_{\text{str}}) = \pi(1 - \pi)/r$. This means that the variance obtained by optimal allocation, which is the smallest variance possible under stratified random sampling, equals the variance that would have been obtained from simple random sampling. Consequently, when there is a possibility that the π_i are all equal, stratified random sampling should be avoided since the effort involved in stratification will not be rewarded by a reduction in variance.

Simple random sampling and stratified random sampling are conceptually and computationally straightforward. There are more complex methods of random sampling such as multistage sampling and cluster sampling. Furthermore, the various methods can be combined to produce even more elaborate sampling strategies. It will come as no surprise that as the method of sampling becomes more complicated so does the corresponding data analysis. In practice, most epidemiologic studies use relatively straightforward sampling procedures. Aside from prevalence studies, which may require complex sampling, the typical epidemiologic study is usually based on simple random sampling or perhaps stratified random sampling, but generally nothing more elaborate.

Most of the procedures in standard statistical packages, such as SAS (1987) and SPSS (1993), assume that data have been collected using simple random sampling or stratified random sampling. For more complicated sampling designs it is necessary to use a statistical package such as SUDAAN (Shah et al., 1996), which is specifically designed to analyze complex survey data. STATA (1999) is a statistical package that has capabilities similar to SAS and SPSS, but with the added feature of being able to analyze data collected using complex sampling. For the remainder of the book it will be assumed that data have been collected using simple random sampling unless stated otherwise.

CHAPTER 2

Measurement Issues in Epidemiology

Unlike laboratory research where experimental conditions can usually be carefully controlled, epidemiologic studies must often contend with circumstances over which the investigator may have little influence. This reality has important implications for the manner in which epidemiologic data are collected, analyzed, and interpreted. This chapter provides an overview of some of the measurement issues that are important in epidemiologic research, an appreciation of which provides a useful perspective on the statistical methods to be discussed in later chapters. There are many references that can be consulted for additional material on measurement issues and study design in epidemiology; in particular, the reader is referred to Rothman and Greenland (1998).

2.1 SYSTEMATIC AND RANDOM ERROR

Virtually any study involving data collection is subject to error, and epidemiologic studies are no exception. The error that occurs in epidemiologic studies is broadly of two types: random and systematic.

Random Error
The defining characteristic of random error is that it is due to "chance" and, as such, is unpredictable. Suppose that a study is conducted on two occasions using identical methods. It is possible for the first replicate to lead to a correct inference about the study hypothesis, and for the second replicate to result in an incorrect inference as a result of random error. For example, consider a study that involves tossing a coin 100 times where the aim is to test the hypothesis that the coin is "fair"—that is, has an equal chance of landing heads or tails. Suppose that unknown to the investigator the coin is indeed fair. In the first replicate, imagine that there are 50 heads and 50 tails, leading to the correct inference that the coin is fair. Now suppose that in the second replicate there are 99 heads and 1 tail, leading to the incorrect inference that the coin is unfair. The erroneous conclusion in the second replicate is due to random error, and this occurs despite the fact that precisely the same study methods were used both times.

31

Since the coin is fair, based on the binomial model, the probability of observing the data in the second replicate is $\binom{100}{99}(1/2)^{99}(1/2)^1 = 7.89 \times 10^{-29}$, an exceedingly small number. Although unlikely, this outcome is possible. The only way to completely eliminate random error in the study is to toss the coin an "infinite" number of times, an obvious impossibility. However, as intuition suggests, tossing the coin a "large" number of times can reduce the probability of random error. Epidemiologic studies are generally based on measurements performed on subjects randomly sampled from a "population." A population can be any well-defined group of individuals, such as the residents of a city, individuals living in the catchment area of a hospital, workers in a manufacturing plant, or patients attending a medical clinic, just to give a few examples. The process of random sampling from a population introduces random error. In theory, such random error could be eliminated by recruiting the entire population into the study. Usually populations of interest are so large or otherwise inaccessible as to make this option a practical impossibility. As a result, random error must be addressed in virtually all epidemiologic studies. Much of the remainder of this book is devoted to methods for analyzing data in the presence of random error.

An epidemiologic study is usually designed with a particular hypothesis in mind, typically having to do with a purported association between a predictor variable and an outcome of interest. For example, in an occupational epidemiologic study it might be hypothesized that exposure to a certain chemical increases the risk of cancer. The classical approach to examining the truth of such a hypothesis is to define the corresponding "null" hypothesis that no association is present. The null hypothesis is then tested using inferential statistical methods and either rejected or not. In the present example, the null hypothesis would be that the chemical is not associated with the risk of cancer. Rejecting the null hypothesis would lead to the inference that the chemical is in fact associated with this risk.

The null hypothesis is either true or not, but due to random error the truth of the matter can never be known with certainty based on statistical methods. The inference drawn from a hypothesis test can be wrong in two ways. If the null hypothesis is rejected when it is true, a type I error has occurred; and if the null hypothesis is not rejected when it is false, there has been a type II error. The probability of a type I error will be denoted by α, and the probability of a type II error will be denoted by β. In a given application the values of α and β are determined by the nature of the study and, as such, are under the control of the investigator. It is desirable to keep α and β to a minimum, but it is not possible to reduce either of them to 0. For a given sample size there is a tradeoff between type I error and type II error, in the sense that α can be reduced by increasing β, and conversely (Chapter 14).

Systematic Error
The cardinal feature of systematic error, and the characteristic that distinguishes it from random error, is that it is reproducible. For the most part, systematic error occurs as a result of problems having to do with study methodology. If these problems are left unattended and if identical methods are used to replicate the study, the same systematic errors will occur. As can be imagined, there are an almost endless number

of possibilities for systematic error in an epidemiologic study. For example, the study sample could be chosen improperly, the questionnaire could be invalid, the statistical analysis could be faulty, and so on. Certain epidemiologic designs are, by their very nature, more prone to systematic error than others. Case-control studies, discussed briefly in Chapter 11, are usually considered to be particularly problematic in this regard due to the reliance on retrospective data collection. With careful attention to study methods it is possible minimize systematic error, at least those sources of systematic error that come to the attention of the investigator. In this chapter we focus on two types of systematic error which are particularly important in epidemiologic studies, namely, confounding and misclassification.

Ordinarily the findings from an epidemiologic study are presented in terms of a parameter estimate based on a probability model. In the coin-tossing example the focus would typically be on the parameter π from a binomial distribution, where π is the (unknown) probability of the coin landing heads. When systematic error is present, the parameter estimate will usually be biased in the sense of Section 1.2, and so it may either over- or underestimate the true parameter value. Epidemiology has borrowed the term "bias" from the statistical literature, using it as a synonym for systematic error. So when an epidemiologic study is subject to systematic error we say that the parameter estimate is biased or, rather more loosely, that the study is biased.

2.2 MEASURES OF EFFECT

In this book we will mostly be concerned with analyzing data from studies in which groups of individuals are compared, the aim being to determine whether a given exposure is related to the occurrence of a particular disease. Here "exposure" and "disease" are used in a generic sense. The term exposure can refer to any characteristic that we wish to investigate as potentially having a health-related impact. Examples are: contact with a toxic substance, treatment with an innovative medical therapy, having a family history of illness, engaging in a certain lifestyle practice, and belonging to a particular sociodemographic group. Likewise, the term disease can refer to the occurrence of any health-related outcome we wish to consider. Examples are: onset of illness, recovery following surgery, and death from a specific cause. In the epidemiologic literature, "risk" is sometimes used synonymously with probability, a convention that tends to equate the term with the probability parameter of a binomial model. Here we use the term risk more generally to connote the propensity toward a particular outcome, whether or not that tendency is modeled using the binomial distribution.

2.2.1 Closed Cohort Study

There are many types of cohort studies, but the common theme is that a group of individuals, collectively termed the cohort, is followed over time and monitored for the occurrence of an outcome of interest. For example, a cohort of breast cancer patients might be followed for 5 years, with death from this disease as the study endpoint. In this example, the cohort is a single sample which is not being contrasted with any comparison group. As another example, suppose that a group of workers in a chemical fabrication plant is followed for 20 years to determine if their risk of leukemia is greater than that in the general population. In this case, the workers are being compared to the population at large.

A reality of cohort studies is that subjects may cease to be under observation prior to either developing the disease or reaching the end of the planned period of follow-up. When this occurs we say that the subject has become "unobservable." This can occur for a variety of reasons, such as the subject being lost to follow-up by the investigator, the subject deciding to withdraw from the study, or the investigator eliminating the subject from further observation due to the development of an inter-current condition which conflicts with the aims of the study. Whatever the reasons, these occurrences pose a methodological challenge to the conduct of a cohort study. For the remainder of this chapter we restrict attention to the least complicated type of cohort study, namely, one in which all subjects have the same maximum observation time and all subjects not developing the disease remain observable throughout the study. A study with this design will be referred to as a closed cohort study.

In a closed cohort study, subjects either develop the disease or not, and all those not developing it necessarily have the same length of follow-up, namely, the maximum observation time. For example, suppose that a cohort of 1000 otherwise healthy middle-aged males are monitored routinely for 5 years to determine which of them develops hypertension (high blood pressure). In order for the cohort to be closed, it is necessary that all those who do not develop hypertension remain under observation for the full 5 years. Once a subject develops hypertension, follow-up for that indi-vidual ceases. In a closed cohort study involving a single sample, the parameter of interest is usually the binomial probability of developing disease. In some of the epi-demiologic literature on closed cohort studies, the probability of disease is referred to as the incidence proportion or the cumulative incidence, but we will avoid this terminology. In most cohort studies, at least a few subjects become unobservable for reasons such as those given above, and so closed cohort studies are rarely encoun-tered in practice. However, the closed cohort design offers a convenient vehicle for introducing a number of ideas that are also important in the context of cohort studies conducted under less restrictive conditions.

Consider a closed cohort study in which the exposure is dichotomous and suppose that at the start of follow-up there are r_1 subjects in the exposed cohort ($E = 1$) and r_2 subjects in the unexposed cohort ($E = 2$). At the end of the period of follow-up each subject will have either developed the disease ($D = 1$) or not ($D = 2$). Some-one who develops the disease will be referred to as a case, otherwise as a noncase. The development of disease in the exposed and unexposed cohorts will be modeled using binomial random variables A_1 and A_2 with parameters (π_1, r_1) and (π_2, r_2), respectively. As discussed in Section 1.2.1, we assume that subjects behave inde-pendently with respect to developing the disease. Tables 2.1(a) and 2.1(b) show the observed counts and expected values for the study, respectively. We do not refer to the entries in Table 2.1(b) as expected counts, for reasons that will be explained in Section 4.1.

TABLE 2.1(a)   Observed Counts:
Closed Cohort Study

D	*Exposed* E	*unexposed*
	1	2
Disease 1	a_1	a_2
No Disease 2	b_1	b_2
	r_1	r_2

TABLE 2.1(b)   Expected Values:
Closed Cohort Study

D		E
	1	2
1	$\pi_1 r_1$	$\pi_2 r_2$
2	$(1 - \pi_1)r_1$	$(1 - \pi_2)r_2$
	r_1	r_2

2.2.2   Risk Difference, Risk Ratio, and Odds Ratio

When an exposure is related to the risk of disease we say that the exposure has an "effect." We now define several measures of effect which quantify the magnitude of the association between exposure and disease in a closed cohort study.

The risk difference, defined by $RD = \pi_1 - \pi_2$, is an intuitively appealing measure of effect. Since $\pi_1 = \pi_2 + RD$, the risk difference measures change on an additive scale. If $RD > 0$, exposure is associated with an increase in the probability of disease; if $RD < 0$, exposure is associated with a decrease in the probability of disease; and if $RD = 0$, exposure is not associated with the disease.

The risk ratio, defined by $RR = \pi_1/\pi_2$, is another intuitively appealing measure of effect. In some of the epidemiologic literature the risk ratio is referred to as the relative risk, but this terminology will not be used in this book. Since $\pi_1 = RR\pi_2$, the risk ratio measures change on a multiplicative scale. Note that RR is undefined when $\pi_2 = 0$, a situation that is theoretically possible but of little interest from an epidemiologic point of view. If $RR > 1$, exposure is associated with an increase in the probability of disease; if $RR < 1$, exposure is associated with a decrease in the probability of disease; and if $RR = 1$, exposure is not associated with the disease. A measure of effect that has both additive and multiplicative features is $(\pi_1 - \pi_2)/\pi_2 = RR - 1$, which is referred to as the excess relative risk (Preston, 2000). A related measure of effect is $(\pi_1 - \pi_2)/\pi_1 = 1 - (1/RR)$, which is called the attributable risk percent (Cole and MacMahon, 1971). These measures of effect are closely related to the risk ratio and will not be considered further.

For a given probability $\pi \neq 1$, the odds ω is defined to be

$$\omega = \frac{\pi}{1 - \pi}.$$

Solving for π gives

$$\pi = \frac{\omega}{1 + \omega}$$

and so probability and odds are equivalent ways of expressing the same information. Although appearing to be somewhat out of place in the context of health-related studies, odds terminology is well established in the setting of games of chance. As an example, the probability of picking an ace at random from a deck of cards is $\pi = 4/52 = 1/13$. The odds is therefore $\omega = (4/52)/(48/52) = 1/12$, which can be written as 1:12 and read as "1 to 12." Despite their nominal equivalence, probability and odds differ in a major respect: π must lie in the interval between 0 and 1, whereas ω can be any nonnegative number. An important characteristic of the odds is that it satisfies a reciprocal property: If $\omega = \pi/(1 - \pi)$ is the odds of a given outcome, then $(1 - \pi)/[1 - (1 - \pi)] = 1/\omega$ is the odds of the opposite outcome. For example, the odds of not picking an ace is $(48/52)/(4/52) = 12$, that is, "12 to 1."

Returning to the discussion of closed cohort studies, let $\omega_1 = \pi_1/(1 - \pi_1)$ and $\omega_2 = \pi_2/(1 - \pi_2)$ be the odds of disease for the exposed and unexposed cohorts, respectively. The odds ratio is defined to be

$$OR = \frac{\omega_1}{\omega_2} = \frac{\pi_1(1 - \pi_2)}{\pi_2(1 - \pi_1)}. \tag{2.1}$$

Since $\omega_1 = OR\omega_2$, the odds ratio is similar to the risk ratio in that change is measured on a multiplicative scale. However, with the odds ratio the scale is calibrated in terms of odds rather than in terms of probability. If $OR > 1$, exposure is associated with an increase in the odds of disease; if $OR < 1$, exposure is associated with a decrease in the odds of disease; and if $OR = 1$, exposure is not associated with the disease. It is easily demonstrated that $\omega_1 > \omega_2, \omega_1 < \omega_2, \omega_1 = \omega_2$ are equivalent to $\pi_1 > \pi_2, \pi_1 < \pi_2, \pi_1 = \pi_2$, respectively, and so statements made in terms of odds are readily translated into corresponding statements about probabilities, and conversely.

When the disease is "rare," $1 - \pi_1$ and $1 - \pi_2$ are close to 1 and so, from (2.1), OR is approximately equal to RR. In some of the older epidemiologic literature the odds ratio was viewed as little more than an approximation to the risk ratio. More recently, some authors have argued against using the odds ratio as a measure of effect in clinical studies on the grounds that it cannot substitute for the clinically more meaningful risk difference and risk ratio (Sinclair and Bracken, 1994). In this book we regard the odds ratio as a measure of effect worthy of consideration in its own right and not merely as a less desirable alternative to the risk ratio. As will be seen shortly, the odds ratio has a number of attractive measurement properties that are not shared by either the risk difference or the risk ratio.

2.2.3   Choosing a Measure of Effect

We now consider which, if any, of the risk difference, risk ratio, or odds ratio is the most desirable measure of effect for closed cohort studies. One of the most contentious issues revolves around the utility of RD and RR as measures of etiology (causation) on the one hand, and measures of population (public health) impact on the other. This is best illustrated with some examples. First, suppose that the probability of developing the disease is small, whether or not there is exposure; for example, $\pi_1 = .0003$ and $\pi_2 = .0001$. Then $RD = .0002$, and so exposure is associated with a small increase in the probability of disease. Unless a large segment of the population has been exposed, the impact of the disease will be small and so, from a public health perspective, this particular exposure is not of major concern. On the other hand, $RR = 3$ and according to usual epidemiologic practice this is large enough to warrant further investigation of the exposure as a possible cause of the disease. Now suppose that $\pi_1 = .06$ and $\pi_2 = .05$, so that $RD = .01$ and $RR = 1.2$. In this example, the risk difference will be of public health importance unless exposure is especially infrequent, while the risk ratio is of relatively little interest from an etiologic point of view.

The above arguments have been expressed in terms of the risk difference and risk ratio, but are in essence a debate over the merits of measuring effect on an additive as opposed to a multiplicative scale. This issue has generated a protracted debate in the epidemiologic literature, with some authors preferring additive models (Rothman, 1974; Berry, 1980) and others preferring the multiplicative approach (Walter and Holford, 1978). Statistical methods have been proposed for deciding whether an additive or multiplicative model provides a better fit to study data. One approach is to compare likelihoods based on best-fitting additive and multiplicative models (Berry, 1980; Gardner and Munford, 1980; Walker and Rothman, 1982). An alternative method is to fit a general model that has additive and multiplicative models as special cases and then decide whether one or the other, or perhaps some intermediate model, fits the data best (Thomas, 1981; Guerrero and Johnson, 1982; Breslow and Storer, 1985; Moolgavkar and Venzon, 1987).

Consider a closed cohort study where $\pi_1 = .6$ and $\pi_2 = .2$, so that $\omega_1 = 1.5$ and $\omega_2 = .25$. Based on these parameters we have the following interpretations: Exposure increases the probability of disease by an increment $RD = .4$; exposure increases the probability of disease by a factor $RR = 3$; and exposure increases the odds of disease by a factor $OR = 6$. This simple example illustrates that the risk difference, risk ratio, and odds ratio are three very different ways of measuring the effect of exposure on the risk of disease. It also illustrates that the risk difference and risk ratio have a straightforward and intuitive interpretation, a feature that is not shared by the odds ratio. Even if $\omega_1 = 1.5$ and $\omega_2 = .25$ are rewritten as "15 to 10" and "1 to 4," these quantities remain less intuitive than $\pi_1 = .6$ and $\pi_2 = .2$. It seems that, from the perspective of ease of interpretation, the risk difference and risk ratio have a distinct advantage over the odds ratio.

Suppose we redefine exposure status so that subjects who were exposed according to the original definition are relabeled as unexposed, and conversely. Denoting the resulting measures of effect with a prime $'$, we have $RD' = \pi_2 - \pi_1$, $RR' = \pi_2/\pi_1$, and $OR' = [\pi_2(1 - \pi_1)]/[\pi_1(1 - \pi_2)]$. It follows that $RD' = -RD$, $RR' = 1/RR$, and $OR' = 1/OR$, and so each of the measures of effect is transformed into a reciprocal quantity on either the additive or multiplicative scale. Now suppose that we redefine disease status so that subjects who were cases according to the original definition are relabeled as noncases, and conversely. Denoting the resulting measures of effect with a double prime $''$, we have $RD'' = (1 - \pi_1) - (1 - \pi_2)$, $RR'' = (1 - \pi_1)/(1 - \pi_2)$, and $OR'' = [(1 - \pi_1)\pi_2]/[(1 - \pi_2)\pi_1]$. It follows that $RD'' = -RD$ and $OR'' = 1/OR$, but $RR'' \neq 1/RR$. The failure of the risk ratio to demonstrate a reciprocal property when disease status is redefined is a distinct shortcoming of this measure of effect. For example, in a randomized controlled trial let "exposure" be active treatment (as compared to placebo) and let "disease" be death from a given cause. With $\pi_1 = .01$ and $\pi_2 = .02$, $RR = .01/.02 = .5$ and so treatment leads to an impressive decrease in the probability of dying. Looked at another way, $RR'' = .99/.98 = 1.01$ and so treatment results in only a modest improvement in the probability of surviving.

Since $0 \leq \pi_1 \leq 1$, there are constraints placed on the values of RD and RR. Specifically, for a given value of π_2, RD and RR must satisfy the inequalities $0 \leq \pi_2 + RD \leq 1$ and $0 \leq RR\pi_2 \leq 1$; or equivalently, $-\pi_2 \leq RD \leq (1 - \pi_2)$ and $0 \leq RR \leq (1/\pi_2)$. In the case of a single 2×2 table, such as being considered here, these constraints do not pose a problem. However, when several tables are being analyzed and an overall measure of effect is being estimated, these constraints have greater implications. First, there is the added complexity of finding an overall measure that satisfies the constraints in each table. Second, and more importantly, the constraint imposed by one of the tables may severely limit the range of possible values for the measure of effect in other tables. The odds ratio has the attractive property of not being subject to this problem. Solving (2.1) for π_1 gives

$$\pi_1 = \frac{OR\pi_2}{OR\pi_2 + (1 - \pi_2)}. \tag{2.2}$$

Since $0 \leq \pi_2 \leq 1$ and $OR \geq 0$, it follows that $0 \leq \pi_1 \leq 1$ for any values of OR and π_2 for which the denominator of (2.2) is nonzero. Figures 2.1(a) and 2.1(b), which

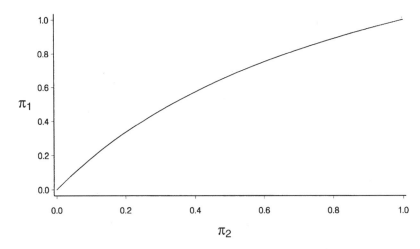

FIGURE 2.1(a) π_1 as a function of π_2, with $OR = 2$

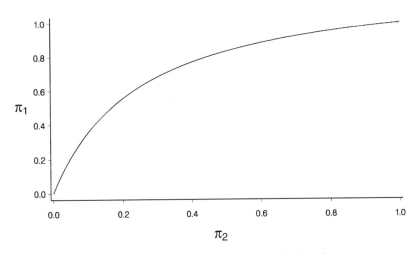

FIGURE 2.1(b) π_1 as a function of π_2, with $OR = 5$

are based on (2.2), show graphs of π_1 as a function of π_2 for $OR = 2$ and $OR = 5$. As can be seen, the curves are concave downward in shape. By contrast, for given values of RD and RR, the graphs of $\pi_1 = \pi_2 + RD$ and $\pi_1 = RR\pi_2$ (not shown) are both linear; the former has a slope of 1 and an intercept of RD, while the latter has a slope of RR and an intercept of 0.

When choosing a measure of effect for a closed cohort study, it is useful to consider the properties discussed above—that is, whether the measure of effect is additive or multiplicative, intuitively appealing, exhibits reciprocal properties, and imposes restrictions on the range of parameter values. However, a more fundamental consideration is whether the measure of effect is consistent with the underlying mechanism of the disease process. For example, if it is known that a set of exposures exert their influence in an additive rather than a multiplicative fashion, it would be appropriate to select the risk difference as a measure of effect in preference to the risk ratio or odds ratio. Unfortunately, in most applications there is insufficient substantive knowledge to help decide such intricate questions. It might be hoped that epidemiologic data could be used to determine whether a set of exposures is operating additively, multiplicatively, or in some other manner. However, the behavior of risk factors at the population level, which is the arena in which epidemiologic research operates, may not accurately reflect the underlying disease process (Siemiatycki and Thomas, 1981; Thompson, 1991).

Walter (2000) has demonstrated that models based on the risk difference, risk ratio, and odds ratio tend to produce similar findings, a phenomenon that will be illustrated later in this book. Currently, in most epidemiologic studies, some form of multiplicative model is used. Perhaps the main reason for this emphasis is a practical consideration: In most epidemiologic research the outcome variable is categorical (discrete) and the majority of statistical methods, along with most of the statistical packages available to analyze such data, are based on the multiplicative approach

(Thomas, 2000). In particular, the majority of regression techniques that are widely used in epidemiology, such as logistic regression and Cox regression, are multiplicative in nature. For this reason the focus of this book will be on techniques that are defined in multiplicative terms.

2.3 CONFOUNDING

One of the defining features of epidemiology as a field of inquiry is the concern (some might say preoccupation) over a particular type of systematic error known as confounding. In many epidemiologic studies the aim is to isolate the causal effect of a particular exposure on the development of a given disease. When there are factors that have the potential to result in a spurious increase or decrease in the observed effect, the possibility of confounding must be considered. Early definitions of confounding were based on the concept of collapsibility, an approach which has considerable intuitive appeal. The current and widely accepted definition of confounding rests on counterfactual arguments that, by contrast, are rather abstract. As will be shown, the collapsibility and counterfactual definitions of confounding have certain features in common. We will develop some preliminary insights into confounding using the collapsibility approach and then proceed to a definition of confounding based on counterfactual arguments (Greenland et al., 1999).

2.3.1 Counterfactuals, Causality, and Risk Factors

The concept of causality has an important place in discussions of confounding (Pearl, 2000, Chapter 6). The idea of what it means for something to "cause" something else is a topic that has engaged philosophers for centuries. Holland (1986) and Greenland et al. (1999) review some of the issues related to causality in the context of inferential statistics. A helpful way of thinking about causality is based on the concept of counterfactuals. Consider the statement "smoking causes lung cancer," which could be given the literal interpretation that everyone who smokes develops this type of tumor. As is well known, there are many people who smoke but do not develop lung cancer and, conversely, there are people who develop lung cancer and yet have never smoked. So there is nothing inevitable about the association between smoking and lung cancer, in either direction. One way of expressing a belief that smoking is causally related to lung cancer is as follows: We imagine that corresponding to an individual who smokes there is an imaginary individual who is identical in all respects, except for being a nonsmoker. We then assert that the risk of lung cancer in the person who smokes is greater than the risk in the imaginary nonsmoker. This type of argument is termed counterfactual (counter to fact) because we are comparing an individual who is a known smoker with the "same" individual minus the history of smoking.

Epidemiologists are usually uncomfortable making claims about causality, generally preferring to discuss whether an exposure and disease are associated or related. The term "risk factor" imparts a sense of causality and at the same time is appropriately conservative for an epidemiologic discussion. So instead of referring to smoking as a cause of lung cancer, it would be usual in an epidemiologic context to say that smoking is a risk factor for this disease. The term risk factor is also used for any condition that forms part of a causal chain connecting an exposure of interest to a given disease. For example, a diet deficient in calcium can lead to osteoporosis, and this can in turn result in hip fractures. We consider both calcium deficiency and osteoporosis to be risk factors for hip fractures. Sometimes the definition of what constitutes a risk factor is broadened to include characteristics that are closely associated with a causal agent but not necessarily causal themselves. In this sense, carrying a lighter can be considered to be a risk factor for lung cancer. We will restrict our use of the term risk factor to those characteristics that have a meaningful etiologic connection with the disease in question.

2.3.2   The Concept of Confounding

The type of problem posed by confounding is best illustrated by an example. Imagine a closed cohort study investigating alcohol consumption as a possible risk factor for lung cancer. The exposed cohort consists of a group of individuals who consume alcohol (drinkers) and the unexposed cohort is a group who do not (nondrinkers). Setting aside the obvious logistical difficulties involved in conducting such a study, suppose that at the end of the period of follow-up the proportion of drinkers who develop lung cancer is greater than the corresponding proportion of nondrinkers. This might be regarded as evidence that alcohol is a risk factor for lung cancer, but before drawing this conclusion we must consider the well-known association between drinking and smoking. Specifically, since smoking is a known cause of lung cancer, and smoking and drinking are lifestyle habits that are often associated, there is the possibility that drinking may only appear to be a risk factor for lung cancer because of the intermediate role played by smoking.

These ideas are captured visually in Figure 2.2(a), which is referred to as a causal diagram. In the diagram we use E, D and F to denote drinking (exposure), lung cancer (disease) and smoking (intermediate factor), respectively. The unidirectional solid arrow between smoking and lung cancer indicates a known causal relationship, the bidirectional solid arrow between drinking and smoking stands for a known non-causal association, and the unidirectional dashed arrow between drinking and lung

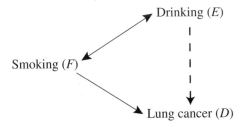

FIGURE 2.2(a)   Causal diagram for drinking as a risk factor for lung cancer

cancer represents an association that results from smoking acting as an intermediate factor.

A quantitative approach to examining whether smoking results in a spurious association between drinking and lung cancer involves stratifying (dividing) the cohort into smokers and nonsmokers, and then reanalyzing the data within strata. Stratification ensures that the subjects in each stratum are identical with respect to smoking status. So if the association between drinking and lung cancer is mediated through smoking, this association will vanish within each of the strata. In a sense, stratifying by smoking status breaks the connection between drinking and lung cancer in each stratum by blocking the route through smoking. In fact, drinking is not a risk factor for lung cancer and so, random error aside, within each smoking stratum the proportion of drinkers who develop lung cancer will be the same as the proportion of nondrinkers. So after accounting (controlling, adjusting) for smoking we conclude that drinking is not a risk factor for this disease. In the crude (unstratified) analysis, drinking appears to be a risk factor for lung cancer due to what we will later refer to as confounding by smoking. The essential feature of smoking which enables it to produce confounding is that it is associated with both drinking and lung cancer.

Now imagine a closed cohort study investigating calcium deficiency (E) as a risk factor for hip fractures (D). We have already noted that calcium deficiency leads to osteoporosis (F) and that both calcium deficiency and osteoporosis cause hip fractures. These associations are depicted in Figure 2.2(b). By analogy with the previous example it is tempting to regard osteoporosis as a source of confounding. However, the situation is different here in that osteoporosis is a step in the causal pathway between calcium deficiency and hip fractures. Consequently, osteoporosis does not induce a spurious risk relationship between calcium deficiency and hip fractures but rather helps to explain a real causal connection. For this reason we do not consider osteoporosis to be a source of confounding.

As with any mathematical construct, the manner in which confounding is operationalized for the purposes of data analysis is a matter of definition; and, as we will see, different definitions are possible. The process of arriving at a definition of confounding is an inductive one, with concrete examples examined for essential features which can then be given a more general formulation. The preceding hypothetical studies illustrate some of the key attributes that should be included as part of a definition of confounding, and these requirements will be adhered to as we explore the

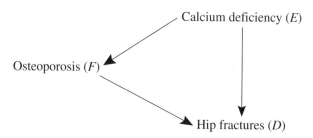

FIGURE 2.2(b)   Causal diagram for calcium deficiency as a risk factor for hip fractures

concept further. Specifically, for a variable F to be a source of confounding (confounder) we require that F satisfy the following conditions: F must be a risk factor for the disease, and F must be associated with the exposure. To these two conditions we add the requirement that F must not be part of the causal pathway between the exposure and the disease.

2.3.3 Some Hypothetical Examples of Closed Cohort Studies

As illustrated in the preceding section, stratification plays an important role in the analysis of epidemiologic data, especially in connection with confounding. In this section we examine a series of hypothetical closed cohort studies in order to develop a sense of how the risk difference, risk ratio, and odds ratio behave in crude and stratified 2×2 tables. This will motivate an analysis that will be useful in the discussion of confounding. In an actual cohort study, subjects are randomly sampled from a population, a process that introduces random error. For the remainder of this chapter it is convenient to avoid issues related to random error by assuming that the entire population has been recruited into the cohort and that, for each individual, the outcome with respect to developing the disease is predetermined (although unknown to the investigator). In this way we replace the earlier probabilistic (stochastic) approach with one that is deterministic. Strictly speaking, we should now refer to π_1 and π_2 in Table 2.1(b) as proportions rather than probabilities because there is no longer a stochastic context. However, for simplicity of exposition we will retain the earlier terminology. In what follows, we continue to make reference to the population, but will now equate it with the cohort at the start of follow-up.

Tables 2.2(a)–2.2(e) give examples of closed cohort studies in which there are three variables: exposure (E), disease (D), and a stratifying variable, (F). We use $E = 1$, $D = 1$, and $F = 1$ to denote the presence of an attribute and use $E = 2$, $D = 2$, and $F = 2$ to indicate its absence. Here, as elsewhere in the book, a dot • denotes summation over all values of an index. We refer to the tables with the

TABLE 2.2(a) Hypothetical Closed Cohort Study: F Is Not a Risk Factor for the Disease and F Is Not Associated with Exposure

	$F = 1$			$F = 2$			$F = \bullet$	
	E			E			E	
D	1	2		1	2		1	2
1	70	40		140	80		210	120
2	30	60		60	120		90	180
	100	100		200	200		300	300
RD	.30			.30			.30	
RR	1.8			1.8			1.8	
OR	3.5			3.5			3.5	

TABLE 2.2(b)   Hypothetical Closed Cohort Study: F Is Not a Risk Factor for the Disease and F Is Not Associated with Exposure

	$F = 1$		$F = 2$		$F = \bullet$	
D	E		E		E	
	1	2	1	2	1	2
1	70	40	160	80	230	120
2	30	60	40	120	70	180
	100	100	200	200	300	300
RD	.30		.40		.37	
RR	1.8		2.0		1.9	
OR	3.5		6.0		4.9	

TABLE 2.2(c)   Hypothetical Closed Cohort Study: F Is Not a Risk Factor for the Disease and F Is Associated with Exposure

	$F = 1$		$F = 2$		$F = \bullet$	
D	E		E		E	
	1	2	1	2	1	2
1	70	80	160	40	230	120
2	30	120	40	60	70	180
	100	200	200	100	300	300
RD	.30		.40		.37	
RR	1.8		2.0		1.9	
OR	3.5		6.0		4.9	

TABLE 2.2(d)   Hypothetical Closed Cohort Study: F Is a Risk Factor for the Disease and F Is Not Associated with Exposure

	$F = 1$		$F = 2$		$F = \bullet$	
D	E		E		E	
	1	2	1	2	1	2
1	90	60	80	20	170	80
2	10	40	120	180	130	220
	100	100	200	200	300	300
RD	.30		.30		.30	
RR	1.5		4.0		2.1	
OR	6.0		6.0		3.6	

TABLE 2.2(e)   Hypothetical Closed Cohort Study: F Is a Risk Factor for the Disease and F Is Associated with Exposure

	$F = 1$			$F = 2$		$F = \bullet$	
D	E			E		E	
	1	2		1	2	1	2
1	90	120		30	10	120	130
2	10	80		170	90	180	170
	100	200		200	100	300	300
RD	.30			.05		−.03	
RR	1.5			1.5		.92	
OR	6.0			1.6		.87	

TABLE 2.2(f)   Hypothetical Closed Cohort Study: F Is a Risk Factor for the Disease and F Is Associated with Exposure

	$F = 1$		$F = 2$		$F = 3$		$F = \bullet$	
D	E		E		E		E	
	1	2	1	2	1	2	1	2
1	140	50	120	20	70	90	330	160
2	60	50	180	180	30	210	270	440
	200	100	300	200	100	300	600	600
RD	.20		.30		.40		.28	
RR	1.4		4.0		2.3		2.1	
OR	2.3		6.0		5.4		3.4	

headings "$F = 1$" and "$F = 2$" as the stratum-specific tables and refer to the table with the heading "$F = \bullet$" as the crude table. The crude table is obtained from the stratum-specific tables by collapsing over F—that is, summing over strata on a cell-by-cell basis. The interpretation of the subheadings of the tables will become clear shortly.

In Table 2.2(a), for each measure of effect, the stratum-specific values are equal to each other and to the crude value. In fact, the entries in stratum 2 are, cell by cell, double those in stratum 1. There would seem to be little reason to retain stratification when analyzing the data in Table 2.2(a). In Tables 2.2(b) and 2.2(c), for each measure of effect, the stratum-specific values increase from stratum 1 to stratum 2. Observe that each of the crude measures of effect falls between the corresponding stratum-specific values.

When some or all of the stratum-specific values of a measure of effect differ (across strata of F) we describe this phenomenon using any of the following synonymous expressions: The measure of effect is heterogeneous (across strata of F),

F is an effect modifier (of the measure of effect), and there is an interaction between E and F. These expressions will be used interchangeably in subsequent discussions. Note that the decision as to whether a measure of effect is heterogeneous is based exclusively on the stratum-specific values and does not involve the crude value. For each of the measures of effect under consideration, when E and F are dichotomous, it can be shown that F is an effect modifier of the E–D association if and only if E is an effect modifier of the F–D association. This means that effect modification is a symmetric relationship between E and F. See Section 2.5.6 for a demonstration of this result for the risk ratio. When heterogeneity is absent—that is, when all the stratum-specific values of the measure of effect are equal—we say there is homogeneity. In Table 2.2(d) there is effect modification of the risk ratio, but not the risk difference or odds ratio. This illustrates that the decision as to whether effect modification is present depends on the measure of effect under consideration.

Surprisingly, it is possible for a crude measure of effect to be either greater or less than any of the stratum-specific values, a phenomenon referred to as Simpson's paradox (Simpson, 1951). In Table 2.2(e), all three measures of effect exhibit Simpson's paradox. Here the crude values not only lie outside the range of the stratum-specific values but, in each instance, point to the opposite risk relationship. The odds ratio in Table 2.2(d) also exhibits Simpson's paradox, a finding that is all the more striking given that there is no effect modification.

2.4   COLLAPSIBILITY APPROACH TO CONFOUNDING

2.4.1   Averageability and Strict Collapsibility in Closed Cohort Studies

In this section we carry out an analysis of the risk difference, risk ratio, and odds ratio in stratified 2×2 tables, where the stratifying variable F has $J \geq 2$ categories. The results of this analysis provide insight into the empirical findings in Tables 2.2(a)–2.2(e), in particular the reason for Simpson's paradox. For a given measure of effect, let M denote the crude value, let μ_j denote the jth stratum-specific value ($j = 1, 2, \ldots, J$), and let $\mu_{\min}$ and $\mu_{\max}$ be the minimum and maximum values of the μ_j. We are particularly interested in determining conditions that ensure that $\mu_{\min} \leq M \leq \mu_{\max}$; that is, conditions that guarantee that Simpson's paradox will not be present.

M is said to be averageable (for a given stratification) if it can be expressed as a weighted average of the μ_j, for some set of weights. We now show that M is averageable if and only if $\mu_{\min} \leq M \leq \mu_{\max}$. A corollary is that if M is averageable, then Simpson's paradox is not present. Suppose that M is averageable and let

$$M = \frac{1}{W} \sum_{j=1}^{J} w_j \mu_j$$

where the w_j are weights and $W = \sum_{j=1}^{J} w_j$. Since $\mu_j \geq \mu_{\min}$ for each j, it follows that

$$M \geq \frac{1}{W} \sum_{j=1}^{J} w_j \mu_{\min} = \mu_{\min}.$$

Likewise, $M \leq \mu_{\max}$ and so $\mu_{\min} \leq M \leq \mu_{\max}$. Conversely, suppose that $\mu_{\min} \leq M \leq \mu_{\max}$. We need to consider two cases. If M equals one of the stratum-specific values, say μ_j, let $w_j = 1$ and set the remaining weights equal to 0. Otherwise, M falls strictly between two of the stratum-specific values, say μ_j and μ_{j+1}. In this case let w_j be the (unique) solution to $M = w_j \mu_j + (1 - w_j)\mu_{j+1}$, let $w_{j+1} = 1 - w_j$, and set the remaining weights equal to 0. In either case, M can be expressed as a weighted average of the μ_j; that is, M is averageable.

M is said to be strictly collapsible (for a given stratification) if $M = \mu_j$ for all j, that is, if the crude and stratum-specific values are all equal (Whittemore, 1978; Ducharme and LePage, 1986; Greenland and Mickey, 1988). For example, the odds ratio in Table 2.2(a) is strictly collapsible, but not the odds ratio in Table 2.2(d). Note that if M is strictly collapsible, then, by definition, the μ_j are homogeneous. Denoting the common stratum-specific value by μ, strict collapsibility means that $M = \mu$ and $\mu_j = \mu$ for all j. We now show that M is strictly collapsible if and only if M is averageable and the μ_j are homogeneous. For arbitrary weights w_j, let

$$M = \frac{1}{W} \sum_{j=1}^{J} w_j \mu_j + \Delta$$

where Δ is, by definition, the difference between M and the weighted average of the μ_j. Suppose that M is strictly collapsible. Then $M = \mu$ and $\mu_j = \mu$ for all j. It follows that $\mu = \mu + \Delta$ and so $\Delta = 0$; that is, M is averageable. Conversely, suppose that M is averageable with weights w_j and suppose that $\mu_j = \mu$ for all j. Then

$$M = \frac{1}{W} \sum_{j=1}^{J} w_j \mu = \mu$$

and so M is strictly collapsible.

With respect to ensuring that Simpson's paradox does not occur, the above results suggest that we search for conditions that are sufficient to guarantee that the risk difference, risk ratio, and risk difference are averageable. Such conditions have been reported by Kupper et al. (1981) for the case of two strata. Tables 2.3(a) and

TABLE 2.3(a) Observed Counts in the jth Stratum: Closed Cohort Study

D	E	
	1	2
1	a_{1j}	a_{2j}
2	b_{1j}	b_{2j}
	r_{1j}	r_{2j}

TABLE 2.3(b)   Expected Values in
the jth Stratum: Closed Cohort Study

D		E	
		1	2
1		$\pi_{1j}r_{1j}$	$\pi_{2j}r_{2j}$
2		$(1 - \pi_{1j})r_{1j}$	$(1 - \pi_{2j})r_{2j}$
		r_{1j}	r_{2j}

2.3(b) give the observed counts and "expected values" for the jth stratum of a set of J (2×2) tables. Since we have assumed a deterministic model, the observed count and expected value for each cell are merely different ways of referring to the same quantity, but the alternative notation and terminology are convenient.

By definition, $\pi_{1j} = a_{1j}/r_{1j}$ and $\pi_{2j} = a_{2j}/r_{2j}$, and so

$$a_{1\bullet} = \sum_{j=1}^{J} a_{1j} = \sum_{j=1}^{J} \pi_{1j}r_{1j}$$

$$a_{2\bullet} = \sum_{j=1}^{J} a_{2j} = \sum_{j=1}^{J} \pi_{2j}r_{2j}$$

$$b_{1\bullet} = \sum_{j=1}^{J} b_{1j} = \sum_{j=1}^{J} (1 - \pi_{1j})r_{1j}$$

$$b_{2\bullet} = \sum_{j=1}^{J} b_{2j} = \sum_{j=1}^{J} (1 - \pi_{2j})r_{2j}. \qquad (2.3)$$

Define

$$p_{1j} = \frac{r_{1j}}{r_{1\bullet}}$$

and

$$p_{2j} = \frac{r_{2j}}{r_{2\bullet}}.$$

Then p_{1j} is the proportion of the exposed cohort in the jth stratum at the start of follow-up and p_{2j} is the corresponding proportion for the unexposed cohort. In other words, the p_{1j} and p_{2j} give the distribution of F in the exposed and unexposed cohorts, respectively. By definition,

$$\pi_1 = \frac{a_{1\bullet}}{r_{1\bullet}} = \sum_{j=1}^{J} \left(\frac{r_{1j}}{r_{1\bullet}} \right) \pi_{1j} = \sum_{j=1}^{J} p_{1j}\pi_{1j} \qquad (2.4)$$

and

$$\pi_2 = \frac{a_{2\bullet}}{r_{2\bullet}} = \sum_{j=1}^{J} \left(\frac{r_{2j}}{r_{2\bullet}} \right) \pi_{2j} = \sum_{j=1}^{J} p_{2j} \pi_{2j}. \qquad (2.5)$$

Since $\sum_{j=1}^{J} p_{1j} = 1$ and $\sum_{j=1}^{J} p_{2j} = 1$, π_1 is a weighted average of the π_{1j} and, likewise, π_2 is a weighted average of the π_{2j}.

2.4.2 Risk Difference

The risk difference for the jth stratum is defined to be $\delta_j = \pi_{1j} - \pi_{2j}$. It follows from (2.4) and (2.5) that

$$\begin{aligned} RD &= \sum_{j=1}^{J} p_{1j} \pi_{1j} - \sum_{j=1}^{J} p_{2j} \pi_{2j} \\ &= \sum_{j=1}^{J} p_{1j} (\pi_{2j} + \delta_j) - \sum_{j=1}^{J} p_{2j} \pi_{2j} \\ &= \sum_{j=1}^{J} p_{1j} \delta_j + \sum_{j=1}^{J} (p_{1j} \pi_{2j} - p_{2j} \pi_{2j}). \end{aligned} \qquad (2.6)$$

If

$$\sum_{j=1}^{J} p_{1j} \pi_{2j} = \sum_{j=1}^{J} p_{2j} \pi_{2j} \qquad (2.7)$$

then

$$RD = \sum_{j=1}^{J} p_{1j} \delta_j. \qquad (2.8)$$

In this case, RD is averageable with weights p_{1j}. Each of the following conditions is sufficient to ensure that (2.7) is true:

(i) $\pi_{2j} = \pi_2$ for all j.
(ii) $p_{1j} = p_{2j}$ for all j.

That is, if either condition (i) or condition (ii) is satisfied then RD is averageable. Condition (i) says that in the unexposed population the probability of disease is the same across strata of F. In other words, F is *not* a risk factor for the disease in the unexposed population. Note that if the π_{2j} are all equal, their common value is π_2, as can be seen from (2.5). Condition (ii) says that F has the same distribution in the exposed and unexposed populations. In other words, F is *not* associated with exposure in the population. Note that conditions (i) and (ii) refer to the cohort at the start of follow-up.

2.4.3 Risk Ratio

The risk ratio for the jth stratum is defined to be $\rho_j = \pi_{1j}/\pi_{2j}$. It follows from (2.4) that $\pi_1 = \sum_{j=1}^{J} p_{1j}\pi_{2j}\rho_j$ and so

$$RR = \frac{\sum_{j=1}^{J} p_{1j}\pi_{2j}\rho_j}{\sum_{j=1}^{J} p_{2j}\pi_{2j}}$$

$$= \frac{\sum_{j=1}^{J} p_{1j}\pi_{2j}}{\sum_{j=1}^{J} p_{2j}\pi_{2j}} \times \frac{\sum_{j=1}^{J} p_{1j}\pi_{2j}\rho_j}{\sum_{j=1}^{J} p_{1j}\pi_{2j}}. \qquad (2.9)$$

If

$$\sum_{j=1}^{J} p_{1j}\pi_{2j} = \sum_{j=1}^{J} p_{2j}\pi_{2j} \qquad (2.10)$$

then

$$RR = \frac{\sum_{j=1}^{J} p_{1j}\pi_{2j}\rho_j}{\sum_{j=1}^{J} p_{1j}\pi_{2j}}. \qquad (2.11)$$

In this case, RR is averageable with weights $p_{1j}\pi_{2j}$. Note that (2.10) is identical to (2.7) and so conditions (i) and (ii) above, as well as the ensuing discussion, apply to the risk ratio.

2.4.4 Odds Ratio

The odds and odds ratio for the jth stratum are defined to be $\omega_{1j} = \pi_{1j}/(1 - \pi_{1j})$, $\omega_{2j} = \pi_{2j}/(1 - \pi_{2j})$, and $\theta_j = \omega_{1j}/\omega_{2j}$. It follows that $\pi_{1j} = \omega_{1j}(1 - \pi_{1j}) = \theta_j\omega_{2j}(1 - \pi_{1j})$ and $\pi_{2j} = \omega_{2j}(1 - \pi_{2j})$, and so, from (2.4) and (2.5),

$$\frac{\pi_1}{1 - \pi_1} = \frac{\sum_{j=1}^{J} p_{1j}\pi_{1j}}{\sum_{j=1}^{J} p_{1j}(1 - \pi_{1j})} = \frac{\sum_{j=1}^{J} p_{1j}(1 - \pi_{1j})\omega_{2j}\theta_j}{\sum_{j=1}^{J} p_{1j}(1 - \pi_{1j})}$$

and

$$\frac{\pi_2}{1 - \pi_2} = \frac{\sum_{j=1}^{J} p_{2j}\pi_{2j}}{\sum_{j=1}^{J} p_{2j}(1 - \pi_{2j})} = \frac{\sum_{j=1}^{J} p_{2j}(1 - \pi_{2j})\omega_{2j}}{\sum_{j=1}^{J} p_{2j}(1 - \pi_{2j})}.$$

Therefore

$$
\begin{aligned}
OR &= \frac{\left[\sum_{j=1}^{J} p_{1j}(1 - \pi_{1j})\omega_{2j}\theta_j\right] \Big/ \left[\sum_{j=1}^{J} p_{1j}(1 - \pi_{1j})\right]}{\left[\sum_{j=1}^{J} p_{2j}(1 - \pi_{2j})\omega_{2j}\right] \Big/ \left[\sum_{j=1}^{J} p_{2j}(1 - \pi_{2j})\right]} \\[2ex]
&= \frac{\left[\sum_{j=1}^{J} p_{1j}(1 - \pi_{1j})\omega_{2j}\right] \Big/ \left[\sum_{j=1}^{J} p_{1j}(1 - \pi_{1j})\right]}{\left[\sum_{j=1}^{J} p_{2j}(1 - \pi_{2j})\omega_{2j}\right] \Big/ \left[\sum_{j=1}^{J} p_{2j}(1 - \pi_{2j})\right]} \\[2ex]
&\quad \times \frac{\sum_{j=1}^{J} p_{1j}(1 - \pi_{1j})\omega_{2j}\theta_j}{\sum_{j=1}^{J} p_{1j}(1 - \pi_{1j})\omega_{2j}}.
\end{aligned}
\tag{2.12}
$$

If

$$
\frac{\sum_{j=1}^{J} p_{1j}(1 - \pi_{1j})\omega_{2j}}{\sum_{j=1}^{J} p_{1j}(1 - \pi_{1j})} = \frac{\sum_{j=1}^{J} p_{2j}(1 - \pi_{2j})\omega_{2j}}{\sum_{j=1}^{J} p_{2j}(1 - \pi_{2j})}
\tag{2.13}
$$

then

$$
OR = \frac{\sum_{j=1}^{J} p_{1j}(1 - \pi_{1j})\omega_{2j}\theta_j}{\sum_{j=1}^{J} p_{1j}(1 - \pi_{1j})\omega_{2j}}.
$$

In this case, OR is averageable with weights $p_{1j}(1 - \pi_{1j})\omega_{2j}$. Identity (2.13) can be written as

$$
\frac{\sum_{j=1}^{J} p_{1j}(1 - \pi_{1j})\omega_{2j}}{1 - \pi_1} = \frac{\sum_{j=1}^{J} p_{2j}(1 - \pi_{2j})\omega_{2j}}{1 - \pi_2}.
\tag{2.14}
$$

Each of the following conditions is sufficient to ensure that (2.14) is true:

(i) $\pi_{2j} = \pi_2$ for all j.

(iii) $\dfrac{p_{1j}(1 - \pi_{1j})}{1 - \pi_1} = \dfrac{p_{2j}(1 - \pi_{2j})}{1 - \pi_2}$ for all j.

Condition (i) is sufficient because the π_{2j} are all equal if and only if the ω_{2j} are all equal. Since $p_{1j} = r_{1j}/r_{1\bullet}$, $1 - \pi_{1j} = b_{1j}/r_{1j}$, and $1 - \pi_1 = b_{1\bullet}/r_{1\bullet}$, it follows that

$$
\frac{p_{1j}(1 - \pi_{1j})}{1 - \pi_1} = \frac{b_{1j}}{b_{1\bullet}}
$$

and likewise

$$\frac{p_{2j}(1 - \pi_{2j})}{1 - \pi_2} = \frac{b_{2j}}{b_{2\bullet}}.$$

Thus condition (iii) is equivalent to

(iii′) $\dfrac{b_{1j}}{b_{1\bullet}} = \dfrac{b_{2j}}{b_{2\bullet}}$ for all j.

Condition (iii′) says that F is *not* associated with exposure in those members of the population who do not develop the disease. This rather unusual condition can only be established once the study has been completed. Indeed, the condition may be satisfied at one point during the course of follow-up and not at another. If the disease is rare in each stratum—that is, if π_{1j} and π_{2j} are small for all j—then condition (iii) is approximately the same as condition (ii).

2.4.5 A Peculiar Property of the Odds Ratio

It was noted in connection with Table 2.2(d) that the decision as to whether there is homogeneity depends on the measure of effect under consideration. When homogeneity is present we denote the common values of δ_j, ρ_j, and θ_j by δ, ρ, and θ, respectively. Suppose that both the risk difference and risk ratio are homogeneous; that is, $\delta_j = \delta$ and $\rho_j = \rho$ for all j. Then $\delta = \pi_{1j} - \pi_{2j} = \rho\pi_{2j} - \pi_{2j}$ and so $\pi_{2j} = \delta/(\rho - 1)$ for all j. Therefore the π_{2j} are all equal and consequently condition (i) is satisfied; that is, F is not a risk factor (Rothman and Greenland, 1998, Chapter 18). A similar conclusion is reached if the starting point is the risk ratio and odds ratio, but not if we start with the risk difference and odds ratio. The explanation for the latter finding is that when $\delta_j = \delta$ and $\theta_j = \theta$ for all j, a quadratic equation results, making possible more than one value of π_{2j}. This is illustrated by Table 2.2(d), where both the risk difference and odds ratio are homogeneous despite the fact that F is a risk factor. For the remainder of this chapter, when effect modification is being examined and F is a risk factor, it will be assumed that each of the measures of effect is being considered separately and in turn.

Suppose that F is a risk factor for the disease and that F is not associated with the exposure. For the moment we also assume that F is not an effect modifier of the risk difference. Since F is not associated with exposure, condition (ii) is satisfied and so RD is averageable. Since F is not an effect modifier of the risk difference, the δ_i are homogeneous. It follows from a result at the beginning of Section 2.4.1 that RD is strictly collapsible with $RD = \delta$. Similarly, if F is not an effect modifier of the risk ratio, then $RR = \rho$. However, unlike the risk difference and risk ratio, condition (ii) does not ensure that the odds ratio is averageable. Therefore, even if F is a risk factor for the disease, F is not associated with exposure, and F is not an effect modifier, it may still be true that $OR \neq \theta$. This explains why in Table 2.2(d) the odds ratio exhibits Simpson's paradox while the risk difference does not. This peculiar property of the odds ratio has been a source of confusion and debate surrounding an appropriate definition of confounding (Grayson, 1987; Greenland et al., 1989).

To formalize the preceding remarks on the odds ratio, suppose that the π_{2j} are not all equal (F is a risk factor for the disease), $p_{1j} = p_{2j}$ for all j (F is not associated with exposure), and $\theta_j = \theta$ for all j (F is not an effect modifier of the odds ratio). It follows from the first line of (2.12) that

$$OR = \theta \frac{\left[\sum_{j=1}^{J} p_{1j}(1 - \pi_{1j})\omega_{2j}\right] \Big/ \left[\sum_{j=1}^{J} p_{1j}(1 - \pi_{1j})\right]}{\left[\sum_{j=1}^{J} p_{1j}(1 - \pi_{2j})\omega_{2j}\right] \Big/ \left[\sum_{j=1}^{J} p_{1j}(1 - \pi_{2j})\right]}. \tag{2.15}$$

About the only obvious condition that ensures that $OR = \theta$ is $\pi_{1j} = \pi_{2j}$ for all j. In this case $\theta_j = 1$ for all j, and hence $OR = \theta = 1$. In Appendix A we show that, when the p_{1j} are all equal, if $\theta > 1$ then $1 < OR < \theta$, and if $\theta < 1$ then $\theta < OR < 1$. Similar inequalities have been demonstrated in the context of matched-pairs case-control studies (Siegel and Greenhouse, 1973; Armitage, 1975), and more general results are available for the logistic regression model (Gail et al., 1984; Gail, 1986; Neuhaus et al., 1991).

2.4.6   Averageability in the Hypothetical Examples of Closed Cohort Studies

We showed at the beginning of Section 2.4.1 that if a measure of effect is averageable, then it is not subject to Simpson's paradox. This is logically equivalent to saying that if Simpson's paradox is present, then the measure of effect is not averageable. For the risk difference and risk ratio, not being averageable means that both conditions (i) and (ii) must fail, and for the odds ratio it means that both conditions (i) and (iii) must fail. For Tables 2.2(a)–2.2(e), Simpson's paradox is exhibited by the risk difference and risk ratio in Table 2.2(e), and by the odds ratio in Tables 2.2(d) and 2.2(e). We now examine these findings in light of conditions (i)–(iii).

Tables 2.2(a)–2.2(e) have only two strata and so condition (i) becomes $\pi_{21} = \pi_{22}$, which is the same as $a_{21}/r_{21} = a_{22}/r_{22}$. Since $p_{11} + p_{12} = 1 = p_{21} + p_{22}$, condition (ii) simplifies to $p_{11} = p_{21}$, which can be expressed as $r_{11}/r_{1\bullet} = r_{21}/r_{2\bullet}$. From Section 2.4.4, condition (iii) is equivalent to $b_{11}/b_{1\bullet} = b_{21}/b_{2\bullet}$. As can be seen from Table 2.4, either condition (i) or condition (ii) is satisfied by all the tables except for Table 2.2(e), and either condition (i) or condition (iii) is satisfied by all the tables except for Tables 2.2(d) and 2.2(e). These findings are consistent with the presence of Simpson's paradox in these same tables.

2.4.7   Collapsibility Definition of Confounding in Closed Cohort Studies

We are now in a position to describe and critique the collapsibility definition of confounding (Yanagawa, 1979; Kupper et al., 1981; Kleinbaum et al., 1982, Chapter 13; Schlesselman, 1982, §2.10; Yanagawa, 1984; Boivin and Wacholder, 1985;

TABLE 2.4 Averageability in the Hypothetical Closed Cohort Studies

Table	Condition (i) a_{21}/r_{21}	a_{22}/r_{22}	Condition (ii) $r_{11}/r_{1\bullet}$	$r_{21}/r_{2\bullet}$	Condition (iii) $b_{11}/b_{1\bullet}$	$b_{21}/b_{2\bullet}$
2.2(a)	.40	.40	.33	.33	.33	.33
2.2(b)	.40	.40	.33	.33	.43	.33
2.2(c)	.40	.40	.33	.67	.43	.67
2.2(d)	.60	.10	.33	.33	.08	.18
2.2(e)	.60	.10	.33	.67	.06	.47

Grayson, 1987). We simplify the discussion by assuming that, aside from E and D, F is the only other variable under consideration and hence the only potential confounder. When there are several potential confounders, the decision as to whether a given variable is a confounder depends on the other risk factors under consideration (Fisher and Patil, 1974). The more general case of several confounders will be considered in Section 2.5.3.

According to the collapsibility definition of confounding, F is a confounder (of a measure of effect) if both of the following conditions are satisfied:

(a) The measure of effect is homogeneous across strata of F.

(b) The common stratum-specific value of the measure of effect does *not* equal the crude value.

When F is a confounder, the common stratum-specific value that is guaranteed by condition (a) is taken to be the "overall" measure of effect for the cohort, and the crude value is said to be confounded (by F). For example, in Table 2.2(d), F is a confounder of the odds ratio but not the risk difference. So, the overall risk difference for the cohort is $RD = \delta = .30$ and the overall odds ratio is $\theta = 6.0$. In Table 2.2(e), F is a confounder of the risk ratio and so the overall risk ratio is $\rho = 1.5$. These examples show that, according to the collapsibility definition, the presence or absence of confounding depends on the measure of effect under consideration. The collapsibility definition of confounding has the attractive feature that it is possible, in theory, to base decisions about confounding entirely on study data. Here we set aside the important issue of random error, which may make the decision about effect modification uncertain, especially when the sample size is small. This topic is discussed at length in subsequent chapters.

For a given measure of effect, suppose that F is not an effect modifier; that is, suppose the measure of effect is homogeneous (across strata of F). Since condition (a) is then satisfied, it follows that F is a confounder if and only if the measure of effect is *not* strictly collapsible. In Section 2.4.1 we showed that a measure of effect is strictly collapsible if and only if it is homogeneous and averageable. Since the measure of effect is assumed to be homogeneous, F is a confounder if and only if the measure of effect is *not* averageable. In Section 2.4.2 and 2.4.3 we described *sufficient* conditions for the risk difference and risk ratio to be averageable; that is, *either* F is *not* a risk factor for the disease in the unexposed population *or* F is *not* associated with exposure in the population. It follows that if the risk difference and risk ratio are *not* averageable, then F must be a risk factor for the disease in the unexposed population *and* F must be associated with exposure in the population. So, given that F is *not* an effect modifier, the following are *necessary* conditions for F to be a confounder of the risk difference and the risk ratio:

1. F is a risk factor for the disease in the unexposed population.
2. F is associated with exposure in the population.

Analogous arguments apply to the odds ratio: Given that F is *not* an effect modifier, the following are *necessary* conditions for F to be a confounder of the odds ratio:

1. F is a risk factor for the disease in the unexposed population.
3. F is associated with exposure among those who do not develop the disease.

At the close of Section 2.3.2 we specified two properties that we felt should form part of any definition of a confounder. In fact, these two properties are basically conditions 1 and 2 above. So, for studies analyzed using the risk difference or the risk ratio, the collapsibility definition of confounding meets our essential requirements. However, a difficulty arises with studies analyzed using the odds ratio. The problem is that, according to the collapsibility definition, a confounder of the odds ratio has to satisfy conditions 1 and 3, but not necessarily condition 2. This means that a variable can be a confounder of the odds ratio even when it is not associated with the exposure. This is a serious shortcoming of the collapsibility definition of confounding. In a sense, the problem lies not so much with the collapsibility approach but rather with the peculiar property of the odds ratio alluded to in Section 2.4.5 and illustrated in Table 2.2(d). However, we wish to use the odds ratio as a measure of effect and so there is no recourse but to search for an alternative definition of confounding.

2.5 COUNTERFACTUAL APPROACH TO CONFOUNDING

2.5.1 Counterfactual Definition of Confounding in Closed Cohort Studies

In Section 2.3.1 we introduced the idea of counterfactual arguments in discussions of causality. The counterfactual approach is well established in the field of philosophy but has only recently been exploited in statistics and epidemiology (Rubin, 1974; Holland, 1986; Holland and Rubin, 1988). Below we present a definition of confounding using counterfactuals (Greenland et al., 1999). The following discussion can be expressed in terms of an arbitrary parameter and an arbitrary measure of effect, but for concreteness we focus on the probability of disease and the risk difference. Continuing with the notation used above, let π_1 be the probability of disease in the exposed cohort, let π_2 the corresponding probability in the unexposed cohort, and let $RD = \pi_1 - \pi_2$ be the risk difference. Consider a closed cohort study where the aim is to determine whether a given exposure is a risk factor for a particular disease. To this end, the exposed and unexposed cohorts are followed over a period of time and the risk difference is estimated. The reason for including the unexposed cohort in the study is to have a comparison group for the exposed cohort. Not surprisingly, the manner in which the unexposed cohort is chosen is crucial to the success of the cohort study.

For each member of the exposed cohort we can imagine an individual, referred to as the counterfactual unexposed individual, who exhibits the risk relationship between exposure and disease that would have been observed in the exposed individual had that person not been exposed. By bringing together the group of counterfactual unexposed individuals, one for each member of the exposed cohort, we obtain what will be referred to as the counterfactual unexposed cohort. The counterfactual unexposed cohort is an imaginary group of individuals, but if such a cohort were available it would constitute the ideal comparison group. Let π_1^* denote the probability of disease in the counterfactual unexposed cohort and let $RD^* = \pi_1 - \pi_1^*$ be the risk difference comparing the exposed cohort to the counterfactual unexposed cohort. In order not to confuse comparison groups, we will refer to the unexposed cohort as the actual unexposed cohort.

Under ideal circumstances the probability of disease in the actual and counterfactual unexposed cohorts would be equal—that is, $\pi_2 = \pi_1^*$—in which case we would have $RD = RD^*$. According to the counterfactual definition, confounding is present when $\pi_2 \neq \pi_1^*$. In this case the risk difference (and other measures of effect) are said to be confounded. In order for confounding to be absent, it is not necessary that the actual unexposed cohort be even remotely similar to the counterfactual unexposed cohort, only that the identity $\pi_2 = \pi_1^*$ is satisfied. For example, a group of females could serve as the actual unexposed cohort in a study of all-cause mortality in prostate cancer patients. The risk difference would be unconfounded provided the probability of death in this comparison group happened to be equal to the probability of death in the counterfactual unexposed cohort. This illustrates a crucial point about the counterfactual definition of confounding: It is based on features of the population at the aggregate level, making no reference to processes at the level of the individual. This distinction is important when interpreting epidemiologic findings with respect to individual risk (Greenland, 1987; Greenland and Robins, 1988; Robins and Greenland, 1989a, 1989b, 1991).

The counterfactual definition of confounding is a useful construct but has the obvious drawback that the counterfactual unexposed cohort is imaginary. However, under certain circumstances the counterfactual unexposed cohort can be reasonably approximated. For example, consider a randomized controlled trial with a crossover design in which a new analgesic is compared to placebo in patients with chronic pain. According to this design, subjects are randomly assigned to receive either analgesic (exposure) or placebo, and after an initial period of observation they are switched (crossed-over) to the other treatment. Suppose that the analgesic is short-acting so that there are no carry-over effects for subjects who receive this medication first. In this case, the counterfactual unexposed cohort is closely approximated by the entire study cohort when it is on placebo. As another example, suppose that a group of workers in a chemical fabricating plant is accidentally exposed to a toxic substance. Due to the accidental nature of the exposure it may be reasonable to assume that the exposed workers do not differ in any systematic way from those who were not involved in the accident. If so, a random sample of unexposed workers would provide a satisfactory approximation to the counterfactual unexposed cohort.

In most epidemiologic studies the actual unexposed cohort does not compare as closely to the counterfactual unexposed cohort as in the above examples. Moreover, in certain instances it is difficult to imagine that a counterfactual unexposed cohort could even exist. For example, suppose that we wish to study the impact of ethnic background on the development of disease. Since ethnic background is closely related to genetics, socioeconomic status, and other fundamental characteristics of the individual, it is almost impossible to conceive of a counterfactual unexposed cohort. Despite these limitations, the counterfactual approach to confounding provides a useful framework for organizing our thinking about causality and risk.

2.5.2   A Model of Population Risk

We now consider a model of population risk due to Greenland and Robins (1986). Suppose that in the exposed and (actual) unexposed cohorts there are four types of individuals as shown in Table 2.5. By definition, exposure has no effect on those who are "doomed" or "immune," and exposure is either "causative" or "preventive" in those who are "susceptible." The distributions of the exposed and unexposed cohorts according to each of the four types are given in Table 2.5, where, by definition, $p_1 + p_2 + p_3 + p_4 = 1$ and $q_1 + q_2 + q_3 + q_4 = 1$. In the exposed cohort, only type 1 and type 2 subjects will develop the disease, and so the probability of disease is $\pi_1 = p_1 + p_2$. In the unexposed cohort the corresponding probability is $\pi_2 = q_1 + q_3$. So $RD = (p_1 + p_2) - (q_1 + q_3)$.

In the counterfactual unexposed cohort the probability of disease is $\pi_1^* = p_1 + p_3$ and so $RD^* = (p_1 + p_2) - (p_1 + p_3) = p_2 - p_3$. By definition, RD will be unconfounded only if $\pi_1^* = \pi_2$, that is, $p_1 + p_3 = q_1 + q_3$, in which case $RD = RD^* = p_2 - p_3$. Note that for the identity $p_1 + p_3 = q_1 + q_3$ to be satisfied it is not necessary that the individual identities $p_1 = q_1$ and $p_3 = q_3$ hold. All that is needed for confounding to be absent is that the net effects be the same in the subcohorts consisting of type 1 and type 3 subjects. This demonstrates the point made earlier

TABLE 2.5 Distribution of Exposed and Unexposed Cohorts According to Type of Outcome

Type	Description	Exposed	Unexposed
1	Exposure has no effect (doomed)	p_1	q_1
2	Exposure is causative (susceptible)	p_2	q_2
3	Exposure is preventive (susceptible)	p_3	q_3
4	Exposure has no effect (immune)	p_4	q_4

that confounding is determined at the population level rather than at the level of individuals.

2.5.3 Counterfactual Definition of a Confounder

As intuition suggests, in order for a measure of effect to be confounded according to the counterfactual definition, the exposed and unexposed cohorts must differ on risk factors for the disease. A variable that is, in whole or in part, "responsible" for confounding is said to be a confounder (of the measure of effect). Note that according to the counterfactual approach, the fundamental concept is confounding, and that confounders are defined secondarily as variables responsible for this phenomenon. This is to be contrasted with the collapsibility definition that first defines confounders and, when these have been identified, declares confounding to be present. As was observed in the previous section, the counterfactual definition of confounding is based on measurements taken at the population level. These usually represent the net effects of many interrelated variables, some of which may play a role in confounding. An important part of the analysis of epidemiologic data involves identifying from among the possibly long list of risk factors those which might be confounders.

Below we discuss the relatively simple case of a single confounder, but in practice there will usually be many such variables to consider. To get a sense of how complicated the interrelationships can be, consider a cohort study investigating hypercholesterolemia (elevated serum cholesterol) as a risk factor for myocardial infarction (heart attack). Myocardial infarction most often results from atherosclerosis (hardening of the arteries). Established risk factors that would typically be considered in such a study are age, sex, family history, hypertension (high blood pressure), and smoking. A number of associations among these risk factors need to be taken into account: Hypertension and hypercholesterolemia tend to increase with age; smoking is related to sex and age; hypercholesterolemia and atherosclerosis are familial; hypercholesterolemia can cause atherosclerosis, which can in turn lead to hypertension; hypertension can damage blood vessels and thereby provide a site for atherosclerosis to develop.

In order to tease out the specific effect, if any, that hypercholesterolemia might have on the risk of myocardial infarction, it is necessary to take account of potential confounding by the other variables mentioned above. As can be imagined, this presents a formidable challenge, in terms of both statistical analysis and pathophysiologic interpretation. Furthermore, the preceding discussion refers only to known risk factors. There may be unknown risk factors that should be considered but that, due to the current state of scientific knowledge, are not included in the study. This example also points out that according to the counterfactual approach (and in contrast to the collapsibility approach) a decision about confounding is virtually never decided solely on the basis of study data. Instead, all available information is utilized—in particular, whatever is known about the underlying disease process and the population being studied (Greenland and Neutra, 1980; Robins and Morgenstern, 1987).

We now formalize the counterfactual definition of a confounder. Let $\mathcal{R}$ be the complete set of risk factors for the disease, both known and unknown, and let S be a subset of $\mathcal{R}$ that does not include E, the exposure of interest. The stratification that results from cross-classifying according to all variables in S will be referred to as stratifying by S, and the resulting strata will be referred to as the strata of S. For example, let $S = \{F_1, F_2\}$, where F_1 is age group (five levels) and F_2 is sex. Then the strata of S are the 10 age group–sex categories obtained by cross-classifying F_1 and F_2. Within each stratum of S we form the 2×2 table obtained by cross-classifying by the exposure of interest and the disease. Associated with the actual exposed cohort in each stratum is a corresponding counterfactual unexposed cohort. We say there is no residual confounding in the strata of S if each stratum is unconfounded; that is, within each stratum the probability of disease in the actual unexposed cohort equals the probability of disease in the counterfactual unexposed cohort.

Suppose that we have constructed the causal diagram relating the risk factors in $\mathcal{R}$ and the exposure E to the disease. Based on the "back-door" criterion, the causal diagram can be used to determine whether, after stratifying by S, there is residual confounding in the strata of S (Pearl, 1993, 1995, 2000, Chapter 3). When there is no residual confounding we say that S is sufficient to control confounding, or simply that S is sufficient. When S is sufficient but no proper subset of S is sufficient, S is said to be minimally sufficient. A minimally sufficient set can be determined by sequentially deleting variables from a sufficient set until no more variables can be dropped without destroying the sufficiency. Depending on the choices made at each step, this process may lead to more than one minimally sufficient set of confounders. This shows that whether we view a risk factor as a confounder depends on the other risk factors under consideration.

It is possible for a minimally sufficient set of confounders to be empty, meaning that the crude measure of effect is unconfounded. A valuable and surprising lesson to be learned from causal diagrams is that confounding can be introduced by enlarging a sufficient set of confounders (Greenland and Robins, 1986; Greenland et al., 1999). The explanation for this seeming paradox is that, as has been observed, confounding is a phenomenon that is determined by net effects at the population level. Just as stratification can prevent confounding by severing the connections between variables that were responsible for a spurious causal relationship, it is equally true that stratification can create confounding by interfering with the paths between variables that were responsible for preventing confounding.

Causal diagrams have the potential drawback of requiring detailed information on the possibly complex interrelationships among risk factors, both known and unknown. However, even if a causal diagram is based on incomplete knowledge, it can be useful for organizing what is known about established and suspected risk factors. In Section 2.5.6 it is demonstrated that for an unknown confounder to produce significant bias it must be highly prevalent in the unexposed population, closely associated with the exposure, and a major risk factor for the disease. It is always possible that such an important risk factor might as yet be unknown, especially when a disease is only beginning to be studied, but for well-researched diseases this seems less likely.

Nevertheless, the impact of an unknown confounder must be kept in mind and so causal diagrams should be interpreted with an appropriate degree of caution. We now specialize to the simple case of a single potential confounder F. The situation with multiple potential confounders is partially subsumed by the present discussion if we consider F to be formed by stratifying on a set of potential confounders. We make the crucial assumption that F is not affected by the exposure being studied. An instance where this assumption would fail is in the study of hypercholesterolemia and myocardial infarction presented earlier, with F taken to be hypertension. As was remarked above, hypercholesterolemia can lead to hypertension, which is a risk factor for myocardial infarction. When risk factors are affected by the exposure under consideration, the analysis of confounding and causality becomes much more complicated, requiring considerations beyond the scope of the present discussion (Rosenbaum, 1984b; Robins, 1989; Robins and Greenland, 1992; Robins et al., 1992; Weinberg, 1993; Robins, 1998; Keiding, 1999). With F assumed to be unaffected by exposure, it follows that F cannot be on the causal pathway between exposure and disease. Recall that this is one of the conditions that was specified in Section 2.3.2 as a requirement for a proper definition of a confounder.

Let π_{1j}^* denote the counterfactual probability of disease in the jth stratum and let p_{1j}^* denote the proportion of the counterfactual unexposed cohort in that stratum ($j = 1, 2, \ldots, J$). Consistent with (2.4), it follows that

$$\pi_1^* = \sum_{j=1}^{J} p_{1j}^* \pi_{1j}^*.$$

We now assume that there is no residual confounding in the strata of F; that is, $\pi_{1j}^* = \pi_{2j}$ for all j. This assumption is related to the assumption of strong ignorability (Rosenbaum and Rubin, 1983; Rosenbaum, 1984a). According to Holland (1989), this is perhaps the most important type of assumption that is made in discussions of causal inference in nonrandomized studies. It was assumed above that F is not affected by E. This implies that if the exposed cohort had in fact been unexposed, the distribution of F would be unchanged; that is, $p_{1j}^* = p_{1j}$ for all j. Consequently,

$$\pi_1^* = \sum_{j=1}^{J} p_{1j} \pi_{2j}. \tag{2.16}$$

From (2.4) and (2.16), the criterion for no confounding, $\pi_1^* = \pi_2$, can be expressed as

$$\sum_{j=1}^{J} p_{1j} \pi_{2j} = \sum_{j=1}^{J} p_{2j} \pi_{2j} \tag{2.17}$$

(Wickramaratne and Holford, 1987; Holland, 1989). Identity (2.17) is the same as identities (2.7) and (2.10), which were shown to be sufficient to guarantee averageability of the risk difference and risk ratio. Consequently, either condition (i) or condition (ii) is sufficient for $\pi_1^* = \pi_2$ to be true. So, provided F is not affected by E and assuming that there is no residual confounding within strata of F, the following are necessary conditions for F to be a confounder according to the counterfactual definition of confounding (Miettinen and Cook, 1981; Rothman and Greenland, 1998):

1. F is a risk factor for the disease in the unexposed population.
2. F is associated with exposure in the population.

These are the same necessary conditions for F to be a confounder of the risk difference and risk ratio that were obtained using the collapsibility definition of confounding. An important observation is that (2.16) was derived without specifying a particular measure of effect. This means that conditions 1 and 2 above are applicable to the odds ratio as well as the risk difference and risk ratio. This avoids the problem related to the odds ratio that was identified as a flaw in the collapsibility definition of confounding.

We return to an examination of confounding and effect modification in the hypothetical cohort studies considered earlier. Based on criterion (2.17), it is readily verified that F is a confounder (according to the counterfactual definition) in Table 2.2(e) but not in Table 2.2(d). We observe that in Table 2.2(d), F is an effect modifier of the risk ratio but not an effect modifier of the risk difference or odds ratio. On the other hand, in Table 2.2(e), F is an effect modifier of the risk difference and the odds ratio but not an effect modifier of the risk ratio. This shows that confounding (according to the counterfactual definition) and effect modification are distinct characteristics that can occur in the presence or absence of one other.

When there are only two strata, (2.17) simplifies to

$$p_{11}\pi_{21} + p_{12}\pi_{22} = p_{21}\pi_{21} + p_{22}\pi_{22}. \tag{2.18}$$

Substituting $p_{12} = 1 - p_{11}$ and $p_{22} = 1 - p_{21}$ in (2.18) and rearranging terms leads to $(\pi_{21} - \pi_{22})(p_{11} - p_{21}) = 0$. This identity is true if and only if either $\pi_{21} = \pi_{22}$ or $p_{11} = p_{21}$. The latter identities are precisely conditions (i) and (ii), respectively. So, when there are only two strata, (2.17) implies and is implied by conditions (i) and (ii). In other words, conditions 1 and 2 completely characterize a dichotomous confounder. In Table 2.2(f), F is a risk factor for the disease and F is associated with exposure, and yet (2.17) is satisfied:

$$\left(\frac{200}{600}\right)\left(\frac{50}{100}\right) + \left(\frac{300}{600}\right)\left(\frac{20}{200}\right) + \left(\frac{100}{600}\right)\left(\frac{90}{300}\right) = \frac{160}{600}$$
$$= \left(\frac{100}{600}\right)\left(\frac{50}{100}\right) + \left(\frac{200}{600}\right)\left(\frac{20}{200}\right) + \left(\frac{300}{600}\right)\left(\frac{90}{300}\right).$$

This illustrates that, when F has three or more strata, even if both conditions 1 and 2 are satisfied, F may not be a confounder. Therefore, when there are three or more strata, conditions 1 and 2 are necessary but not sufficient for confounding.

2.5.4   Standardized Measures of Effect

For the following discussion, the number of subjects in the exposed cohort who develop disease will be denoted by O, a quantity we refer to as the observed count. So we have $O = a_{1\bullet} = \pi_1 r_{1\bullet}$. From (2.4) and $p_{1j} r_{1\bullet} = r_{1j}$, it follows that

$$O = \left(\sum_{j=1}^{J} p_{1j} \pi_{1j} \right) r_{1\bullet} = \sum_{j=1}^{J} \pi_{1j} r_{1j}. \tag{2.19}$$

Based on the probability of disease in the actual unexposed cohort, the number of subjects in the exposed cohort expected to develop disease in the absence of exposure is $cE = \pi_2 r_{1\bullet}$, a quantity we refer to as the crude expected count. From (2.5) it follows that

$$cE = \left(\sum_{j=1}^{J} p_{2j} \pi_{2j} \right) r_{1\bullet}.$$

Since $\pi_1 = O/r_{1\bullet}$ and $\pi_2 = cE/r_{1\bullet}$, the risk difference, risk ratio, and odds ratio can be expressed as

$$cRD = \pi_1 - \pi_2 = \frac{O - cE}{r_{1\bullet}}$$

$$cRR = \frac{\pi_1}{\pi_2} = \frac{O}{cE}$$

and

$$cOR = \frac{\pi_1(1 - \pi_2)}{\pi_2(1 - \pi_1)} = \frac{O(r_{1\bullet} - cE)}{cE(r_{1\bullet} - O)}$$

which we refer to as the crude measures of effect.

Based on the probability of disease in the counterfactual unexposed cohort, the number of subjects in the exposed cohort expected to develop disease in the absence of exposure is $sE = \pi_1^* r_{1\bullet}$, a quantity we refer to as the standardized expected count. It follows that the criterion for no confounding, $\pi_2 = \pi_1^*$, can be expressed as $cE = sE$. Assume that F is the only potential confounder and that F is not affected by exposure. Also assume that there is no residual confounding in the strata of F. From (2.16) and $p_{1j} r_{1\bullet} = r_{1j}$ it follows that

$$sE = \left(\sum_{j=1}^{J} p_{1j} \pi_{2j} \right) r_{1\bullet} = \sum_{j=1}^{J} \pi_{2j} r_{1j}. \tag{2.20}$$

Note that sE can be estimated from study data.

The standardized measures of effect are defined to be

$$sRD = \pi_1 - \pi_1^* = \frac{O - sE}{r_{1\bullet}}$$

$$sRR = \frac{\pi_1}{\pi_1^*} = \frac{O}{sE}$$

and

$$sOR = \frac{\pi_1(1 - \pi_1^*)}{\pi_1^*(1 - \pi_1)} = \frac{O(r_{1\bullet} - sE)}{sE(r_{1\bullet} - O)}.$$

Note that sRD was denoted by RD^* in previous sections. When F is not a confounder, $cE = sE$ and so the crude and standardized measures of effect are equal. When F is a confounder, the standardized measures of effect can be thought of as overall measures of effect for the cohort after controlling for confounding due to F.

It is readily verified that

$$sRD = \frac{\sum_{j=1}^{J} \pi_{1j}r_{1j} - \sum_{j=1}^{J} \pi_{2j}r_{1j}}{r_{1\bullet}} = \sum_{j=1}^{J} p_{1j}\delta_j \tag{2.21}$$

$$sRR = \frac{\sum_{j=1}^{J} \pi_{1j}r_{1j}}{\sum_{j=1}^{J} \pi_{2j}r_{1j}} = \frac{\sum_{j=1}^{J} p_{1j}\pi_{2j}\rho_j}{\sum_{j=1}^{J} p_{1j}\pi_{2j}} \tag{2.22}$$

and

$$sOR = \frac{\left(\sum_{j=1}^{J} p_{1j}\pi_{1j}\right)\left(1 - \sum_{j=1}^{J} p_{1j}\pi_{2j}\right)}{\left(\sum_{j=1}^{J} p_{1j}\pi_{2j}\right)\left(1 - \sum_{j=1}^{J} p_{1j}\pi_{1j}\right)}$$

$$= \frac{\left[\sum_{j=1}^{J} p_{1j}(1 - \pi_{1j})\omega_{2j}\theta_j\right]\left[\sum_{j=1}^{J} p_{1j}(1 - \pi_{2j})\right]}{\left[\sum_{j=1}^{J} p_{1j}(1 - \pi_{2j})\omega_{2j}\right]\left[\sum_{j=1}^{J} p_{1j}(1 - \pi_{1j})\right]}. \tag{2.23}$$

The second equality in (2.23) follows from identities established in Section 2.4.4. When the risk difference, risk ratio, and odds ratio are homogeneous, it follows from (2.21)–(2.23) that $sRD = \delta$, $sRR = \rho$, and

$$sOR = \theta \frac{\left[\sum_{j=1}^{J} p_{1j}(1 - \pi_{1j})\omega_{2j}\right] \Big/ \left[\sum_{j=1}^{J} p_{1j}(1 - \pi_{1j})\right]}{\left[\sum_{j=1}^{J} p_{1j}(1 - \pi_{2j})\omega_{2j}\right] \Big/ \left[\sum_{j=1}^{J} p_{1j}(1 - \pi_{2j})\right]}.$$

Condition (i) is sufficient to ensure that $sOR = \theta$; but in general, $sOR \neq \theta$.

In an actual study the stratum-specific values of a measure of effect may be numerically close but are virtually never exactly equal. Once it is determined that (after accounting for random error) there is no effect modification, the stratum-specific estimates can be combined to create what is referred to as a summarized or summary measure of effect. Usually this takes the form of a weighted average of stratum-specific estimates where the weights are chosen in a manner that reflects the amount of information contributed by each stratum. Numerous examples of this approach to combining stratum-specific measures of effect will be encountered in later chapters. A summarized measure of effect may be interpreted as an estimate of the common stratum-specific value of the measure of effect. Since we have used a deterministic approach here, the interpretation of Tables 2.2(a)–2.2(f) is that there is no effect modification only if stratum-specific values are precisely equal. Accordingly, in Tables 2.2(a)–2.2(f), when there is no effect modification (in this sense) we take the summarized value to be the common stratum-specific value. For example, in Table 2.2(d) the summary odds ratio is 6.0.

When reporting the results of a study, a decision must be made as to which of the crude, standardized, summarized, and stratum-specific values should be presented. Table 2.6(a) offers some guidelines in this regard with respect to the risk difference and risk ratio, and Table 2.6(b) does the same for the odds ratio. Here we assume that summarization is carried out by forming a weighted average of stratum-specific values. When there is no confounding, the crude value should be reported because it represents the overall measure of effect for the cohort. On the other hand, when confounding is present, the crude value is, by definition, a biased estimate of the overall measure of effect and so the standardized value should be reported instead. When there is no effect modification, the summarized value should be reported because it represents the common stratum-specific value of the measure of effect. However, when effect modification is present, the stratum-specific values should be given individually because the pattern across strata may be of epidemiologic interest.

Under certain conditions there will be equalities among the crude, standardized, summarized, and stratum-specific values of a measure of effect. When there is no effect modification, the summarized measure of effect equals the common stratum-

TABLE 2.6(a)   Guidelines for Reporting Risk Difference and Risk Ratio Results

	Effect modification	
Confounding	No	Yes
No	crude = summarized	crude and stratum-specific
Yes	standardized = summarized	standardized and stratum-specific

TABLE 2.6(b)   Guidelines for Reporting Odds Ratio Results

	Effect modification	
Confounding	No	Yes
No	crude and summarized	crude and stratum-specific
Yes	standardized and summarized	standardized and stratum-specific

specific value. When there is no confounding, the crude and standardized values of a measure of effect are equal (by definition). Also, when there is no confounding, identity (2.17) is satisfied and therefore so are (2.7) and (2.10). In this case the risk difference and risk ratio are averageable. If, in addition, there is no effect modification, these measures of effect are strictly collapsible. This justifies the equality in the upper left cell of Table 2.6(a). The equality in the lower left cell follows from remarks made in connection with (2.21) and (2.22). In general, the preceding equalities do not hold for the odds ratio and so they have not been included in Table 2.6(b). This means that when there is no effect modification, two values of the odds ratio should be reported—the crude and summarized values when there is no confounding, and the standardized and summarized values when confounding is present.

We now illustrate some of the above considerations with specific examples. From Table 2.2(c) we have $O = 230$, $cE = (120/300)300 = 120$, and

$$sE = \left(\frac{80}{200} \right) 100 + \left(\frac{40}{100} \right) 200 = 120.$$

Since $cE = sE$ there is no confounding and so the crude value of each measure of effect should be reported. For each measure of effect there is effect modification and so the stratum-specific values should be given individually rather than summarized.

Now consider Table 2.2(d), where $O = 170$, $cE = (80/300)300 = 80$, and

$$sE = \left(\frac{60}{100} \right) 100 + \left(\frac{20}{200} \right) 200 = 80.$$

Since $cE = sE$ there is no confounding and so the crude value of each measure of effect should be reported. For the risk ratio, effect modification is present and so the stratum-specific values should be given separately. For the risk difference, there is no effect modification and, consistent with Table 2.6(a), the crude and summarized values are equal. For the odds ratio, effect modification is absent. Consistent with Table 2.6(b), the crude value, $OR = 3.6$, does not equal the summarized value, $\theta = 6.0$, and so both should be reported. We view the crude odds ratio as the overall odds ratio for the cohort, and we regard the summarized odds ratio as the common value of the stratum-specific odds ratios.

The fact that two odds ratios are needed to characterize the exposure–disease relationship in Table 2.2(d) creates frustrating difficulties with respect to interpretation, as we now illustrate. Suppose that the strata have been formed by categorizing subjects according to sex. So for males and females considered separately the odds ratio is $\theta = 6.0$, whereas for the population as a whole it is $OR = 3.6$. This means that, despite effect modification being absent, there is no single answer to the question "What is the odds ratio for the exposure–disease relationship?" Intuitively it is difficult to accept the idea that even though the odds ratios for males and females are the same, this common value is nevertheless different from the odds ratio for males and females combined. Furthermore, there is the frustration that the difference cannot be blamed on confounding.

As has been noted previously, the source of the difficulty is that condition (ii) is not sufficient to ensure that the odds ratio is averageable. This drawback of the odds ratio has led Greenland (1987) to argue that this measure of effect is epidemiologically meaningful only insofar as it approximates the risk ratio or hazard ratio (defined in Chapter 8). As noted in Section 2.2.2, the odds ratio is approximately equal to the risk ratio when the disease is rare, and so using the odds ratio is justified when this condition is met. Alternatively, the failure of the odds ratio to be averageable can be acknowledged and the necessity of having to report two odds ratios accepted as an idiosyncrasy of this measure of effect.

It is instructive to apply the above methods to data from the University Group Diabetes Program (1970), a study that was quite controversial when first published. Rothman and Greenland (1998, Chapter 15) analyzed these data using an approach that is slightly different from what follows. The UGDP study was a randomized controlled trial comparing tolbutamide (a blood sugar-lowering drug) to placebo in patients with diabetes. Long-standing diabetes can cause cardiovascular complications, and this increases the risk of such potentially fatal conditions as myocardial infarction (heart attack), stroke, and renal failure. Tolbutamide helps to normalize blood sugar and would therefore be expected to reduce mortality in diabetic patients. Table 2.7 gives data from the UGDP study stratified by age at enrollment, with death from all causes as the study endpoint. The following analysis is based on the risk difference.

Since $cRD = .045$ it appears that, contrary to expectation, tolbutamide increases mortality. Note also that Simpson's paradox is present. As will be discussed in Section 2.5.5, randomization is expected to produce treatment arms with similar patient characteristics. But this can only be guaranteed over the course of many replications of a study, not in any particular instance. From Table 2.7, $p_{12} = 98/204 = .48$ and $p_{22} = 85/205 = .41$. So the proportion of subjects in the 55+ age group is greater in the tolbutamide arm than in the placebo arm. This raises the possibility that the excess mortality observed in patients receiving tolbutamide might be a consequence of their being older. Since age is associated with exposure (type of treatment) and also increases mortality risk, age meets the two necessary conditions to be a confounder.

TABLE 2.7   UGDP Study Data

		Age <55			Age 55+			All ages		
	Survival	Tolbutamide			Tolbutamide			Tolbutamide		
		yes	no		yes	no		yes	no	
	dead	8	5		22	16		30	21	
	alive	98	115		76	69		174	184	
		106	120		98	85		204	205	
RD		.034			.036			.045		
RR		1.81			1.19			1.44		
OR		1.88			1.25			1.51		

Before employing the techniques developed above, we need to verify two assumptions, namely, that age is not in the causal pathway between tolbutamide and all-cause mortality, and there is no residual confounding in each of the age-specific strata. Since tolbutamide does not cause aging, the first assumption is clearly satisfied. There is evidence in the UGDP data (not shown) that variables other than age were distributed unequally in the two treatment arms. However, for the sake of illustration we assume that there is no residual confounding in each stratum. Then $O = 30$, $cE = (21/205)204 = 20.90$, and $sE = (5/120)106 + (16/85)98 = 22.86$. The difference between cE and sE is not large, but there is enough of a disparity to suggest that age is a confounder. On these grounds we take $sRD = (30 - 22.86)/204 = .035$ to be the overall risk difference as opposed to the somewhat larger $cRD = (30 - 20.90)/204 = .045$. So, even after accounting for age, tolbutamide still appears to increase mortality risk in diabetic patients.

At the beginning of this chapter we introduced the concept of confounding as a type of systematic error. The confounding in the UGDP data has its origins in the uneven manner in which randomization allocated subjects to the tolbutamide and placebo arms. The apparent conflict in terminology between confounding (systematic error) and randomization (random error) is resolved once it is realized that confounding is a property of allocation (Greenland, 1990). Therefore, given (conditional on) the observed allocation in the UGDP study, it is appropriate to consider age as a source of confounding.

2.6 METHODS TO CONTROL CONFOUNDING

Confounding is of concern in virtually every epidemiologic study. The methods that are commonly used to control (adjust for) confounding are randomization, stratification, restriction, matching, and regression.

Of all the methods used to control confounding, randomization comes the closest to satisfying the counterfactual ideal. Randomization is the defining feature of randomized controlled trials, but it is rarely, if ever, used in other types of epidemiologic studies. Consider a randomized controlled trial in which a new (experimental) treatment is compared to a conventional (control) treatment. As in any epidemiologic study, subjects must meet certain eligibility criteria before being enrolled in the study. Randomization is carried out by randomly assigning subjects to either the experimental or control arms. As a result of randomization, the treatment and control arms will "on average" have identical distributions (balance) with respect to all confounders, both known and unknown. This last property, the ability to control unknown confounders, is an important feature of randomization that is not shared by other methods used to control confounding.

The phrase "on average" is an important caveat. Unfortunately, randomization does not guarantee balance in any particular randomized controlled trial, as is illustrated by the UGDP study. In a randomized controlled trial it is usual to check if randomization "worked" by comparing the distribution of known risk factors in the two treatment arms. As intuition suggests, the larger the sample size the greater the chance that the treatment arms will be balanced (Greenland, 1990). In the ideal situation where randomization has resulted in perfect balance (for both known and unknown confounders), the control arm is equivalent to the counterfactual unexposed cohort. When the treatment arms are not balanced on important confounders, methods such as stratification and regression should be employed to control confounding.

As might be imagined from earlier discussions, stratification is one of the cornerstones of epidemiologic data analysis. After stratification by a sufficient set of confounders, the subjects in each stratum have the same confounder values and so there is no longer a pathway from exposure to disease through the confounders. Methods based on stratification will be considered throughout the remainder of the book. A drawback of stratification is that it can result in tables with small or even zero cell counts, especially when a large number of confounders have to be controlled simultaneously. Nevertheless, stratification is almost always used in the exploratory stages of an epidemiologic data analysis.

When there are many confounders to control, and especially when some of them are continuous variables, the question arises as to how strata should be created. One approach is based on the propensity score (Rosenbaum and Rubin, 1983; Joffe and Rosenbaum, 1999). The propensity score is a function defined in terms of known risk factors (other than the exposure of interest) which gives the probability that an individual belongs to the exposed population. If strata are created by grouping together individuals with the same propensity score, the exposed and unexposed subjects in each stratum will be balanced on known risk factors (Rosenbaum and Rubin, 1983; Rosenbaum, 1995, Chapter 9). The propensity score can be estimated from study data using, for example, logistic regression (Rosenbaum and Rubin, 1984; Rosenbaum and Rubin, 1985).

Restriction is another method used to control confounding. According to this approach, only those subjects who have a given value of the confounder are eligible for the study. This mechanism acts to control confounding in much the same way as stratification. For example, if smoking is a confounder, then restricting the study to nonsmokers prevents this variable from providing a pathway between exposure and disease. A drawback of restriction is that study findings will have a correspondingly limited generalizability.

Matching is seldom used in cohort studies but, as discussed in Chapter 11, has an important role in case-control studies. In a matched-pairs cohort study each exposed subject is matched to an unexposed subject on values of the matching variables. For example, matching might be based on age, sex, socioeconomic status, and medical history. As a result of matching, the distribution of matching variables is the same in the exposed and unexposed cohorts, and consequently these variables are eliminated as sources of confounding.

Regression techniques control confounding by including confounders as independent variables in the regression equation. When outcomes are measured on a continuous scale, methods such as linear regression and analysis of variance are used. Often, an epidemiologic study is concerned with categorical (discrete) outcomes. The regression methods that are most widely used in epidemiology for the analysis of categorical outcomes are logistic regression for case-control data and Cox regression for censored survival data, both of which are discussed in Chapter 15.

In an epidemiologic study it is not unusual for data to be collected on a wide range of variables, some of which may not be well understood as risk factors. As has been pointed out, the counterfactual definition of confounding relies heavily on an understanding of causal relationships as they exist in the population. When detailed knowledge on causation is lacking, there is little recourse but to rely on study data for clues to potential confounding. A problem with this approach is that the data may not accurately reflect the situation in the population as a result of random and systematic errors. A number of different strategies have been proposed for identifying confounders based on study data. To a greater or lesser extent, these methods involve a comparison of crude and summarized measures of effect, which is to say they are based on the collapsibility approach to confounding. Mickey and Greenland (1989) and Maldonado and Greenland (1993) evaluate a number of data-based strategies for confounder selection and offer guidelines for their use. According to one such strategy, a variable is designated a confounder if the relative increase or decrease in the adjusted compared to the crude measure of effect exceeds some fairly small magnitude such as 10%. Another strategy involves making this decision on the basis of a formal statistical test where the cutoff for the p-value is set at a relatively large value such as .20. According to both strategies, the idea is to set a relatively low threshold for treating a variable as a confounder. The former strategy will be used (in a somewhat informal manner) when commenting on the numerical examples presented in this book.

2.7 BIAS DUE TO AN UNKNOWN CONFOUNDER

In this section we investigate the extent to which the risk ratio can be biased due to an unknown confounder. We assume that E is the exposure of interest and that F is the unknown confounder. Tables 2.8(a) and 2.8(b) give the observed counts for a closed cohort study after stratification by F and E, respectively. In order to distinguish risk ratios arising from the two tables, we use subscripted notation such as $\rho_{ED|F=1} = (a_{11}/r_{11})/(a_{21}/r_{21})$.

From (2.4) and $p_{11} + p_{12} = 1$, we have

$$\pi_1 = p_{11}\pi_{11} + p_{12}\pi_{12}$$

$$= \pi_{12}\left[\frac{p_{11}\pi_{11}}{\pi_{12}} + (1 - p_{11})\right]$$

$$= \pi_{12}[p_{11}\rho_{FD|E=1} + (1 - p_{11})]$$

$$= \pi_{12}[1 + (\rho_{FD|E=1} - 1)p_{11}].$$

Likewise, from (2.5) and $p_{21} + p_{22} = 1$, we have

$$\pi_2 = p_{21}\pi_{21} + p_{22}\pi_{22}$$

TABLE 2.8(a)   Observed Counts: Closed
Cohort Study Stratified by F

	$F = 1$			$F = 2$	
D	E			E	
	1	2		1	2
1	a_{11}	a_{21}		a_{12}	a_{22}
2	b_{11}	b_{21}		b_{12}	b_{22}
	r_{11}	r_{21}		r_{12}	r_{22}

TABLE 2.8(b)   Observed Counts: Closed
Cohort Study Stratified by E

	$E = 1$			$E = 2$	
D	F			F	
	1	2		1	2
1	a_{11}	a_{12}		a_{21}	a_{22}
2	b_{11}	b_{12}		b_{21}	b_{22}
	r_{11}	r_{12}		r_{21}	r_{22}

TABLE 2.8(c)   Observed Counts:
Closed Cohort Study

F	E	
	1	2
1	r_{11}	r_{21}
2	r_{12}	r_{22}
	$r_{1\bullet}$	$r_{2\bullet}$

$$= \pi_{22}\left[\frac{p_{21}\pi_{21}}{\pi_{22}} + (1 - p_{21})\right]$$
$$= \pi_{22}[p_{21}\rho_{FD|E=2} + (1 - p_{21})]$$
$$= \pi_{22}[1 + (\rho_{FD|E=2} - 1)p_{21}].$$

With $RR_{ED} = \pi_1/\pi_2$ and $\rho_{ED|F=2} = \pi_{12}/\pi_{22}$ it follows that

$$RR_{ED} = \rho_{ED|F=2}\left[\frac{1 + (\rho_{FD|E=1} - 1)p_{11}}{1 + (\rho_{FD|E=2} - 1)p_{21}}\right].$$

We now assume that the risk ratio for the E–D relationship is homogeneous across strata of F, that is, $\rho_{ED|F=1} = \rho_{ED|F=2}$ $(= \rho_{ED})$. Since

$$\rho_{ED|F=1} = \frac{a_{11}/r_{11}}{a_{21}/r_{21}}$$

and

$$\rho_{ED|F=2} = \frac{a_{12}/r_{12}}{a_{22}/r_{22}}$$

it follows that $\rho_{FD|E=1} = \rho_{FD|E=2}$ $(= \rho_{FD})$ and, consequently, that

$$RR_{ED} = \rho_{ED} \left[\frac{1 + (\rho_{FD} - 1)p_{11}}{1 + (\rho_{FD} - 1)p_{21}} \right]$$

(Cornfield et al., 1959; Schlesselman, 1978; Simon, 1980; Gastwirth et al., 2000). From Table 2.8(c), which can be derived from either Table 2.8(a) or 2.8(b), the risk ratio for the E–F relationship is $RR_{EF} = p_{11}/p_{21}$. Note that p_{11} is the prevalence rate of the confounder in the exposed population, and p_{21} is the corresponding prevalence rate in the unexposed population. Writing $p_{11} = p_{21}RR_{EF}$, we have

$$RR_{ED} = \rho_{ED} \left[\frac{1 + (\rho_{FD} - 1)p_{21}RR_{EF}}{1 + (\rho_{FD} - 1)p_{21}} \right]. \tag{2.24}$$

According to terminology introduced earlier, RR_{ED} is the crude risk ratio for the E–D relationship and ρ_{ED} is the summary risk ratio. Since F is a confounder but not an effect modifier, Table 2.6(a) specifies that ρ_{ED} should be reported for the study. When F is an unknown confounder, RR_{ED} will be reported instead. In most applications, F will be positively associated with both D and E, that is, $\rho_{FD} > 1$ and $RR_{EF} > 1$. Consequently RR_{ED}/ρ_{ED}, which is a measure of the bias due to F being an unknown confounder, will usually be greater than 1.

Table 2.9 gives values of RR_{ED}/ρ_{ED} for selected values of p_{21}, RR_{EF}, and ρ_{FD}. As can be seen, provided the unknown confounder has a low prevalence in the unexposed population, is not closely associated with exposure, and is not a major risk fac-

TABLE 2.9 RR_{ED}/ρ_{ED} for Selected Values of p_{21}, RR_{EF}, and ρ_{FD}

		ρ_{FD}		
p_{21}	RR_{EF}	2	5	10
.01	2	1.01	1.04	1.08
.01	5	1.04	1.15	1.33
.05	2	1.05	1.17	1.31
.05	5	1.19	1.67	2.24
.10	2	1.09	1.29	1.47
.10	5	1.36	2.14	2.89

tor for the disease, the degree of bias will be relatively small. Analogous arguments can be used to determine the bias due to an unknown confounder of the odds ratio in a case-control study (Greenland, 1996a; Rothman and Greenland, 1998, Chapter 19).

2.8   MISCLASSIFICATION

In addition to confounding, there are other important types of systematic error that can arise in epidemiologic studies, one of which is misclassification. Misclassification, which can be either random or systematic, is said to have occurred when subjects are assigned incorrectly to exposure–disease categories. For example, due to chance coding errors, a case might be labeled incorrectly as a noncase (random misclassification). On the other hand, a subject who is actually free of disease might deliberately misrepresent symptoms and be diagnosed incorrectly as having an illness (systematic misclassification). In what follows we discuss only systematic misclassification.

Consider a closed cohort study in which the assessment of disease status, but not the measurement of exposure, is prone to misclassification. We assume that each subject has a "true" disease state that is unknown to the investigator. The true disease state of each subject could be determined by appealing to a "gold standard," but this is not part of the study (otherwise there would be no misclassification). Table 2.1(a) will be used to represent the true cross-classification of subjects in the study.

Let α_1 be the proportion of the exposed cohort who truly develop the disease and are diagnosed correctly as having the disease, and let β_1 be the proportion of the exposed cohort who truly do not develop the disease and are diagnosed correctly as not having the disease. We refer to α_1 and β_1 as the sensitivity and specificity of diagnosis for the exposed cohort, respectively. Table 2.10(a) gives the cross-classification of the r_1 exposed subjects in terms of observed (misclassified) and true disease status. For example, of the a_1 subjects who are exposed and truly develop the disease, $\alpha_1 a_1$ are diagnosed correctly as having the disease and $(1 - \alpha_1)a_1$ are misclassified. A corresponding interpretation applies to $\beta_1 b_1$ and $(1 - \beta_1)b_1$. In a similar manner we define α_2 and β_2, the sensitivity and specificity of diagnosis for the unexposed cohort, and derive Table 2.10(b). From the far right columns of Tables 2.10(a) and 2.10(b) we obtain Table 2.11, which gives the observed counts for the cohort study in the presence of misclassification.

TABLE 2.10(a)   Observed and True Counts: Exposed Cohort

Observed	True		
	$D = 1$	$D = 2$	
$D = 1$	$\alpha_1 a_1$	$(1 - \beta_1)b_1$	$\alpha_1 a_1 + (1 - \beta_1)b_1$
$D = 2$	$(1 - \alpha_1)a_1$	$\beta_1 b_1$	$(1 - \alpha_1)a_1 + \beta_1 b_1$
	a_1	b_1	r_1

TABLE 2.10(b) Observed and True Counts: Unexposed Cohort

Observed	True $D = 1$	$D = 2$	
$D = 1$	$\alpha_2 a_2$	$(1 - \beta_2)b_2$	$\alpha_2 a_2 + (1 - \beta_2)b_2$
$D = 2$	$(1 - \alpha_2)a_2$	$\beta_2 b_2$	$(1 - \alpha_2)a_2 + \beta_2 b_2$
	a_2	b_2	r_2

TABLE 2.11 Observed Counts: Closed Cohort Study

D	E 1	2
1	$\alpha_1 a_1 + (1 - \beta_1)b_1$	$\alpha_2 a_2 + (1 - \beta_2)b_2$
2	$(1 - \alpha_1)a_1 + \beta_1 b_1$	$(1 - \alpha_2)a_2 + \beta_2 b_2$
	r_1	r_2

Misclassification is said to be nondifferential when the sensitivities and specificities do not depend on exposure status—that is, when $\alpha_1 = \alpha_2 \ (= \alpha)$ and $\beta_1 = \beta_2$ $(= \beta)$. When these conditions are not satisfied, the term differential misclassification is used. In the nondifferential case, the observed odds ratio is

$$OR^* = \frac{[\alpha a_1 + (1 - \beta)b_1][(1 - \alpha)a_2 + \beta b_2]}{[\alpha a_2 + (1 - \beta)b_2][(1 - \alpha)a_1 + \beta b_1]}.$$

Expanding the numerator and denominator of OR^*, factoring out $\alpha \beta a_2 b_1$, and noting that $\omega_1 = a_1/b_1$ and $\omega_2 = a_2/b_2$, we find that

$$OR^* = \frac{OR + \phi + \psi}{1 + \phi OR + \psi} \tag{2.25}$$

where

$$\phi = \frac{(1 - \alpha)(1 - \beta)}{\alpha \beta}$$

$$\psi = \frac{(1 - \alpha)\omega_1}{\beta} + \frac{(1 - \beta)\omega_2}{\alpha} \tag{2.26}$$

and

$$OR = \frac{a_1 b_2}{a_2 b_1}$$

is the true odds ratio.

Suppose that $OR > 1$ and that $0 < \alpha < 1$ and $0 < \beta < 1$. Since $\phi > 0$ and $\psi > 0$, it follows that $(1 + \phi + \psi) < (1 + \phi OR + \psi)$. So, from (2.25), we have

$$OR^* < \frac{OR + \phi + \psi}{1 + \phi + \psi} < OR. \tag{2.27}$$

Using (2.25) it is readily demonstrated that $OR^* > 1$ if and only if $\phi < 1$, the latter inequality being equivalent to $\alpha + \beta > 1$. A diagnostic process that is so prone to misclassification that $\alpha + \beta < 1$ is unlikely to be used in an epidemiologic study. So, in practice, we usually have $OR^* > 1$. This can be combined with (2.27) to give $1 < OR^* < OR$. Likewise, when $OR < 1$ we find that $OR < OR^* < 1$. This shows that, provided $\alpha + \beta > 1$, nondifferential misclassification biases the observed odds ratio toward the "null"—that is, toward 1 (Copeland et al., 1977). When misclassification is differential, no such general statement can be made about the direction of bias.

As before, let π_2 be the probability that a subject in the unexposed cohort truly develops the disease. Then $\omega_2 = \pi_2/(1 - \pi_2)$ and $\omega_1 = OR\omega_2 = OR\pi_2/(1 - \pi_2)$. Substituting in (2.26), we have

$$\psi = \left[\frac{(1 - \alpha)OR}{\beta} + \frac{(1 - \beta)}{\alpha} \right] \left(\frac{\pi_2}{1 - \pi_2} \right)$$

TABLE 2.12   Values of OR^*/OR for Selected Values of α, β, π_2, and OR

			OR		
α	β	π_2	2	5	10
.95	.95	.25	.92	.83	.76
.90	.95	.25	.90	.78	.67
.95	.90	.25	.86	.75	.67
.90	.90	.25	.84	.70	.59
.95	.95	.10	.83	.73	.68
.90	.95	.10	.82	.70	.63
.95	.90	.10	.75	.60	.53
.90	.90	.10	.74	.57	.50
.95	.95	.05	.75	.59	.54
.90	.95	.05	.74	.58	.51
.95	.90	.05	.66	.46	.39
.90	.90	.05	.66	.45	.37

and so OR^* is a function of α, β, π_2, and OR. Table 2.12 gives values of OR^*/OR for selected values of α, β, π_2, and OR. As can be seen, even when the sensitivity and specificity are quite high, nondifferential misclassification can lead to severe underestimation of the true odds ratio, especially when the probability of disease in the unexposed population is small.

A similar analysis can be carried out for case-control studies, except that it is exposure rather than disease which is assumed to be prone to misclassification. When the exposure variable is polychotomous—that is, has more than two categories—the observed odds ratios may be biased away from the null or may even have values on the other side of the null (Dosemeci et al., 1990; Birkett, 1992). Misclassification can also affect confounding variables. When this occurs, odds ratios may be biased away from the null, spurious heterogeneity may appear, and true heterogeneity may be masked (Greenland, 1980; Brenner, 1993). Under these conditions the usual methods to control confounding are not effective (Greenland, 1980; Greenland and Robins, 1985a).

2.9 SCOPE OF THIS BOOK

We conclude this chapter with a few remarks on the scope of this book and the role of statistics in epidemiology. There is no question that statisticians have had, and continue to have, an enormous impact on epidemiology as a field of scientific inquiry. Probability models have been used to clarify fundamental methodologic issues, and innovative statistical techniques have been developed to accommodate features of epidemiologic study designs. Examples of the latter are logistic regression for case-control studies and Cox regression for cohort studies, but these are just two of many achievements that could be cited.

This book is mostly concerned with the technical aspects of statistical methods as applied to epidemiologic data. However, there are deeper issues that need to be considered when examining the role of statistics in epidemiology. We began this chapter with the observation that there are two types of error in epidemiologic studies, namely, random and systematic. Much of the effort expended in an epidemiologic study is devoted to ensuring that systematic error is kept to a minimum. In the ideal situation where this type of error has been eliminated, only random error remains. Under these circumstances, inferential statistical methods, such as hypothesis tests and confidence intervals, are appropriate. Outside of the specialized setting of randomized controlled trials it is difficult to ensure that systematic error, in particular confounding, has been satisfactorily addressed. This raises important issues about the role of inferential statistical methods in epidemiologic research (Greenland, 1990).

In this book we survey a number of nonregression methods that have been developed to analyze data from epidemiologic studies. With the increasing availability of sophisticated statistical packages it is now easy to fit complicated regression models and produce masses of computer output. However, it is well to remember that elaborate statistical methods cannot compensate for a badly designed study and poorly collected data (Freedman, 1999). An advantage of nonregression techniques is that they bring the investigator into close contact with data in a way that regression methods typically do not. In addition, nonregression techniques are often conceptually more accessible than their regression counterparts, a feature that is helpful when explaining the results of a data analysis to those with a limited background in statistical methods. One of the aims of this book is to identify a select number of nonregression techniques that are computationally convenient and which can be used to explore epidemiologic data prior to a more elaborate regression analysis.

CHAPTER 3

Binomial Methods for Single Sample Closed Cohort Data

In Chapter 2 a number of measurement issues that are important in epidemiology were discussed. For expository purposes a deterministic approach was used, thereby eliminating the need to consider random error. We now return to the stochastic setting and describe methods for analyzing data from a closed cohort study. Recall from Section 2.2.1 that in a closed cohort study, subjects either develop the disease or do not, and those not developing it necessarily have the same length of follow-up. This is the least complicated cohort design, but it nevertheless provides a convenient vehicle for presenting some basic methods of data analysis. In this chapter we consider only a single sample—that is, one where the cohort is considered in its entirety, with no comparisons made across exposure categories. For example, the following methods could be used to analyze data from a cohort study in which a group of cancer patients is followed for 5 years with the aim of estimating the 5-year mortality rate. The methods to be described are based on the binomial distribution,

$$P(A = a|\pi) = \binom{r}{a} \pi^a (1 - \pi)^{r-a}$$

where π is the probability of developing the disease, r is the number of individuals in the cohort, and a is the number of cases that occur during the course of follow-up.

3.1 EXACT METHODS

As remarked in Section 1.1.3, in statistics the term "exact" means that an actual probability function is being used to perform calculations, as opposed to a normal approximation. The advantage of exact methods is that they do not rely on asymptotic properties and hence are valid regardless of sample size. The drawback, as will soon become clear, is that exact methods can be computationally intensive, especially when the sample size is large. Fortunately, this is precisely the situation where a normal approximation is appropriate, and so between the two approaches it is usually possible to perform a satisfactory data analysis.

3.1.1 Hypothesis Test

Suppose that we wish to test the null hypothesis $H_0 : \pi = \pi_0$, where π_0 is a given value of the probability parameter. If H_0 is rejected, we conclude that π does not equal π_0. To properly interpret this finding it is necessary to have an explicit alternative hypothesis H_1. For example, it may be that there are only two possible values of π, namely, π_0 and π_1. In this case the alternative hypothesis is necessarily $H_1 : \pi = \pi_1$. If we believe that π cannot be less than π_0 we are led to consider the one-sided alternative hypothesis $H_1 : \pi > \pi_0$. Before proceeding with a one-sided test it is important to ensure that the one-sided assumption is valid. For example, in a randomized controlled trial of a new drug compared to usual therapy, it may be safe to assume that the innovative drug is at least as beneficial as standard treatment. The two-sided alternative hypothesis corresponding to $H_0 : \pi = \pi_0$ is $H_1 : \pi \neq \pi_0$. Ordinarily it is difficult to justify a one-sided alternative hypothesis, and so in most applications a two-sided test is used. Except for portions of this chapter, in this book we consider only two-sided tests. In particular, all chi-square tests are two-sided.

To test $H_0 : \pi = \pi_0$ we need to decide whether the observed outcome is likely or unlikely under the assumption that π_0 is the true value of π. With a as the outcome, the lower and upper tail probabilities for the binomial distribution with parameters (π_0, r) are defined to be

$$P(A \leq a | \pi_0) = \sum_{x=0}^{a} \binom{r}{x} \pi_0^x (1 - \pi_0)^{r-x}$$

$$= 1 - \sum_{x=a+1}^{r} \binom{r}{x} \pi_0^x (1 - \pi_0)^{r-x} \tag{3.1}$$

and

$$P(A \geq a | \pi_0) = \sum_{x=a}^{r} \binom{r}{x} \pi_0^x (1 - \pi_0)^{r-x}$$

$$= 1 - \sum_{x=0}^{a-1} \binom{r}{x} \pi_0^x (1 - \pi_0)^{r-x} \tag{3.2}$$

respectively.

Let $p_{\min}$ be the smaller of $P(A \leq a | \pi_0)$ and $P(A \geq a | \pi_0)$. Then $p_{\min}$ is the probability of observing an outcome at least as "extreme" as a at that end of the distribution. For a one-sided alternative hypothesis, $p_{\min}$ is defined to be the p-value of the test. To compute the two-sided p-value we need a method of determining a corresponding probability at the "other end" of the distribution. The two-sided p-value is defined to be the sum of these two probabilities. One possibility is to define the second probability to be the largest tail probability at the other end of the distribution which does not exceed $p_{\min}$. We refer to this approach as the cumulative method. An alternative is to define the second probability to be equal to $p_{\min}$, in which case the two-sided p-value is simply $2 \times p_{\min}$. We refer to this approach as the doubling method. Evidently the doubling method produces a two-sided p-value at least as large as the one obtained using the cumulative approach. When the distribution is approximately symmetrical, the two methods produce similar results. For the binomial distribution this will be the case when π_0 is near .5. There does not appear to be a consensus as to whether the cumulative method or the doubling method is the best approach to calculating two-sided p-values (Yates, 1984).

In order to make a decision about whether or not to reject H_0, we need to select a value for α, the probability of a type I error (Section 2.1). According to the classical approach to hypothesis testing, when the p-value is less than α, the null hypothesis is rejected. An undesirable practice is to simply report a hypothesis test as either "statistically significant" or "not statistically significant," according to whether the p-value is less than α or not, respectively. A more informative way of presenting results is to give the actual p-value. This avoids confusion when the value of α has not been made explicit, and it gives the reader the option of interpreting the hypothesis test according to other choices of α. In this book we avoid any reference to "statistical significance," preferring instead to comment on the "evidence" provided by a p-value (relative to a given α). For descriptive purposes we adopt the current convention of setting $\alpha = .05$. So, for example, when the p-value is "much smaller" than .05 we comment on this finding by saying that the data provide "little evidence" for H_0.

Referring to the hypothesis test based on (3.1) and (3.2) as "exact" is apt to leave the impression that, when the null hypothesis is true and $\alpha = .05$, over the course of many replications of the study the null hypothesis will be rejected 5% of the time. In most applications, an exact hypothesis test based on a discrete distribution will reject the null hypothesis less frequently than is indicated by the nominal value of α. The reason is that the tail probabilities of a discrete distribution do not assume all possible values between 0 and 1. Borrowing an example from Yates (1984), consider a study in which a coin is tossed 10 times. Under the null hypothesis that the coin is fair, the study can be modeled using the binomial distribution with parameters (.5, 10). In this case, $P(A \geq 8|.5) = 5.5\%$ and $P(A \geq 9|.5) = 1.1\%$. It follows that, based on a one-sided test with $\alpha = .05$, the null hypothesis will be rejected 1.1%, not 5%, of the time. For this reason, exact tests are said to be conservative.

In the examples presented in this book we routinely use more decimal places than would ordinarily be justified by the sample size under consideration. The reason is that we often wish to compare findings based on several statistical techniques, and in many instances the results are so close in value that a large number of decimal places is needed in order to demonstrate a difference. Most of the calculations in this book were performed on a computer. In many of the examples, one or more intermediate steps have been included rather than just the final answer. The numbers in the intermediate steps have necessarily been rounded and so may not lead to precisely the final answer given in the example.

Example 3.1   Let $a = 2$ and $r = 10$, and consider $H_0 : \pi_0 = .4$. The binomial distribution with parameters (.4, 10) is given in Table 3.1. Since $P(A \leq 2|.4) = .167$ and $P(A \geq 2|.4) = .954$, it follows that $p_{\min} = .167$. At the other end of the

TABLE 3.1   Probability Function (%) for the Binomial
Distribution with Parameters (.4, 10)

a	$P(A = a\|.4)$	$P(A \leq a\|.4)$	$P(A \geq a\|.4)$
0	.60	.60	100
1	4.03	4.64	99.40
2	12.09	16.73	95.36
3	21.50	38.23	83.27
4	25.08	63.31	61.77
5	20.07	83.38	36.69
6	11.15	94.52	16.62
7	4.25	98.77	5.48
8	1.06	99.83	1.23
9	.16	99.99	.17
10	.01	100	.01

distribution the largest tail probability not exceeding p_{min} is $P(A \geq 6|.4) = .166$.
So the two-sided p-value based on the cumulative method is $p = .167 + .166 = .334$.
According to the doubling approach the two-sided p-value is also $p = 2(.167) = .334$. In view of these results there is little evidence to reject H_0.

3.1.2   A Critique of p-values and Hypothesis Tests

A "small" p-value means that, under the assumption that H_0 is true, an outcome as
extreme as, or more extreme than, the one that was observed is unlikely. We interpret
such a finding as evidence against H_0, but this is not the same as saying that evidence
has been found in favor of H_1. The reason for making this distinction is that the p-
value is defined exclusively in terms of H_0 and so does not explicitly contrast H_0 with
H_1. Intuitively it seems that a decision about whether an outcome should be regarded
as likely or unlikely ought to depend on a direct comparison with other possible
outcomes. This comparative feature is missing from the p-value. For this reason and
others, the p-value is considered by some authors to be a poor measure of "evidence"
(Goodman and Royall, 1988; Goodman, 1993; Schervish, 1996). An approach that
avoids this problem is to base inferential procedures on the likelihood ratio. This
quantity is defined to be the quotient of the likelihood of the observed data under
the assumption that H_0 is true, divided by the corresponding likelihood under the
assumption that H_1 is true (Edwards, 1972; Clayton and Hills, 1993; Royall, 1997).
Likelihood ratio methods involve considerations beyond the scope of this book and
so, except for likelihood ratio tests, this approach to statistical inference will not be
considered further.

From the epidemiologic perspective, another problem with the classical use of
p-values is that they are traditionally geared toward all-or-nothing decisions: Based
on the magnitude of the p-value and the agreed-upon α, H_0 is either rejected or not.
In epidemiology this approach to data analysis is usually unwarranted because the

findings from a single epidemiologic study are rarely, if ever, definitive. Rather, the advancement of knowledge based on epidemiologic research tends to be cumulative, with each additional study contributing incrementally to our understanding. Not infrequently, epidemiologic studies produce conflicting results, making the evaluation of research findings that much more challenging. Given these uncertainties, in epidemiology we are usually interested not only in whether the true value of a parameter is equal to some hypothesized value but, more importantly, what is the range of plausible values for the parameter. Confidence intervals provide this kind of information. In recent years the epidemiologic literature has become critical of *p*-values and has placed increasingly greater emphasis on the use of confidence intervals (Rothman, 1978; Gardner and Altman, 1986; Poole, 1987). Despite these concerns, hypothesis tests and *p*-values are in common use in the analysis of epidemiologic data, and so they are given due consideration in this book.

3.1.3 Confidence Interval

Let α be the probability of a type I error and consider a given method of testing the null hypothesis $H_0 : \pi = \pi_0$. A $(1 - \alpha) \times 100\%$ confidence interval for π is defined to be the set of all parameter values π_0 such that H_0 is not rejected. In other words, the confidence interval is the set of all π_0 that are consistent with the study data (for a given choice of α). Note that according to this definition, different methods of testing the null hypothesis, such as exact and asymptotic methods, will usually lead to somewhat different confidence intervals. The process of obtaining a confidence interval in this manner is referred to as inverting the hypothesis test. In Example 3.1 the "data" consist of $a = 2$ and $r = 10$. Based on the doubling method, the exact test of $H_0 : \pi_0 = .4$ resulted in a two-sided *p*-value of .334. By definition, .4 is in the exact $(1 - \alpha) \times 100\%$ confidence interval for π when the null hypothesis $H_0 : \pi = .4$ is not rejected, and this occurs when the *p*-value is greater than α—that is, when $\alpha \leq .334$.

A $(1 - \alpha) \times 100\%$ confidence interval for π will be denoted by $[\underline{\pi}, \overline{\pi}]$. We refer to $\underline{\pi}$ and $\overline{\pi}$ as the lower and upper bounds of the confidence interval, respectively. In keeping with established mathematical notation, square brackets are used to indicate that the confidence interval contains all possible values of the parameter that are greater than or equal to $\underline{\pi}$ and less than or equal to $\overline{\pi}$. It is sometimes said (incorrectly) that $1 - \alpha$ is the probability that π is in $[\underline{\pi}, \overline{\pi}]$. This manner of speaking suggests that the confidence interval is fixed and that, for a given study, π is a random quantity which is either in the confidence interval or not. In fact, precisely the opposite is true. Both $\underline{\pi}$ and $\overline{\pi}$ are random variables and so $[\underline{\pi}, \overline{\pi}]$ is actually a random interval. This explains why a confidence interval is sometimes referred to as an interval estimate. An appropriate interpretation of $[\underline{\pi}, \overline{\pi}]$ is as follows: Over the course of many replications of the study, $(1 - \alpha) \times 100\%$ of the "realizations" of the confidence interval will contain π.

An exact $(1 - \alpha) \times 100\%$ confidence interval for π is obtained by solving the equations

$$\frac{\alpha}{2} = P(A \geq a|\underline{\pi}) = \sum_{x=a}^{r} \binom{r}{x} \underline{\pi}^x (1 - \underline{\pi})^{r-x}$$

$$= 1 - \sum_{x=0}^{a-1} \binom{r}{x} \underline{\pi}^x (1 - \underline{\pi})^{r-x} \tag{3.3}$$

and

$$\frac{\alpha}{2} = P(A \leq a|\overline{\pi}) = \sum_{x=0}^{a} \binom{r}{x} \overline{\pi}^x (1 - \overline{\pi})^{r-x}$$

$$= 1 - \sum_{x=a+1}^{r} \binom{r}{x} \overline{\pi}^x (1 - \overline{\pi})^{r-x} \tag{3.4}$$

for $\underline{\pi}$ and $\overline{\pi}$. Since there is only one unknown in each equation, the solutions can be found by trial and error. If $a = 0$, which sometimes happens when the probability of disease is low, we define $\underline{\pi} = 0$ and $\overline{\pi} = 1 - \alpha^{1/r}$ (Louis, 1981; Jovanovic, 1998). In StatXact (1998, §12.3), a statistical package designed to perform exact calculations, the upper bound is defined to be $\overline{\pi} = 1 - (\alpha/2)^{1/r}$. Since an exact confidence interval is obtained by inverting an exact test, when the distribution is discrete, the resulting exact confidence interval will be conservative—that is, wider than is indicated by the nominal value of α (Armitage and Berry, 1994, p.123).

Example 3.2   Let $a = 2$ and $r = 10$. From

$$.025 = 1 - \sum_{x=0}^{1} \binom{10}{x} \underline{\pi}^x (1 - \underline{\pi})^{10-x}$$

$$= 1 - \left[(1 - \underline{\pi})^{10} + 10\underline{\pi}(1 - \underline{\pi})^9 \right]$$

and

$$.025 = \sum_{x=0}^{2} \binom{10}{x} \overline{\pi}^x (1 - \overline{\pi})^{10-x}$$

$$= (1 - \overline{\pi})^{10} + 10\overline{\pi}(1 - \overline{\pi})^9 + 45\overline{\pi}^2 (1 - \overline{\pi})^8$$

a 95% confidence interval for π is [.025, .556]. Note that the confidence interval includes $\pi_0 = .4$, a finding that is consistent with the results of Example 3.1.

3.2   ASYMPTOTIC METHODS

When r is large the calculations required by exact methods can be prohibitive. Under these conditions an asymptotic approach based on a normal approximation provides a practical alternative. Sometimes, by transforming a random variable the normal approximation can be improved. We first discuss methods where no transformation is involved and then consider the odds and log-odds transformations.

3.2.1 No Transformation

Point Estimate
The maximum likelihood estimates of π and var($\hat{\pi}$) are

$$\hat{\pi} = \frac{a}{r}$$

and

$$\widehat{\text{var}}(\hat{\pi}) = \frac{\hat{\pi}(1-\hat{\pi})}{r} = \frac{a(r-a)}{r^3}.$$

Confidence Interval
As defined in Section 1.1.2, for $0 < \gamma < 1$, z_γ is the number that cuts off the upper γ-tail probability of the standard normal distribution. That is, $P(Z \geq z_\gamma) = \gamma$ where Z is standard normal. According to (3.3), we need to solve the equation $P(A \geq a|\underline{\pi}) = \alpha/2$ for $\underline{\pi}$. This equation can be written in the equivalent form,

$$P\left(\frac{A - \underline{\pi}r}{\sqrt{\underline{\pi}(1-\underline{\pi})r}} \geq \frac{a - \underline{\pi}r}{\sqrt{\underline{\pi}(1-\underline{\pi})r}} \middle| \underline{\pi}\right) = \frac{\alpha}{2}. \tag{3.5}$$

From Section 1.1.3 the random variable $Z = (A - \underline{\pi}r)/\sqrt{\underline{\pi}(1-\underline{\pi})r}$ is asymptotically standard normal. It follows that, when r is large, (3.5) is approximately the same probability statement as

$$\frac{a - \underline{\pi}r}{\sqrt{\underline{\pi}(1-\underline{\pi})r}} = z_{\alpha/2}. \tag{3.6}$$

An analogous argument leads to

$$\frac{a - \overline{\pi}r}{\sqrt{\overline{\pi}(1-\overline{\pi})r}} = -z_{\alpha/2}. \tag{3.7}$$

We can combine (3.6) and (3.7) into the single identity

$$\frac{(a - \pi r)^2}{\pi(1-\pi)r} = (z_{\alpha/2})^2$$

where, for the moment, we treat π as a continuous variable. This is a second-degree polynomial in π which can be solved using the "quadratic formula" to give the $(1-\alpha) \times 100\%$ confidence interval

$$[\underline{\pi}, \overline{\pi}] = \frac{-u \pm \sqrt{u^2 - 4tv}}{2t}$$

where

$$t = r\left[r + (z_{\alpha/2})^2\right]$$

$$u = -r\left[2a + (z_{\alpha/2})^2\right]$$

$$v = a^2.$$

This will be referred to as the implicit method of estimating the confidence interval since $\underline{\pi}$ and $\overline{\pi}$ are present in the variance terms of (3.6) and (3.7). An alternative approach, which we refer to as the explicit method, is to replace $\underline{\pi}$ and $\overline{\pi}$ in the variance terms with the point estimate $\hat{\pi} = a/r$. This gives

$$\frac{a - \underline{\pi}r}{\sqrt{\hat{\pi}(1 - \hat{\pi})r}} = z_{\alpha/2}$$

and

$$\frac{a - \overline{\pi}r}{\sqrt{\hat{\pi}(1 - \hat{\pi})r}} = -z_{\alpha/2}$$

or, equivalently,

$$\underline{\pi} = \hat{\pi} - z_{\alpha/2}\sqrt{\frac{\hat{\pi}(1 - \hat{\pi})}{r}}$$

and

$$\overline{\pi} = \hat{\pi} + z_{\alpha/2}\sqrt{\frac{\hat{\pi}(1 - \hat{\pi})}{r}}.$$

In a more compact notation we can write

$$[\underline{\pi}, \overline{\pi}] = \hat{\pi} \pm z_{\alpha/2}\sqrt{\frac{\hat{\pi}(1 - \hat{\pi})}{r}}. \tag{3.8}$$

A potential problem with the explicit method is that one or both of the bounds may fall outside the range of 0 to 1. As illustrated below, this is especially likely to occur when $\hat{\pi}$ is close to 0 or 1, and r is small.

Continuity corrections were included in the calculations in Example 1.6, and in a similar fashion they could have been incorporated into the above asymptotic formulas. The question of whether continuity corrections should be used has been debated at length in the statistical literature with no clear resolution of the issue (Grizzle, 1967; Mantel and Greenhouse, 1968; Conover, 1974). When sample sizes are moderately large, the effect of a continuity correction is usually negligible. In order to simplify formulas, continuity corrections will not be used in this book.

Hypothesis Test

Under the null hypothesis $H_0 : \pi = \pi_0$, the maximum likelihood estimates of the mean and variance of $\hat{\pi}$ are $E_0(\hat{\pi}) = \pi_0$ and $\mathrm{var}_0(\hat{\pi}) = \pi_0(1 - \pi_0)/r$. A subscript 0 has been added to the notation to indicate that calculations are being performed under the null hypothesis. A test of H_0 is

$$X^2 = \frac{(\hat{\pi} - \pi_0)^2}{\pi_0(1 - \pi_0)/r} = \frac{(a - \pi_0 r)^2}{\pi_0(1 - \pi_0)r} \qquad (\mathrm{df} = 1). \qquad (3.9)$$

The notation in (3.9) is meant to indicate that X^2 is asymptotically chi-square with 1 degree of freedom. This convention will be adhered to throughout the book because virtually all random variables denoted by the X^2 notation are asymptotically, rather than exactly, chi-square.

3.2.2 Odds and Log-Odds Transformations

Point Estimate

Recall from Section 2.2.2 that for $\pi \neq 1$ the odds is defined to be $\omega = \pi/(1 - \pi)$. For $0 < \pi < 1$, we define the log-odds to be $\log(\omega) = \log[\pi/(1 - \pi)]$. In this book the only logarithm considered is the logarithm to the base e. The maximum likelihood estimates of ω and $\log(\omega)$ are

$$\hat{\omega} = \frac{\hat{\pi}}{1 - \hat{\pi}} = \frac{a}{r - a} \qquad (3.10)$$

and

$$\log(\hat{\omega}) = \log\left(\frac{\hat{\pi}}{1 - \hat{\pi}}\right) = \log\left(\frac{a}{r - a}\right). \qquad (3.11)$$

If either a or $r - a$ equals 0, we replace (3.10) and (3.11) with

$$\hat{\omega} = \frac{a + .5}{r - a + .5}$$

and

$$\log(\hat{\omega}) = \log\left(\frac{a + .5}{r - a + .5}\right).$$

Haldane (1955) and Anscombe (1956) showed that $\log(\hat{\omega})$ is less biased when .5 is added to a and $r - a$, whether they are 0 or not. This practice does not appear to be in widespread use and so it will not be followed here.

Figures 1.1(b)–1.5(b) and Figures 1.1(c)–1.5(c) show graphs of the distributions of $\hat{\omega}$ and $\log(\hat{\omega})$, respectively, corresponding to the binomial distributions in Figures

1.1(a)–1.5(a). Evidently, $\hat{\omega}$ can be highly skewed, especially when r is small. On the other hand, $\log(\hat{\omega})$ is relatively symmetric, but no more so than the untransformed distribution. On the basis of these findings, there seems to be little incentive to consider either the odds or log-odds transformations in preference to the untransformed distribution when analyzing single sample binomial data. As will be demonstrated in Chapter 4, the log-odds ratio transformation has an important role to play when analyzing data using odds ratio methods.

Confidence Interval

The maximum likelihood estimate of var[$\log(\hat{\omega})$] is

$$\hat{\mathrm{var}}[\log(\hat{\omega})] = \frac{1}{\hat{\pi}(1 - \hat{\pi})r} = \frac{1}{a} + \frac{1}{r - a}. \tag{3.12}$$

If either a or $r - a$ equals 0, we replace (3.12) with

$$\hat{\mathrm{var}}[\log(\hat{\omega})] = \frac{1}{a + .5} + \frac{1}{r - a + .5}.$$

Gart and Zweifel (1967) showed that $\hat{\mathrm{var}}[\log(\hat{\omega})]$ is less biased when .5 is added to a and $r - a$, whether they are 0 or not. Similar to the situation with $\log(\hat{\omega})$, this convention does not appear to be widely accepted and so it will not be adopted in this book. A $(1 - \alpha) \times 100\%$ confidence interval for $\log(\omega)$ is

$$[\log(\underline{\omega}), \log(\overline{\omega})] = \left[\log\left(\frac{\hat{\pi}}{1 - \hat{\pi}} \right) \right] \pm \frac{z_{\alpha/2}}{\sqrt{\hat{\pi}(1 - \hat{\pi})r}}. \tag{3.13}$$

To obtain $[\underline{\pi}, \overline{\pi}]$ we first exponentiate (3.13) to get $[\underline{\omega}, \overline{\omega}]$, and then use

$$\underline{\pi} = \frac{\underline{\omega}}{1 + \underline{\omega}} \tag{3.14}$$

and

$$\overline{\pi} = \frac{\overline{\omega}}{1 + \overline{\omega}} \tag{3.15}$$

to determine $\underline{\pi}$ and $\overline{\pi}$. Since the exponential function is nonnegative, it follows from (3.13) that $\underline{\omega}$ and $\overline{\omega}$ are always nonnegative, and hence that $\underline{\pi}$ and $\overline{\pi}$ are always between 0 and 1.

Hypothesis Test

Under the null hypothesis $H_0 : \pi = \pi_0$, the maximum likelihood estimates of the mean and variance of $\log(\hat{\omega})$ are $E_0[\log(\hat{\omega})] = \log[\pi_0/(1 - \pi_0)]$ and $\mathrm{var}_0[\log(\hat{\omega})] =$

$1/[\pi_0(1 - \pi_0)r]$. A test of H_0 is

$$X^2 = \left[\log\left(\frac{\hat{\pi}}{1 - \hat{\pi}} \right) - \log\left(\frac{\pi_0}{1 - \pi_0} \right) \right]^2 [\pi_0(1 - \pi_0)r] \qquad (\text{df} = 1). \qquad (3.16)$$

Example 3.3 Let $a = 10$ and $r = 50$. From (3.8) and (3.13), 95% confidence intervals are

$$[\underline{\pi}, \overline{\pi}] = .2 \pm 1.96\sqrt{\frac{.2(.8)}{50}} = [.089, .311]$$

and

$$[\log(\underline{\omega}), \log(\overline{\omega})] = \log\left(\frac{.2}{.8} \right) \pm \frac{1.96}{\sqrt{.2(.8)(50)}} = [-2.08, -.693]. \qquad (3.17)$$

Exponentiating (3.17) results in $[\underline{\omega}, \overline{\omega}] = [.125, .500]$, and applying (3.14) and (3.15) gives $[\underline{\pi}, \overline{\pi}] = [.111, .333]$.

An approach to determining whether a method of estimation is likely to produce satisfactory results is to perform a simulation study, also called a Monte-Carlo study. This proceeds by programming a random number generator to create a large number of replicates of a hypothetical study. From these "data," results based on different methods of estimation are compared to quantities that were used to program the random number generator. In most simulation studies, exact methods tend to perform better than asymptotic methods, especially when the sample size in each replicate is small. Consequently it is useful to compare asymptotic and exact estimates as in the following examples, with exact results used as the benchmark.

Example 3.4 Table 3.2 gives 95% confidence intervals for π, where, in each case, $\hat{\pi} = .2$. When $a = 10$, the implicit and log-odds methods perform quite well compared to the exact approach. To a lesser extent this is true for $a = 2$ and $a = 5$. The explicit method does not compare as favorably, especially for $a = 2$, where the lower bound is a negative number.

TABLE 3.2 95% Confidence Intervals (%) for π

Method	$a = 2$ $r = 10$		$a = 5$ $r = 25$		$a = 10$ $r = 50$	
	$\underline{\pi}$	$\overline{\pi}$	$\underline{\pi}$	$\overline{\pi}$	$\underline{\pi}$	$\overline{\pi}$
Exact	2.52	55.61	6.83	40.70	10.03	33.72
Implicit (3.6, 3.7)	5.67	50.98	8.86	39.13	11.24	33.04
Explicit (3.8)	−4.79	44.79	4.32	35.68	8.91	31.09
Log-odds (3.13)	5.04	54.07	8.58	39.98	11.11	33.33

TABLE 3.3 p-Values for Hypothesis Tests of $H_0 : \pi = .4$

Method	$a = 2$ $r = 10$	$a = 5$ $r = 25$	$a = 10$ $r = 50$
Exact (cumulative)	.334	.043	.004
No-transformation (3.9)	.197	.041	.004
Log-odds (3.16)	.129	.016	<.001

Example 3.5   Table 3.3 gives p-values for hypothesis tests of $H_0 : \pi = .4$, where, in each case, $\hat{\pi} = .2$. With the exact p-value based on the cumulative method as the benchmark, the no-transformation method performs somewhat better than the log-odds approach.

Example 3.6   In Chapter 4, data are presented from a closed cohort study in which 192 female breast cancer patients were followed for up to 5 years with death from breast cancer as the endpoint of interest. There were $a = 54$ deaths and so $\hat{\pi} = 54/192 = .281$. The 95% confidence interval based on the implicit method is [.222, .349].

Odds Ratio Methods for Unstratified Closed Cohort Data

In Chapter 2 we compared the measurement properties of the odds ratio, risk ratio and risk difference. None of these measures of effect was found to be superior to the other two in every respect. In this chapter we discuss odds ratio methods for analyzing data from a closed cohort study (Section 2.2.1). The reason for giving precedence to the odds ratio is that there is a wider range of statistical techniques available for this measure of effect than for either the risk ratio or risk difference. Thus the initial focus on the odds ratio reflects an organizational approach and is not meant to imply that the odds ratio is somehow "better" than the risk ratio or risk difference for analyzing closed cohort data. However, compared to the risk ratio and risk difference, it is true that methods based on the odds ratio are more readily applied to other epidemiologic study designs. As shown in Chapter 9, odds ratio methods for closed cohort studies can be used to analyze censored survival data; and, as discussed in Chapter 11, these same techniques can be adapted to the case-control setting.

For the most part, the material in this chapter has been organized according to whether methods are exact or asymptotic on the one hand, and unconditional or conditional on the other. This produces four broad categories: exact unconditional, asymptotic unconditional, exact conditional, and asymptotic conditional. Not all odds ratio methods fit neatly into this scheme, but the classification is useful. Within each of the categories we focus primarily on three topics: point estimation, (confidence) interval estimation, and hypothesis testing. For certain categories some of these topics will not be covered because the corresponding methods are not in wide use or their exposition requires a level of mathematical sophistication beyond the scope of this book. Exact unconditional methods will not be considered at all for several reasons: They can be intensely computational; they offer few, if any, advantages over the other techniques to be described; and they are rarely, if ever, used in practice. In Sections 4.1–4.5 we discuss odds ratio methods for tables in which the exposure is dichotomous, and in Section 4.6 we consider the case of a polychotomous exposure variable. General references for this chapter and the next are Breslow and Day (1980), Fleiss (1981), Sahai and Khurshid (1996), and Lachin (2000).

4.1 ASYMPTOTIC UNCONDITIONAL METHODS FOR A SINGLE 2 × 2 TABLE

The methods of this section are referred to as unconditional, a term that will be understood once the conditional approach has been introduced in Section 4.2. Table 4.1, which is similar to Table 2.1(a), gives the observed counts for a closed cohort study in which exposure is dichotomous. We assume either that the cohort is a simple random sample that has been split into exposed and unexposed cohorts or that the exposed and unexposed cohorts are distinct simple random samples. In either case we treat r_1 and r_2 as known constants.

We refer to r_1 and r_2 as the column marginal totals, and we refer to m_1 and m_2 as the row marginal totals. Taken together these four quantities are termed the marginal totals. As in Section 2.2.1, we assume that the development of disease in the exposed and unexposed cohorts is governed by binomial random variables A_1 and A_2 with parameters (π_1, r_1) and (π_2, r_2), respectively. As discussed in Section 2.2.1, it is assumed that subjects behave independently with respect to the development of disease. It follows that A_1 and A_2 are independent, and so their joint probability function is the product of the individual probability functions,

$$P(A_1 = a_1, A_2 = a_2 | \pi_1, \pi_2)$$

$$= \binom{r_1}{a_1} \pi_1^{a_1} (1 - \pi_1)^{r_1 - a_1} \times \binom{r_2}{a_2} \pi_2^{a_2} (1 - \pi_2)^{r_2 - a_2}. \tag{4.1}$$

Recall from Section 2.2 that $\omega_1 = \pi_1/(1 - \pi_1)$, $\omega_2 = \pi_2/(1 - \pi_2)$,

$$OR = \frac{\omega_1}{\omega_2} = \frac{\pi_1(1 - \pi_2)}{\pi_2(1 - \pi_1)}$$

and

$$\pi_1 = \frac{OR\pi_2}{OR\pi_2 + (1 - \pi_2)}. \tag{4.2}$$

In order to make the role of OR explicit, we substitute (4.2) in (4.1), which reparameterizes the joint probability function in terms of OR and π_2,

TABLE 4.1   Observed Counts: Closed Cohort Study

Disease	Exposure		
	yes	no	
yes	a_1	a_2	m_1
no	b_1	b_2	m_2
	r_1	r_2	r

$$P(A_1 = a_1, A_2 = a_2 | OR, \pi_2)$$

$$= \binom{r_1}{a_1} \left[\frac{OR\pi_2}{OR\pi_2 + (1 - \pi_2)} \right]^{a_1} \left[\frac{1 - \pi_2}{OR\pi_2 + (1 - \pi_2)} \right]^{r_1 - a_1}$$

$$\times \binom{r_2}{a_2} \pi_2^{a_2} (1 - \pi_2)^{r_2 - a_2}. \tag{4.3}$$

Following Section 1.2.1, we view (4.3) as a likelihood that is a function of the parameters OR and π_2.

Point Estimate

The unconditional maximum likelihood equations are

$$a_1 = \frac{\widehat{OR}_u \hat{\pi}_2 r_1}{\widehat{OR}_u \hat{\pi}_2 + (1 - \hat{\pi}_2)} \tag{4.4}$$

and

$$m_1 = \frac{\widehat{OR}_u \hat{\pi}_2 r_1}{\widehat{OR}_u \hat{\pi}_2 + (1 - \hat{\pi}_2)} + \hat{\pi}_2 r_2 \tag{4.5}$$

where $\widehat{OR}_u$ denotes the unconditional maximum likelihood estimate of OR. This is a system of two equations in the two unknowns $\widehat{OR}_u$ and $\hat{\pi}_2$, which can be solved to give

$$\widehat{OR}_u = \frac{\hat{\omega}_1}{\hat{\omega}_2} = \frac{\hat{\pi}_1(1 - \hat{\pi}_2)}{\hat{\pi}_2(1 - \hat{\pi}_1)} = \frac{a_1 b_2}{a_2 b_1} \tag{4.6}$$

and

$$\hat{\pi}_2 = \frac{a_2}{r_2}.$$

The estimates of π_1, ω_1, and ω_2 which appear in (4.6) are given by

$$\hat{\pi}_1 = \frac{\widehat{OR}_u \hat{\pi}_2}{\widehat{OR}_u \hat{\pi}_2 + (1 - \hat{\pi}_2)} = \frac{a_1}{r_1}$$

$$\hat{\omega}_1 = \frac{\hat{\pi}_1}{1 - \hat{\pi}_1} = \frac{a_1}{b_1}$$

and

$$\hat{\omega}_2 = \frac{\hat{\pi}_2}{1 - \hat{\pi}_2} = \frac{a_2}{b_2}.$$

If any of a_1, a_2, b_1, or b_2 equals 0, we replace (4.6) with

$$\widehat{OR}_u = \frac{(a_1 + .5)(b_2 + .5)}{(a_2 + .5)(b_1 + .5)}.$$

Other approaches to the problem of zero cells are available (Walter, 1987). It can be shown that $\widehat{OR}_u$ is less biased when .5 is added to all the interior cells, whether they are zero or not (Walter, 1985). However, as in Chapter 3, this practice will not be followed here.

Log-Odds Ratio Transformation

The log-odds ratio $\log(OR)$ plays an important role in the analysis of data from closed cohort studies. It can be shown that the unconditional maximum likelihood estimate of $\log(OR)$ is $\log(\widehat{OR}_u)$. For convenience of notation we sometimes write $\log \widehat{OR}_u$ instead of $\log(\widehat{OR}_u)$. According to the observations made in Section 3.2.2, $\hat{\omega}$ can be rather skewed, while $\log(\hat{\omega})$ is generally more or less symmetric. It is therefore not surprising that $\widehat{OR}_u = \hat{\omega}_1/\hat{\omega}_2$ can also be quite skewed and that $\log(\widehat{OR}_u) = \log(\hat{\omega}_1) - \log(\hat{\omega}_2)$ is usually relatively symmetric. We illustrate this with examples.

Consider the binomial distributions with parameters $(\pi_1, r_1) = (.4, 10)$ and $(\pi_2, r_2) = (.2, 25)$. Then $\widehat{OR}_u = [a_1(25 - a_2)]/[a_2(10 - a_1)]$ and $\log(\widehat{OR}_u) = \log[a_1(25 - a_2)] - \log[a_2(10 - a_1)]$. The sample space of $\widehat{OR}_u$ extends from 9.34×10^{-4} to 1071, but the distribution is extremely skewed with odds ratios less than or equal to 12.25 accounting for 95.6% of the probability. Figure 4.1(a) shows the distribution of $\widehat{OR}_u$ after truncation on the right at 12.25. As in Figure 1, magnitudes are not shown on the axes because we are primarily concerned with the shapes of distributions. The data points for Figure 4.1(a) were constructed by dividing the truncated sample space into 10 equally spaced intervals and then summing the probability elements within each interval. The distribution of $\log(\widehat{OR}_u)$ is shown in Figure 4.1(b). The horizontal axis has been truncated on the left and

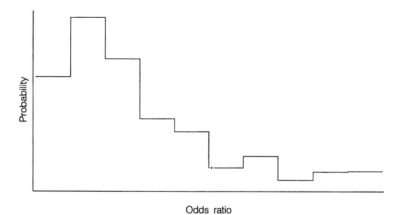

FIGURE 4.1(a)   Distribution of odds ratio for binomial distributions with parameters (.4, 10) and (.2, 25)

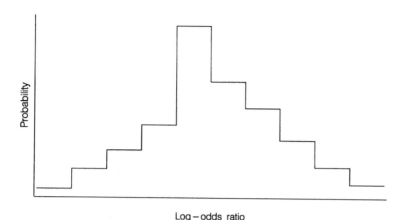

FIGURE 4.1(b)   Distribution of log-odds ratio for binomial distributions with parameters (.4, 10) and (.2, 25)

on the right, in both instances corresponding to a tail probability of 1%. As can be seen, $\log(\widehat{OR}_u)$ is far more symmetric than $\widehat{OR}_u$ and so, with respect to normal approximations, it is preferable to base calculations on $\log(\widehat{OR}_u)$ rather than $\widehat{OR}_u$. Figures 4.2(a) and 4.2(b) show the distributions of $\widehat{OR}_u$ and $\log(\widehat{OR}_u)$ based on binomial distributions with parameters $(\pi_1, r_1) = (.4, 25)$ and $(\pi_2, r_2) = (.2, 50)$. Even though both binomial distributions have a mean of 10, $\widehat{OR}_u$ is quite skewed, while $\log(\widehat{OR}_u)$ is relatively symmetric. Based on empirical evidence such as this, $\log(\widehat{OR}_u)$ should be reasonably symmetric provided the means of the component binomial distributions are 5 or more, while much larger means are required to ensure that $\widehat{OR}_u$ is symmetric.

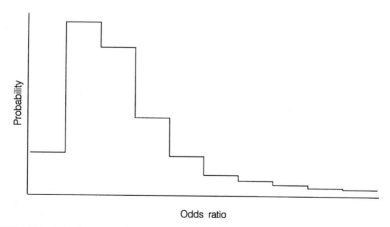

FIGURE 4.2(a)   Distribution of odds ratio for binomial distributions with parameters (.4, 25) and (.2, 50)

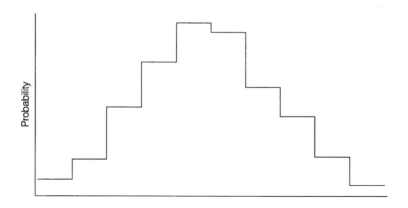

FIGURE 4.2(b)   Distribution of log-odds ratio for binomial distributions with parameters (.4, 25) and (.2, 50)

Confidence Interval

The maximum likelihood estimate of var(log $\widehat{OR}_u$) is

$$\widehat{\text{var}}(\log \widehat{OR}_u) = \frac{1}{a_1} + \frac{1}{a_2} + \frac{1}{b_1} + \frac{1}{b_2}. \tag{4.7}$$

Note that $\widehat{\text{var}}(\log \widehat{OR}_u)$, like $\widehat{OR}_u$, is expressed entirely in terms of the interior cell entries of the 2×2 table. A $(1 - \alpha) \times 100\%$ confidence interval for $\log(OR)$ is

$$\left[\log \underline{OR}_u, \log \overline{OR}_u\right] = \log(\widehat{OR}_u) \pm z_{\alpha/2}\sqrt{\frac{1}{a_1} + \frac{1}{a_2} + \frac{1}{b_1} + \frac{1}{b_2}}$$

which can be exponentiated to give a confidence interval for OR,

$$\left[\underline{OR}_u, \overline{OR}_u\right] = \widehat{OR}_u \exp\left(\pm z_{\alpha/2}\sqrt{\frac{1}{a_1} + \frac{1}{a_2} + \frac{1}{b_1} + \frac{1}{b_2}}\right).$$

If any of a_1, a_2, b_1, or b_2 equals 0, we replace (4.7) with

$$\widehat{\text{var}}(\log \widehat{OR}_u) = \frac{1}{a_1 + .5} + \frac{1}{a_2 + .5} + \frac{1}{b_1 + .5} + \frac{1}{b_2 + .5}.$$

The convention of adding .5 when there are zero cells applies to $\widehat{OR}_u$ and $\widehat{\text{var}}(\log \widehat{OR}_u)$, but not to the other formulas discussed in this section.

Pearson's Test of Association

Pearson's test of association does not have any particular connection to the odds ratio, being equally applicable to analyses based on the risk ratio and risk difference. It is introduced here as a matter of convenience. We say there is no association between exposure and disease when the probability of disease is the same in the exposed and unexposed cohorts, that is, $\pi_1 = \pi_2$. Under the hypothesis of no association $H_0 : \pi_1 = \pi_2$, the expected counts are defined to be

$$\hat{e}_1 = \frac{r_1 m_1}{r} \qquad \hat{e}_2 = \frac{r_2 m_1}{r}$$

$$\hat{f}_1 = \frac{r_1 m_2}{r} \qquad \hat{f}_2 = \frac{r_2 m_2}{r}$$

Using the term "expected" in this context is potentially confusing because these quantities are not expected values (constants). This is because m_1, the number of cases, is unknown until the study has been completed, and hence is a random variable. It would be preferable to refer to the expected counts as "fitted counts under the hypothesis of no association"; however, the term "expected counts" is well established by convention. Note that the expected count for a given interior cell is calculated by multiplying together the corresponding marginal totals and then dividing by r. It is easily shown that the observed and expected marginal totals agree—for example, $a_1 + a_2 = m_1 = \hat{e}_1 + \hat{e}_2$—and so the expected counts can be displayed as in Table 4.2.

Large differences between observed and expected counts provide evidence that the hypothesis of no association may be false. This idea is embodied in Pearson's test of association,

$$X_{\mathrm{p}}^2 = \frac{(a_1 - \hat{e}_1)^2}{\hat{e}_1} + \frac{(a_2 - \hat{e}_2)^2}{\hat{e}_2} + \frac{(b_1 - \hat{f}_1)^2}{\hat{f}_1} + \frac{(b_2 - \hat{f}_2)^2}{\hat{f}_2} \qquad (\mathrm{df} = 1). \quad (4.8)$$

Observe the similarity in form to (1.14) and (1.15). The normal approximation underlying Pearson's test should be satisfactory provided all the expected counts are greater than or equal to 5. According to Yates (1984), this "rule of 5" originated with Fisher (1925). From $a_1 + a_2 = \hat{e}_1 + \hat{e}_2$ it follows that $(a_1 - \hat{e}_1)^2 = (a_2 - \hat{e}_2)^2$. There are similar identities for the other rows and columns, and this allows (4.8) to be expressed in any of the following equivalent forms:

TABLE 4.2   Expected Counts: Closed Cohort Study

Disease	Exposure		
	yes	no	
yes	$\hat{e}_1$	$\hat{e}_2$	m_1
no	$\hat{f}_1$	$\hat{f}_2$	m_2
	r_1	r_2	r

$$X_{\mathrm{p}}^2 = (a_1 - \hat{e}_1)^2 \left(\frac{1}{\hat{e}_1} + \frac{1}{\hat{e}_2} + \frac{1}{\hat{f}_1} + \frac{1}{\hat{f}_2} \right) \tag{4.9}$$

$$X_{\mathrm{p}}^2 = \frac{(a_1 b_2 - a_2 b_1)^2 r}{r_1 r_2 m_1 m_2} \tag{4.10}$$

and

$$X_{\mathrm{p}}^2 = \frac{r}{m_2} \left[\frac{(a_1 - \hat{e}_1)^2}{\hat{e}_1} + \frac{(a_2 - \hat{e}_2)^2}{\hat{e}_2} \right]. \tag{4.11}$$

Wald and Likelihood Ratio Tests of Association

Since $\pi_1 = \pi_2$ is equivalent to $OR = 1$, which in turn is equivalent to $\log(OR) = 0$, the hypothesis of no association can be expressed as $H_0 : \log(OR) = 0$. Under H_0 an estimate of $\mathrm{var}(\log \widehat{OR}_{\mathrm{u}})$ is

$$\widehat{\mathrm{var}}_0(\log \widehat{OR}_{\mathrm{u}}) = \frac{1}{\hat{e}_1} + \frac{1}{\hat{e}_2} + \frac{1}{\hat{f}_1} + \frac{1}{\hat{f}_2}$$

$$= \frac{r^3}{r_1 r_2 m_1 m_2}$$

which is obtained from (4.7) by replacing the observed with expected counts. The Wald test and likelihood ratio tests of association are

$$X_{\mathrm{w}}^2 = (\log \widehat{OR}_{\mathrm{u}})^2 \left(\frac{1}{\hat{e}_1} + \frac{1}{\hat{e}_2} + \frac{1}{\hat{f}_1} + \frac{1}{\hat{f}_2} \right)^{-1}$$

$$= \frac{(\log \widehat{OR}_{\mathrm{u}})^2 r_1 r_2 m_1 m_2}{r^3} \qquad (\mathrm{df} = 1)$$

and

$$X_{\mathrm{lr}}^2 = 2 \left[a_1 \log \left(\frac{a_1}{\hat{e}_1} \right) + a_2 \log \left(\frac{a_2}{\hat{e}_2} \right) + b_1 \log \left(\frac{b_1}{\hat{f}_1} \right) + b_2 \log \left(\frac{b_2}{\hat{f}_2} \right) \right] \qquad (\mathrm{df} = 1) \tag{4.12}$$

respectively. As x approaches 0, the limiting value of $x \log(x)$ is 0. If any of the observed counts is 0, the corresponding term in X_{lr}^2 is assigned a value of 0.

Provided the sample size is large, and sometimes even when it is not so large, Wald, score, and likelihood ratio tests (which can be shown to be asymptotically equivalent) tend to produce similar findings. When there is a meaningful difference among test results the question arises as to which of the tests is to be preferred. Based on asymptotic properties, likelihood ratio tests are generally the first choice, followed by score tests and then Wald tests (Kalbfleisch and Prentice, 1980, p. 48; Lachin, 2000, p. 482). Problems can arise with Wald tests when the variance is not estimated under the null hypothesis (Mantel, 1987). A major disparity among test

TABLE 4.3   Observed Counts:
Antibody–Diarrhea

Diarrhea	Antibody		
	low	high	
yes	12	7	19
no	2	9	11
	14	16	30

results may be an indication that the sample size is too small for the asymptotic approach and that exact methods should be considered.

Example 4.1 (Antibody–Diarrhea)   Table 4.3 gives a portion of the data from a cohort study conducted in Bangladesh which investigated whether antibodies present in breast milk protect infants from diarrhea due to cholera (Glass et al., 1983). These data have been analyzed by Rothman (1986, p. 169).

We first analyze the exposed and unexposed cohorts separately using the methods of Chapter 3. The estimates $\hat{\pi}_1 = 12/14 = .86$ and $\hat{\pi}_2 = 7/16 = .44$ suggest that low antibody level increases the risk of diarrhea. Exact 95% confidence intervals for π_1 and π_2 are [.57, .98] and [.20, .70], respectively. The degree of overlap in the confidence intervals suggests that π_1 and π_2 may be equal, but this impression needs to be formally evaluated using a test of association.

The odds ratio estimate is $\widehat{OR}_u = (12 \times 9)/(7 \times 2) = 7.71$, and so once again it appears that low antibody level increases the risk of diarrhea. To be technically correct we should express this observation by saying that low antibody level seems to increase the odds of developing diarrhea. From

$$\widehat{var}(\log \widehat{OR}_u) = \frac{1}{12} + \frac{1}{7} + \frac{1}{2} + \frac{1}{9} = .84$$

the 95% confidence interval for $\log(OR)$ is $\log(7.71) \pm 1.96\sqrt{.84} = [.25, 3.84]$. Exponentiating, the 95% confidence interval for OR is $[1.28, 46.37]$. With a sample size as small as the one in this study, it is not surprising that the confidence interval is extremely wide. Our impression is that OR may be larger than 1, but how much larger is difficult to say. The expected counts, shown in Table 4.4, are all greater

TABLE 4.4   Expected Counts:
Antibody–Diarrhea

Diarrhea	Antibody		
	low	high	
yes	8.87	10.13	19
no	5.13	5.87	11
	14	16	30

than 5. The Pearson, Wald, and likelihood ratio tests are similar in value and provide considerable evidence that low antibody level is associated with the development of diarrhea,

$$X_p^2 = \frac{(12 - 8.87)^2}{8.87} + \frac{(7 - 10.13)^2}{10.13} + \frac{(2 - 5.13)^2}{5.13} + \frac{(9 - 5.87)^2}{5.87}$$
$$= 5.66 \ (p = .02)$$

$$X_w^2 = (\log 7.71)^2 \left(\frac{1}{8.87} + \frac{1}{10.13} + \frac{1}{5.13} + \frac{1}{5.87} \right)^{-1} = 7.24 \ (p = .01)$$

and

$$X_{lr}^2 = 2 \left[12 \log \left(\frac{12}{8.87} \right) + 7 \log \left(\frac{7}{10.13} \right) + 2 \log \left(\frac{2}{5.13} \right) + 9 \log \left(\frac{9}{5.87} \right) \right]$$
$$= 6.02 \ (p = .01).$$

Example 4.2 (Receptor Level–Breast Cancer)   The data for this example were kindly provided by the Northern Alberta Breast Cancer Registry. This is a population-based registry that collects information on all cases of breast cancer treated in the northern half of the province of Alberta, Canada. After initial treatment, patients are reviewed on an annual basis, or more frequently if necessary. When an annual follow-up appointment is missed, an attempt is made to obtain current informa-tion on the patient by corresponding with the patient and the treating physicians. When this fails, a search is made of provincial and national vital statistics records to determine if the patient has died and, if so, of what cause. Due to the intensive methods that are used to ensure follow-up of registrants, it is reasonable to assume that patients who are not known to have died are still alive.

The cohort for this example was assembled by selecting a random sample of 199 female breast cancer patients who registered during 1985. Entry into the cohort was restricted to women with either stage I, II, or III disease, thereby excluding cases of disseminated cancer (stage IV). It has been well documented that breast cancer mortality increases as stage of disease becomes more advanced. Another predictor of survival from breast cancer is the amount of estrogen receptor that is present in breast tissue. Published reports show that patients with higher levels of estrogen receptor generally have a better prognosis. Receptor level is measured on a continuous scale, but for the present analysis this variable has been dichotomized into low and high levels using a conventional cutoff value.

For this example the maximum length of follow-up was taken to be 5 years and the endpoint was defined to be death from breast cancer. Of the 199 subjects in the cohort, seven died of a cause other than breast cancer. These individuals were dropped from the analysis, leaving a cohort of 192 subjects. Summarily dropping subjects in this manner is methodologically incorrect, but for purposes of illustration this issue will be ignored. Methods for analyzing cohort data when there are losses to follow-up are presented in later chapters.

TABLE 4.5(a)   Observed Counts:
Receptor Level–Breast Cancer

Survival	Receptor Level		
	low	high	
dead	23	31	54
alive	25	113	138
	48	144	192

Table 4.5(a) gives the breast cancer data with receptor level as the exposure variable. The estimates $\hat{\pi}_1 = 23/48 = .479$ and $\hat{\pi}_2 = 31/144 = .215$ suggest that low receptor level increases the mortality risk from breast cancer. Based on the explicit method, the 95% confidence intervals for π_1 and π_2 are [.338, .620] and [.148, .282], respectively. The confidence intervals are far from overlapping which suggests that π_1 and π_2 are likely unequal. The odds ratio estimate is $\widehat{OR}_u = (23 \times 113)/(31 \times 25) = 3.35$. From $\widehat{\text{var}}(\log \widehat{OR}_u) = .125$, the 95% confidence interval for OR is [1.68, 6.70]. The confidence interval is not especially narrow but does suggest that receptor level is meaningfully associated with breast cancer mortality.

At this point it is appropriate to consider the potential impact of misclassification on the odds ratio estimate. Let a'_1, a'_2, b'_1, and b'_2 denote what would have been the observed counts in the absence of misclassification. From Tables 2.11 and 4.5(a), the following linear equations must be satisfied:

$$\alpha_1 a'_1 + (1 - \beta_1)b'_1 = 23$$
$$\alpha_2 a'_2 + (1 - \beta_2)b'_2 = 31$$
$$(1 - \alpha_1)a'_1 + \beta_1 b'_1 = 25$$
$$(1 - \alpha_2)a'_2 + \beta_2 b'_2 = 113$$

where α_1 and α_2 are the sensitivities, and β_1 and β_2 are the specificities (Section 2.6). One potential source of misclassification is that Registry staff may have failed to identify all the deaths in the cohort. For purposes of illustration we set $\alpha_1 = \alpha_2 = .90$; that is, we assume that only 90% of deaths were ascertained. It seems unlikely that someone who survived would have been recorded as having died, and so we set $\beta_1 = \beta_2 = .99$. The above equations become

$$(.90a'_1) + (.01b'_1) = 23$$
$$(.90a'_2) + (.01b'_2) = 31$$
$$(.10a'_1) + (.99b'_1) = 25$$
$$(.10a'_2) + (.99b'_2) = 113$$

TABLE 4.5(b) Observed Counts after Adjusting for
Misclassification: Receptor Level–Breast Cancer

Survival	Receptor Level		
	low	high	
dead	25.30	33.21	58.51
alive	22.70	110.79	133.49
	48	144	192

which have the solutions given in Table 4.5(b). After accounting for misclassification, the estimated odds ratio is $\widehat{OR}_u' = (25.30 \times 110.79)/(33.21 \times 22.70) = 3.72$, which is only slightly larger than the estimate based on the (possibly) misclassified data. This shows that misclassification is unlikely to be a major source of bias in the present study.

Returning to an analysis of the data in Table 4.5(a), the expected counts, given in Table 4.6, are all much greater than 5. The Pearson, Wald, and likelihood ratio tests are similar in value and provide considerable evidence that low receptor level is associated with an increased risk of dying of breast cancer,

$$X_p^2 = \frac{(23 - 13.5)^2}{13.5} + \frac{(31 - 40.5)^2}{40.5} + \frac{(25 - 34.5)^2}{34.5} + \frac{(113 - 103.5)^2}{103.5}$$
$$= 12.40 \quad (p < .001)$$

$$X_w^2 = (\log 3.35)^2 \left(\frac{1}{13.5} + \frac{1}{40.5} + \frac{1}{34.5} + \frac{1}{103.5} \right)^{-1}$$
$$= 10.66 \quad (p = .001).$$

$$X_{lr}^2 = 2\left[23 \log \left(\frac{23}{13.5} \right) + 31 \log \left(\frac{31}{40.5} \right) + 25 \log \left(\frac{25}{34.5} \right) + 113 \log \left(\frac{113}{103.5} \right) \right]$$
$$= 11.68 \quad (p = .001).$$

TABLE 4.6 Expected Counts: Receptor
Level–Breast Cancer

Survival	Receptor Level		
	low	high	
dead	13.5	40.5	54
alive	34.5	103.5	138
	48	144	192

4.2 EXACT CONDITIONAL METHODS FOR A SINGLE 2 × 2 TABLE

The methods presented in the preceding section are computationally convenient but have the drawback of being valid only under asymptotic conditions. Provided binomial means are 5 or more, the asymptotic methods are likely to produce reasonable results; but when the sample size is very small, the asymptotic approach cannot be relied upon. In this case there is little alternative other than to resort to exact calculations, despite the inevitable increase in computational burden. EGRET (1999) has procedures for calculating the exact confidence interval and hypothesis test presented below.

The asymptotic unconditional methods described above involve two parameters, OR and π_2. We are primarily interested in OR, but based on the unconditional approach it is necessary to estimate both OR and π_2. In a sense, we are using data to estimate π_2 that could be better utilized estimating OR. For this reason, π_2 is referred to as a nuisance parameter. We now describe exact conditional methods for analyzing 2 × 2 tables. These techniques have the desirable feature of eliminating the nuisance parameter π_2 so that only OR, the parameter of interest, remains to be estimated. It was pointed out above that m_1, the total number of cases, is a random variable. The conditional approach proceeds by assuming that m_1 is a known constant. This is certainly true when the study has been completed, but the same can be said for all of the interior cell counts and marginal totals. When the random variable m_1 is treated as a known constant, we say that we have conditioned on m_1. An informal justification for the conditional assumption is that, from the point of view of comparing risk across cohorts, it is not the absolute numbers of cases in the exposed and unexposed cohorts that are important but rather their relative magnitudes. From this perspective, the total number of cases gives little information about the parameter of interest and so we are free to treat m_1 as if it had been fixed by study design (Yates, 1984; Clayton and Hills, 1993, §13.3). More formal arguments for adopting the conditional approach have been provided (Yates, 1984; Little, 1989; Greenland, 1991).

Since r_1 and r_2 are constants, once we have conditioned on m_1, it follows that m_2 is also a known constant. With all of the marginal totals fixed, knowledge of any one of the four interior cell counts determines the remaining three. We refer to a particular choice of interior cell as the index cell. With the upper left cell taken to be the index cell we can display the table of observed counts as in Table 4.7. The choice of which interior cell to use as the index cell is a matter of convenience and does not affect inferences made using the conditional approach.

TABLE 4.7 Observed Counts with Fixed Marginal Totals: Closed Cohort Study

Disease	Exposure		
	yes	no	
yes	a_1	$m_1 - a_1$	m_1
no	$r_1 - a_1$	$r_2 - m_1 + a_1$	m_2
	r_1	r_2	r

Hypergeometric Distribution

Once we have conditioned on m_1, the random variables A_1 and A_2 are no longer independent. Specifically, we have the constraint $A_1 + A_2 = m_1$, and so A_2 is completely determined by A_1 (and *vice versa*). As a result of conditioning on m_1 we have gone from two independent binomial random variables to a single random variable corresponding to the index cell. We continue to denote the random variable in question by A_1, allowing the context to make clear which probability model is being considered. As shown in Appendix C, conditioning on m_1 results in a (noncentral) hypergeometric distribution. The probability function is

$$P(A_1 = a_1 | OR) = \frac{1}{C} \binom{r_1}{a_1} \binom{r_2}{m_1 - a_1} OR^{a_1} \tag{4.13}$$

where

$$C = \sum_{x=l}^{u} \binom{r_1}{x} \binom{r_2}{m_1 - x} OR^x.$$

Viewed as a hypergeometric random variable, A_1 has the sample space $\{l, l + 1, \ldots, u\}$, where $l = \max(0, r_1 - m_2)$ and $u = \min(r_1, m_1)$. Here max and min mean that l is the maximum of 0 and $r_1 - m_2$, and u is the minimum of r_1 and m_1. Since $r_1 - m_2 = (r - r_2) - (r - m_1) = m_1 - r_2$, l is sometimes written as $\max(0, m_1 - r_2)$. Evidently, $l \geq 0$ and $u \leq r_1$, and so the hypergeometric sample space of A_1 is contained in the binomial sample space. For a given set of marginal totals, the hypergeometric distribution is completely determined by the parameter OR. Therefore, by conditioning on m_1 we have eliminated the nuisance parameter π_2. The numerator of (4.13) gives the distribution its basic shape, and the denominator C ensures that (1.1) is satisfied. From (1.2) and (1.3), the hypergeometric mean and variance are

$$E(A_1 | OR) = \frac{1}{C} \sum_{x=l}^{u} x \binom{r_1}{x} \binom{r_2}{m_1 - x} OR^x \tag{4.14}$$

and

$$\mathrm{var}(A_1 | OR) = \frac{1}{C} \sum_{x=l}^{u} [x - E(A_1 | OR)]^2 \binom{r_1}{x} \binom{r_2}{m_1 - x} OR^x. \tag{4.15}$$

Unfortunately, (4.13), (4.14), and (4.15) do not usually simplify to less complicated expressions. An instance where simplification does occur is when $OR = 1$. In this case we say that A_1 has a central hypergeometric distribution. For the central hypergeometric distribution,

$$P_0(A_1 = a_1) = \frac{\binom{r_1}{a_1} \binom{r_2}{m_1 - a_1}}{\binom{r}{m_1}} = \frac{r_1! \, r_2! \, m_1! \, m_2!}{a_1! \, (m_1 - a_1)! \, (r_1 - a_1)! \, (r_2 - m_1 + a_1)! \, r!} \tag{4.16}$$

$$e_1 = E_0(A_1) = \frac{r_1 m_1}{r} \qquad (4.17)$$

and

$$v_0 = \mathrm{var}_0(A_1) = \frac{r_1 r_2 m_1 m_2}{r^2(r-1)}. \qquad (4.18)$$

Since m_1 is now being treated as a constant, e_1 and v_0 are the exact mean and variance rather than just estimates. However, for the sake of uniformity of notation, we will denote these quantities by $\hat{e}_1$ and $\hat{v}_0$ in what follows. Observe that, other than $r!$, the denominator of the final expression in (4.16) is the product of factorials defined in terms of the interior cells of Table 4.7. A convenient method of tabulating a central hypergeometric probability function is to form each of the possible 2×2 tables and calculate probability elements using (4.16).

Confidence Interval

Since the hypergeometric distribution involves the single parameter OR, the approach to exact interval estimation and hypothesis testing is a straightforward adaptation of the techniques described for the binomial distribution in Sections 3.1.1 and 3.1.2. An exact $(1-\alpha) \times 100\%$ confidence interval for OR is obtained by solving the equations

$$\frac{\alpha}{2} = P(A_1 \geq a_1 | \underline{OR}_{\mathrm{c}}) = \frac{1}{\underline{C}_{\mathrm{c}}} \sum_{x=a_1}^{u} \binom{r_1}{x}\binom{r_2}{m_1 - x}(\underline{OR}_{\mathrm{c}})^x$$

$$= 1 - \frac{1}{\underline{C}_{\mathrm{c}}} \sum_{x=l}^{a_1 - 1} \binom{r_1}{x}\binom{r_2}{m_1 - x}(\underline{OR}_{\mathrm{c}})^x$$

and

$$\frac{\alpha}{2} = P(A_1 \leq a_1 | \overline{OR}_{\mathrm{c}}) = \frac{1}{\overline{C}_{\mathrm{c}}} \sum_{x=l}^{a_1} \binom{r_1}{x}\binom{r_2}{m_1 - x}(\overline{OR}_{\mathrm{c}})^x$$

$$= 1 - \frac{1}{\overline{C}_{\mathrm{c}}} \sum_{x=a_1 + 1}^{u} \binom{r_1}{x}\binom{r_2}{m_1 - x}(\overline{OR}_{\mathrm{c}})^x$$

for $\underline{OR}_{\mathrm{c}}$ and $\overline{OR}_{\mathrm{c}}$, where $\underline{C}_{\mathrm{c}}$ and $\overline{C}_{\mathrm{c}}$ stand for C with $\underline{OR}_{\mathrm{c}}$ and $\overline{OR}_{\mathrm{c}}$ substituted for OR, respectively.

Fisher's Exact Test

It is possible to test hypotheses of the form $H_0: OR = OR_0$ for an arbitrary choice of OR_0 but, in practice, interest is mainly in the hypothesis of no association $H_0 : OR = 1$. The exact test of association based on the central hypergeometric distribution is referred to as Fisher's (exact) test (Fisher, 1936; §21.02). The tail probabilities are

$$P_0(A_1 \geq a_1) = \sum_{x=a_1}^{u} \frac{\binom{r_1}{x}\binom{r_2}{m_1-x}}{\binom{r}{m_1}} = 1 - \sum_{x=l}^{a_1-1} \frac{\binom{r_1}{x}\binom{r_2}{m_1-x}}{\binom{r}{m_1}}$$

and

$$P_0(A_1 \leq a_1) = \sum_{x=l}^{a_1} \frac{\binom{r_1}{x}\binom{r_2}{m_1-x}}{\binom{r}{m_1}} = 1 - \sum_{x=a_1+1}^{u} \frac{\binom{r_1}{x}\binom{r_2}{m_1-x}}{\binom{r}{m_1}}.$$

Calculation of the two-sided p-value using either the cumulative or doubling method follows precisely the steps described for the binomial distribution in Section 3.1.1. Recall the discussion in Chapter 3 regarding the conservative nature of an exact test when the distribution is discrete. This conservatism, which is a feature of Fisher's test, is more pronounced when the sample size is small. This is precisely the condition under which an asymptotic test, such as Pearson's test, becomes invalid. These issues have led to a protracted debate regarding the relative merits of these two tests when the sample size is small. Currently, Fisher's test appears to be regarded more favorably (Yates, 1984; Little, 1989).

Example 4.3 (Hypothetical Data)   Data from a hypothetical cohort study are given in Table 4.8. For these data, $l = 1$ and $u = 3$. Note that 0, which is an element of the binomial sample space of A_1, cannot be an element of the hypergeometric sample space since that would force the lower right cell count to be -1.

The central hypergeometric probability function is given in Table 4.9. The mean and variance are $\hat{e}_1 = 1.80$ and $\hat{v}_0 = .36$.

The noncentral hypergeometric probability function corresponding to Table 4.8 is

$$P(A_1 = a_1 | OR) = \frac{1}{C} \binom{3}{a_1} \binom{2}{3 - a_1} OR^{a_1}$$

where

$$C = \sum_{x=1}^{3} \binom{3}{x} \binom{2}{3 - x} OR^x = 3OR + 6OR^2 + OR^3.$$

TABLE 4.8   Observed Counts:
Hypothetical Cohort Study

Disease	Exposure		
	yes	no	
yes	2	1	3
no	1	1	2
	3	2	5

TABLE 4.9 Central Hypergeometric Probability Function: Hypothetical Cohort Study

a_1	$P_0(A_1 = a_1)$
1	$\dfrac{3!\,2!\,3!\,2!}{1!\,2!\,2!\,0!\,5!} = .3$
2	$\dfrac{3!\,2!\,3!\,2!}{2!\,1!\,1!\,1!\,5!} = .6$
3	$\dfrac{3!\,2!\,3!\,2!}{3!\,0!\,0!\,2!\,5!} = .1$

The exact conditional 95% confidence interval for OR is $[.013, 234.5]$, which is obtained by solving the equations

$$.025 = \sum_{x=2}^{3} P(A_1 = x \,|\, \underline{OR}_c) = \frac{6(\underline{OR}_c)^2 + (\underline{OR}_c)^3}{3\underline{OR}_c + 6(\underline{OR}_c)^2 + (\underline{OR}_c)^3}$$

and

$$.025 = \sum_{x=1}^{2} P(A_1 = x \,|\, \overline{OR}_c) = \frac{3\overline{OR}_c + 6(\overline{OR}_c)^2}{3(\overline{OR}_c) + 6(\overline{OR}_c)^2 + (\overline{OR}_c)^3}$$

for $\underline{OR}_c$ and $\overline{OR}_c$.

Example 4.4 (Antibody–Diarrhea) For the data in Table 4.3, $l = 3$ and $u = 14$. The central hypergeometric distribution is given in Table 4.10.

The exact conditional 95% confidence interval for OR is $[1.05, 86.94]$ which is quite wide and just misses containing 1. The p-value for Fisher's test based on the

TABLE 4.10 Central Hypergeometric Probability Function (%): Antibody–Diarrhea

a_1	$P_0(A_1 = a_1)$	$P_0(A_1 \leq a_1)$	$P_0(A \geq a_1)$
3	<.01	<.01	100
4	.03	.03	99.99
5	.44	.47	99.97
6	3.08	3.55	99.53
7	11.43	14.98	96.45
8	24.01	38.99	85.02
9	29.35	68.34	61.01
10	20.96	89.31	31.66
11	8.58	97.88	10.69
12	1.91	99.79	2.12
13	.21	99.99	.21
14	.01	100	.01

TABLE 4.11   Central Hypergeometric Probability
Function (%): Receptor Level–Breast Cancer

a_1	$P_0(A_1 = a_1)$	$P_0(A_1 \leq a_1)$	$P_0(A_1 \geq a_1)$
⋮	⋮	⋮	⋮
3	<.01	<.01	100
4	.01	.02	99.99
5	.07	.09	99.98
⋮	⋮	⋮	⋮
11	9.91	23.13	86.78
12	12.88	36.01	76.87
13	14.54	50.55	63.99
14	14.33	64.88	49.45
15	12.37	77.25	35.12
16	9.39	86.64	22.75
⋮	⋮	⋮	⋮
22	.13	99.94	.19
23	.04	99.98	.06
24	.01	99.99	.02
⋮	⋮	⋮	⋮

cumulative method is $P_0(A_1 \geq 12) + P_0(A_1 \leq 5) = .026$, and based on the doubling method is $2 \times P_0(A_1 \geq 12) = .042$. For these data, there is a noticeable difference between the cumulative and doubling results, but in either case we infer that low antibody level is associated with an increased risk of diarrhea. A comparison of the preceding results with those of Example 4.1 illustrates that exact confidence intervals tend to be wider than asymptotic ones, and exact p-values are generally larger than their asymptotic counterparts.

Example 4.5 (Receptor Level–Breast Cancer)   For Table 4.5(a), $l = 0$ and $u = 48$. The central hypergeometric distribution is given, in part, in Table 4.11.

The exact conditional 95% confidence interval for OR is [1.58, 7.07], and the p-value for Fisher's test based on the cumulative method is $P_0(A_1 \geq 23) + P_0(A_1 \leq 4) = .08\%$. The remark made in Example 4.4 about exact results being conservative holds here (except for Pearson's test), as may be seen from a comparison with Example 4.2. However, when the sample size is large, the differences between exact and asymptotic findings are often of little practical importance, as is the case here.

4.3   ASYMPTOTIC CONDITIONAL METHODS
FOR A SINGLE 2 × 2 TABLE

The exact conditional methods described in the preceding section have the desirable feature of eliminating the nuisance parameter π_1, but there is the drawback that they involve extensive calculations. Asymptotic conditional methods make it possible to reduce the computational burden, at least in the case of the test of association.

Point Estimate

For the asymptotic conditional analysis we consider (4.13) to be a likelihood that is a function of the parameter OR. The conditional maximum likelihood equation is

$$a_1 = E(A_1|\widehat{OR}_c) = \frac{1}{\hat{C}} \sum_{x=l}^{u} x \binom{r_1}{x} \binom{r_2}{m_1 - x} (\widehat{OR}_c)^x \qquad (4.19)$$

where

$$\hat{C} = \sum_{x=l}^{u} \binom{r_1}{x} \binom{r_2}{m_1 - x} (\widehat{OR}_c)^x$$

and $\widehat{OR}_c$ denotes the conditional maximum likelihood estimate of OR. Equation (4.19) is usually a polynomial of high degree, but it can be solved for the single unknown $\widehat{OR}_c$ by trial and error. It can be shown that for a given 2 × 2 table, $\widehat{OR}_c$ is closer to 1 than $\widehat{OR}_u$ (Mantel and Hankey, 1975).

Confidence Interval

We present two methods of interval estimation, one implicit and the other explicit. As in the binomial case discussed in Section 3.2.1, the difference between the two approaches is that the explicit method specifies a particular point estimate of the variance, while the implicit method does not. Analogous to (3.6) and (3.7), an implicit $(1 - \alpha) \times 100\%$ confidence interval for OR is obtained by solving the equations

$$\frac{a_1 - E(A_1|\underline{OR}_c)}{\sqrt{\text{var}(A_1|\underline{OR}_c)}} = z_{\alpha/2} \qquad (4.20)$$

and

$$\frac{a_1 - E(A_1|\overline{OR}_c)}{\sqrt{\text{var}(A_1|\overline{OR}_c)}} = -z_{\alpha/2} \qquad (4.21)$$

for $\underline{OR}_c$ and $\overline{OR}_c$ (Mantel, 1977). The mean and variance terms in (4.20) and (4.21) are defined by (4.14) and (4.15), and the equations are solved by trial and error. These may be equations of high degree with multiple solutions. The bounds for the confidence interval are defined to be those solutions which give the widest confidence interval—that is, the one which is most conservative.

By definition, $\widehat{OR}_c$ satisfies the equation $a_1 = E(A_1|\widehat{OR}_c)$. It follows from (4.15) that an estimate of $\text{var}(A_1|OR)$ is

$$\hat{v} = \widehat{\text{var}}(A_1|\widehat{OR}_c) = \frac{1}{\hat{C}} \sum_{x=l}^{u} (x - a_1)^2 \binom{r_1}{x} \binom{r_2}{m_1 - x} (\widehat{OR}_c)^x. \qquad (4.22)$$

As shown in Appendix C, an estimate of var($\log \widehat{OR}_c$) is

$$\widehat{\text{var}}(\log \widehat{OR}_c) = \frac{1}{\hat{v}} \tag{4.23}$$

(Birch, 1964). The reciprocal relationship in (4.23) between the estimated variance of the log-odds ratio and the estimated variance of the index cell count is an example of a phenomenon that will appear in other contexts. An explicit $(1 - \alpha) \times 100\%$ confidence interval for OR is obtained by exponentiating

$$\left[\log \underline{OR}_c, \log \overline{OR}_c\right] = \log(\widehat{OR}_c) \pm \frac{z_{\alpha/2}}{\sqrt{\hat{v}}}.$$

Mantel–Haenszel Test of Association
The mean and variance of the central hypergeometric distribution are given by (4.17) and (4.18). Perhaps the most widely used test of association in epidemiology, especially in its stratified form (Section 5.2), is due to Mantel and Haenszel (1959)

$$X_{mh}^2 = \frac{(a_1 - \hat{e}_1)^2}{\hat{v}_0} \quad (\text{df} = 1). \tag{4.24}$$

It is readily shown that X_{mh}^2 can be expressed as

$$X_{mh}^2 = \frac{(a_1 b_2 - a_2 b_1)^2 (r - 1)}{r_1 r_2 m_1 m_2} \tag{4.25}$$

and so, from (4.10), we have

$$X_{mh}^2 = \left(\frac{r-1}{r}\right) X_p^2. \tag{4.26}$$

This shows that $X_{mh}^2 < X_p^2$, and so the Mantel–Haenszel test is conservative compared to Pearson's test, in the sense that the p-value for X_{mh}^2 is always larger than the p-value for X_p^2. Evidently, when r is large the difference between X_{mh}^2 and X_p^2 will be negligible. It follows from (4.11) and (4.26) that

$$X_{mh}^2 = \left(\frac{r-1}{m_2}\right)\left[\frac{(a_1 - \hat{e}_1)^2}{\hat{e}_1} + \frac{(a_2 - \hat{e}_2)^2}{\hat{e}_2}\right]. \tag{4.27}$$

Example 4.6 (Hypothetical Data)   For the hypothetical data, the odds ratio estimate is $\widehat{OR}_c = 1.73$, which is obtained by solving

$$2 = \frac{\sum_{x=1}^{3} x \binom{3}{x}\binom{2}{3-x}(\widehat{OR}_c)^x}{\sum_{x=1}^{3} \binom{3}{x}\binom{2}{3-x}(\widehat{OR}_c)^x} = \frac{3\widehat{OR}_c + 12(\widehat{OR}_c)^2 + 3(\widehat{OR}_c)^3}{3\widehat{OR}_c + 6(\widehat{OR}_c)^2 + (\widehat{OR}_c)^3}.$$

Since $\widehat{OR}_u = 2.0$, $\widehat{OR}_c$ is closer to 1 than $\widehat{OR}_u$.

Example 4.7 (Antibody–Diarrhea)   For the antibody–diarrhea data, the odds ratio estimate is $\widehat{OR}_c = 7.17$, which is obtained by solving

$$12 = \frac{\sum_{x=3}^{14} x \binom{14}{x}\binom{16}{19-x}(\widehat{OR}_c)^x}{\sum_{x=3}^{14} \binom{14}{x}\binom{16}{19-x}(\widehat{OR}_c)^x}.$$

The implicit 95% confidence interval for OR is $[1.34, 36.21]$. From

$$\hat{v} = \widehat{var}(A_1|7.17) = \frac{\sum_{x=3}^{14}(x-12)^2\binom{14}{x}\binom{16}{19-x}(7.17)^x}{\sum_{x=3}^{14}\binom{14}{x}\binom{16}{19-x}(7.17)^x} = 1.25$$

we have $\widehat{var}(\log \widehat{OR}_c) = 1/1.25 = .80$. Exponentiating $\log(7.17) \pm 1.96\sqrt{.80} = [.21, 3.73]$, the explicit 95% confidence interval for OR is $[1.24, 41.52]$. The Mantel–Haenszel test is

$$X_{mh}^2 = \frac{(12 - 8.87)^2}{1.79} = 5.47 \ (p = .02).$$

Example 4.8 (Receptor Level–Breast Cancer)   The odds ratio estimate is $\widehat{OR}_c = 3.33$, the implicit and explicit 95% confidence intervals for OR are $[1.68, 6.60]$ and $[1.67, 6.63]$, respectively, and $X_{mh}^2 = 12.34 \ (p < .001)$.

4.4   CORNFIELD'S APPROXIMATION

As in the previous section, let A_1 denote a hypergeometric random variable with parameter OR. Cornfield (1956) describes a normal approximation to the exact distribution of A_1. The mean of the approximation will be denoted by $E^*(A_1|OR)$ or a_1^*, and the variance will be denoted by v^*. For a given value of OR, the Cornfield approximation to $E(A_1|OR)$, the exact hypergeometric mean of A_1, is defined to be the value of a_1^* which solves the equation

$$OR = \frac{a_1^*(r_2 - m_1 + a_1^*)}{(m_1 - a_1^*)(r_1 - a_1^*)} \tag{4.28}$$

and also satisfies $l \le a_1^* \le u$, where, as before, $l = \max(0, r_1 - m_2)$ and $u = \min(r_1, m_1)$. Since we are considering a normal approximation, a_1^* is not required to be a nonnegative integer. It is easily verified that as a_1^* varies between l and u, OR ranges over all nonnegative numbers. Conversely, for any value of OR, there is a corresponding value of a_1^* between l and u which satisfies (4.28). For a given value of OR, we can view (4.28) as a second-degree polynomial in the unknown a_1^*. For $OR \neq 1$, the quadratic formula gives the solution

$$a_1^* = \frac{-y - \sqrt{y^2 - 4xz}}{2x} \tag{4.29}$$

where

$$x = OR - 1$$
$$y = -[(m_1 + r_1)OR - m_1 + r_2]$$
$$z = ORm_1r_1$$

(Fleiss, 1979). When the quadratic formula is used to solve an equation, there is the choice of a positive or negative root. A justification for choosing the negative root in (4.29) is given in Appendix D. When $OR = 1$, the above approach fails since $x = 0$ in the denominator. In this case, (4.28) can be solved directly to give

$$a_1^* = \frac{r_1 m_1}{r} = \hat{e}_1. \tag{4.30}$$

Once a_1^* has been determined, the remaining cell entries are defined as in Table 4.12, thereby ensuring that the estimated counts agree with the original marginal totals.

The Cornfield approximation to $\text{var}(A_1|OR)$, the exact hypergeometric variance of A_1, is defined to be

$$v^* = \left(\frac{1}{a_1^*} + \frac{1}{m_1 - a_1^*} + \frac{1}{r_1 - a_1^*} + \frac{1}{r_2 - m_1 + a_1^*} \right)^{-1}. \tag{4.31}$$

Note that, in contrast to (4.7), there is an exponent -1 in (4.31). This is another example of the reciprocal relationship referred to in connection with (4.23). When $OR = 1$, (4.31) simplifies to

$$v_0^* = \frac{r_1 r_2 m_1 m_2}{r^3}. \tag{4.32}$$

TABLE 4.12   Estimated Counts Based on Cornfield's Approximation: Closed Cohort Study

Disease	Exposure		
	yes	no	
yes	a_1^*	$m_1 - a_1^*$	m_1
no	$r_1 - a_1^*$	$r_2 - m_1 + a_1^*$	m_2
	r_1	r_2	r

Note the similarity between (4.32) and (4.18). From (4.30) and (4.32), a test of H_0 : $OR = 1$ based on Cornfield's approximation is $X^2 = (a_1 - a_1^*)^2/v_0^*$, which is identical to X_p^2 (4.10) and almost identical to X_{mh}^2 (4.25).

Analogous to (4.20) and (4.21), an implicit $(1 - \alpha) \times 100\%$ confidence interval for a_1^* is found by solving the equations

$$(a_1 - \underline{a}_1^*)\sqrt{\frac{1}{\underline{a}_1^*} + \frac{1}{m_1 - \underline{a}_1^*} + \frac{1}{r_1 - \underline{a}_1^*} + \frac{1}{r_2 - m_1 + \underline{a}_1^*}} = z_{\alpha/2} \qquad (4.33)$$

and

$$(a_1 - \overline{a}_1^*)\sqrt{\frac{1}{\overline{a}_1^*} + \frac{1}{m_1 - \overline{a}_1^*} + \frac{1}{r_1 - \overline{a}_1^*} + \frac{1}{r_2 - m_1 + \overline{a}_1^*}} = -z_{\alpha/2} \qquad (4.34)$$

for $\underline{a}_1^*$ and $\overline{a}_1^*$ (Cornfield, 1956; Gart, 1971). Equations (4.33) and (4.34) are fourth-degree polynomials in $\underline{a}_1$ and $\overline{a}_1$, and may have more than one set of solutions. The solutions that fall within the bounds l and u and give the widest confidence interval are the ones that are chosen. Once the estimates $\underline{a}_1^*$ and $\overline{a}_1^*$ have been determined, the estimates $\underline{OR}^*$ and $\overline{OR}^*$ are obtained using

$$\underline{OR}^* = \frac{\underline{a}_1^*(r_2 - m_1 + \underline{a}_1^*)}{(m_1 - \underline{a}_1^*)(r_1 - \underline{a}_1^*)}$$

and

$$\overline{OR}^* = \frac{\overline{a}_1^*(r_2 - m_1 + \overline{a}_1^*)}{(m_1 - \overline{a}_1^*)(r_1 - \overline{a}_1^*)}.$$

Example 4.9 (Antibody–Diarrhea)　The solutions to

$$(12 - \underline{a}_1^*)\sqrt{\frac{1}{\underline{a}_1^*} + \frac{1}{19 - \underline{a}_1^*} + \frac{1}{14 - \underline{a}_1^*} + \frac{1}{\underline{a}_1^* - 3}} = 1.96$$

and

$$(12 - \overline{a}_1^*)\sqrt{\frac{1}{\overline{a}_1^*} + \frac{1}{19 - \overline{a}_1^*} + \frac{1}{14 - \overline{a}_1^*} + \frac{1}{\overline{a}_1^* - 3}} = -1.96$$

are $\underline{a}_1^* = 9.44$ and $\overline{a}_1^* = 13.39$, and the 95% confidence interval for OR is [1.39, 40.65].

Example 4.10 (Receptor Level–Breast Cancer)　From $\underline{a}_1^* = 17.49$ and $\overline{a}_1^* = 28.36$, the 95% confidence interval for OR is [1.69, 6.67].

TABLE 4.13   Summary of Antibody–Diarrhea Results

Result	AU	EC	AC	CF
$\widehat{OR}$	7.71	—	7.17	—
$[\underline{OR}, \overline{OR}]$	[1.28, 46.37]	[1.05, 86.94]	$[1.24, 41.52]^a$	[1.39, 40.65]
Association p-value	$.01^b$	$.03^c$	.02	—

^{a}Explicit
^{b}Wald
^{c}Cumulative

TABLE 4.14   Summary of Receptor Level–Breast Cancer Results

Result	AU	EC	AC	CF
$\widehat{OR}$	3.35	—	3.33	—
$[\underline{OR}, \overline{OR}]$	[1.68, 6.70]	[1.58, 7.07]	$[1.67, 6.63]^a$	[1.69, 6.67]
Association p-value	$.001^b$	<.001	<.001	—

^{a}Explicit
^{b}Wald

4.5   SUMMARY OF EXAMPLES AND RECOMMENDATIONS

Table 4.13 summarizes the results of the antibody–diarrhea analyses based on the asymptotic unconditional (AU), exact conditional (EC), asymptotic conditional (AC), and Cornfield (CF) methods. Despite the small sample size involved, the four methods give reasonably similar results and lead to the conclusion that low antibody level is associated with an increased risk of diarrhea.

Table 4.14 summarizes the results of the receptor level–breast cancer analyses. For these data the four approaches produce results that are, for practical purposes, identical.

Walter (1987) and Walter and Cook (1991) recommend using the estimate $\log(\widehat{OR}_u)$, with .5 added to all cells, in preference to $\log(\widehat{OR}_c)$. Research on interval estimation reveals that, of the approximate techniques studied, Cornfield's method is the most accurate (Gart and Thomas, 1972; Brown, 1981; Gart and Thomas, 1982). In practice, as long as the sample size is reasonably large, the asymptotic methods generally give similar results. When there is concern that the sample size may be too small for an asymptotic analysis, exact methods should be used.

4.6   ASYMPTOTIC METHODS FOR A SINGLE 2 × I TABLE

To this point we have considered only dichotomous exposure variables. When there are several exposure categories (polychotomous) it is of interest to search for dose–response relationships and other patterns in the data, options that do not exist when exposure is dichotomous. In this section we describe asymptotic unconditional and asymptotic conditional methods for the analysis 2 × *I* tables, where $I \geq 2$.

The manner in which exposure categories are defined in a given study depends on a number of considerations—in particular, whether the exposure variable is continuous, discrete, or ordinal. An ordinal variable is one that is qualitative and where there is an implicit ordering of categories. For example, arthritis pain might be rated as mild, moderate, or severe. Stage of breast cancer is also ordinal, even though integers are used to designate the different stages. Discrete and ordinal variables are automatically in categorized form. In certain settings it may be reasonable to regard a discrete variable with many categories as continuous. For example, the number of cigarettes smoked per day is, strictly speaking, discrete, but in many applications it would be treated as a continuous variable.

When the exposure variable is continuous, categories can be created by selecting cutpoints to partition the range of exposures. To the extent possible, it is desirable to have categories that are consistent with the published literature. For instance, in Example 4.2, the continuous variable receptor level was dichotomized using a conventional cutpoint. The sample size of the study and the distribution of the exposure variable in the data also have implications for the choice of cutpoints, and hence for the number and width of categories. In particular, if a predetermined set of cutpoints results in categories that have few or even no subjects, it may be necessary to collapse over categories so as to avoid sparse data problems. When categories are created, it is implicitly assumed that, within each category, the association between exposure and disease is relatively uniform. This assumption may be violated when the categories are made too wide. It sometimes happens that neither substantive knowledge nor study data suggest a method of creating categories, making the choice of cutpoints somewhat arbitrary. In this situation, one option is to use percentiles as cutpoints. For example, quartiles can be formed using the 25th, 50th, and 75th percentiles. This results in four ordered categories consisting of the same (or nearly the same) numbers of subjects.

The data layout for the case of $I \geq 2$ exposure categories is given in Table 4.15. It is usual to order the categories from low to high exposure so that $i = 1$ corresponds to the lowest exposure. Thus the orientation of categories in Table 4.15 is the opposite of the 2 × 2 case. We model the ith exposure category using the binomial distribution with parameters (π_i, r_i) $(i = 1, 2, \ldots, I)$. The odds for the ith exposure category is

TABLE 4.15   Observed Counts: Closed Cohort Study

Disease	Exposure category						
	1	2	$\cdots$	i	$\cdots$	I	
yes	a_1	a_2	$\cdots$	a_i	$\cdots$	a_I	m_1
no	b_1	b_2	$\cdots$	b_i	$\cdots$	b_I	m_2
	r_1	r_2	$\cdots$	r_i	$\cdots$	r_I	r

$\omega_i = \pi_i/(1 - \pi_i)$. With $i = 1$ as the reference category, the odds ratio is

$$OR_i = \frac{\pi_i(1 - \pi_1)}{\pi_1(1 - \pi_i)}.$$

Point Estimates, Confidence Intervals, and Pearson and Mantel–Haenszel Tests of Association

The unconditional maximum likelihood estimates of ω_i and OR_i are $\hat{\omega}_i = a_i/b_i$ and

$$\widehat{OR}_{ui} = \frac{a_i b_1}{a_1 b_i}$$

where we note that $\widehat{OR}_{u1} = 1$. A confidence interval for OR_i can be estimated using (4.7). We say there is no association between exposure and disease if $\pi_1 = \pi_2 = \cdots = \pi_I$. The expected counts for the ith exposure category are

$$\hat{e}_i = \frac{r_i m_1}{r} \quad \text{and} \quad \hat{f}_i = \frac{r_i m_2}{r}.$$

It is readily verified that $\hat{e}_\bullet = a_\bullet = m_1$. It is possible to test each pair of categories for association using any of the tests for 2×2 tables described above. This involves $\binom{I}{2} = I(I-1)/2$ separate tests and, if I is at all large, several of the tests may provide evidence for association even when it is absent, purely on the basis of chance (type I error). For example, with $I = 10$ there would be 45 hypothesis tests. With $\alpha = .05 = 1/20$, even if there is no association between exposure and disease, on average, at least two of the 45 tests would provide evidence in favor of association. This is an example of the problem of multiple comparisons, an issue that has received quite a lot of attention in the epidemiologic literature (Rothman and Greenland, 1998). An approach that avoids this difficulty is to perform tests of association which consider all I exposure categories simultaneously, as we now describe.

The Pearson test of association for a $2 \times I$ table is

$$X_p^2 = \sum_{i=1}^{I} \left[\frac{(a_i - \hat{e}_i)^2}{\hat{e}_i} + \frac{(b_i - \hat{f}_i)^2}{\hat{f}_i} \right] \quad (\text{df} = I - 1). \tag{4.35}$$

Note that there are $I - 1$ degrees of freedom. Using earlier arguments it can be shown that

$$X_p^2 = \left(\frac{r}{m_2} \right) \sum_{i=1}^{I} \frac{(a_i - \hat{e}_i)^2}{\hat{e}_i} \quad (\text{df} = I - 1). \tag{4.36}$$

Conditioning on the total number of cases m_1 results in the multidimensional hypergeometric distribution (Appendix E). The Mantel–Haenszel test of association for a $2 \times I$ table is

$$X_{mh}^2 = \left(\frac{r-1}{m_2}\right) \sum_{i=1}^{I} \frac{(a_i - \hat{e}_i)^2}{\hat{e}_i} \qquad (\text{df} = I - 1). \qquad (4.37)$$

Observe that (4.35), (4.36), and (4.37) are generalizations of (4.8), (4.11), and (4.27), respectively. From (4.36) and (4.37), we have

$$X_{mh}^2 = \left(\frac{r-1}{r}\right) X_p^2$$

just as in the dichotomous case. When the null hypothesis is rejected by either of the above tests, the interpretation is that overall there is evidence for an association between exposure and disease. This does not mean that each of the pairwise tests necessarily has a small p-value. Indeed, it is possible for the pairwise tests to individually provide little evidence for association and yet for the simultaneous test to indicate that an association is present.

Test for Linear Trend

The Pearson and Mantel–Haenszel tests of association are designed to detect whether the probability of disease differs across exposure categories. These are rather nonspecific tests in that they fail to take into account patterns that may exist in the data. We now describe a test designed to detect linear trend. In order to apply this test, it is necessary to assign an exposure level (dose, score) to each category. For a continuous exposure variable, a reasonable approach is to define the exposure level for each category to be the midpoint of the corresponding cutpoints. As an illustration, for age groups 65–69, 70–74, and 75–79, the midpoints are 67.5, 72.5, and 77.5. A problem arises when there is an open-ended category since, in this case, the midpoint is undefined. For example, there is no obvious way of defining a midpoint for an age group such as 80+. An alternative that avoids this problem is to define the exposure level for each category to be the mean or median exposure based on study data.

When the exposure variable is ordinal, the assignment of exposure levels is more complicated. For example, in the breast cancer study described in Example 4.2, there are three stages of disease: Stage I is less serious than stage II, which in turn is less serious than stage III. However, it is not clear how exposure levels should be assigned. In a case like this, it is usual to simply define the exposure levels to be the consecutive integers 1, 2, and 3. Defining exposure levels in this way implicitly assumes that the "distance" between stage I and stage II is the same as that between stage II and stage III. An assumption such as this ultimately depends on some notion of "severity" of disease, and therefore needs to be justified.

Let s_i be the exposure level for the ith category with $s_1 < s_2 < \cdots < s_I$. The ω_i are unknown parameters, but we can imagine the scatter plot of $\log(\hat{\omega}_i)$ against s_i ($i = 1, 2, \ldots, I$). Let $\log(\hat{\omega}_i) = \hat{\alpha} + \hat{\beta} s_i$ be the "best-fitting" straight line for these points, where α and β are constants. We are interested in testing the hypothesis $H_0 : \beta = 0$. When $\beta \neq 0$ we say there is a linear trend in the log-odds, in which case the best-fitting straight line has a nonzero slope. As shown in Appendix E, the score test of $H_0 : \beta = 0$, which will be referred to as the test for linear trend (in log-odds), is

$$X_t^2 = \left(\frac{r-1}{m_2} \right) \frac{\left[\sum_{i=1}^{I} s_i (a_i - \hat{e}_i) \right]^2}{\sum_{i=1}^{I} s_i^2 \hat{e}_i - \left(\sum_{i=1}^{I} s_i \hat{e}_i \right)^2 \Big/ \hat{e}_\bullet} \qquad (\text{df} = 1) \qquad (4.38)$$

(Cochran, 1954; Armitage, 1955). Large values of X_t^2 provide evidence in favor of a linear trend. Although X_t^2 has been presented in terms of log-odds, it can be interpreted as a test for linear trend in probabilities, odds, or odds ratios. Accordingly, we can examine study data for the presence of linear trend using any of the corresponding category-specific parameter estimates.

It is important to appreciate that if H_0 is rejected—that is, if it is decided that a linear trend is present—it does not follow that the log-odds is a linear function of exposure (Rothman, 1986, p. 347; Maclure and Greenland, 1992). Instead the much more limited inference can be drawn that the "linear component" of the functional relationship relating log-odds to exposure has a nonzero slope. In many applications, especially when toxic exposures are being considered, it is reasonable to assume that, as exposure increases, there will be a corresponding increase in the risk of disease. However, more complicated risk relationships are possible. For example, the risk of having a stroke is elevated when blood pressure is either too high or too low. Consequently the functional relationship between blood pressure and stroke has something of a J-shape. The best-fitting straight line to such a curve has a positive slope and so the hypothesis of no linear trend would be rejected, even though the underlying functional relationship is far from linear.

Example 4.11 (Stage–Breast Cancer)   Table 4.16 gives the observed counts for the breast cancer data introduced in Example 4.2, but now with stage of disease as the exposure variable.

A useful place to begin the analysis is to compare stages II and III to stage I using 2×2 methods. Table 4.17 gives odds ratio estimates and 95% confidence intervals, with stage I as the reference category. As can be seen, there is an increasing trend in odds ratios across stages I, II, and III (where $\widehat{OR}_{u1} = 1$).

The expected counts, given in Table 4.18, are all greater than 5. The Pearson and Mantel–Haenszel tests are $X_p^2 = 38.55$ ($p < .001$) and $X_{mh}^2 = 38.35$ ($p < .001$),

TABLE 4.16   Observed Counts: Stage–Breast Cancer

Survival	Stage			
	I	II	III	
dead	7	26	21	54
alive	60	70	8	138
	67	96	29	192

TABLE 4.17   Odds Ratio Estimates and 95% Confidence
Intervals: Stage–Breast Cancer

Stage	$\widehat{OR}_{ui}$	$\underline{OR}_{ui}$	$\overline{OR}_{ui}$
II	3.18	1.29	7.85
III	22.50	7.27	69.62

TABLE 4.18   Expected Counts:
Stage–Breast Cancer

Survival	Stage 1	2	3	
dead	18.84	27.00	8.16	54
alive	48.16	69.00	20.84	138
	67	96	29	192

both of which provide considerable evidence for an association between stage of disease and breast cancer mortality. Setting $s_1 = 1$, $s_2 = 2$, and $s_3 = 3$, the test for linear trend is

$$X_t^2 = \left(\frac{192 - 1}{138} \right) \left[\frac{(24.69)^2}{200.25 - (97.31)^2/54} \right] = 33.90 \ (p < .001)$$

which is consistent with the observation made above. For the sake of illustration, suppose that the "severity" of stage III compared to stage II is regarded as three times the "severity" of stage II compared to stage I. For example, this determination might be based on an assessment of quality of life or projected mortality. With $s_1 = 1$, $s_2 = 2$, and $s_3 = 5$, the test for linear trend is $X_t^2 = 38.32 \ (p < .001)$, a finding that is close to the earlier result.

CHAPTER 5

Odds Ratio Methods for Stratified Closed Cohort Data

In most epidemiologic studies it is necessary to consider confounding and effect modification, and usually this involves some form of stratified analysis. In this chapter we discuss odds ratio methods for closed cohort studies in which there is stratification. The asymptotic unconditional and asymptotic conditional methods presented here are generalizations of those given in Chapter 4. Exact conditional methods are not discussed because they involve especially detailed computations. Appendix B gives the derivations of many of the asymptotic unconditional formulas that appear in this chapter and in Chapters 6 and 7.

5.1 ASYMPTOTIC UNCONDITIONAL METHODS FOR J (2 × 2) TABLES

Asymptotic methods require large sample sizes to be valid. The asymptotic unconditional techniques presented in this section work best when there are relatively few strata, and within each stratum the number of subjects in each exposure category is large (Breslow, 1981). These conditions, which will be referred to as the large-strata conditions, ensure that a large amount of data is available to estimate relatively few parameters, a situation that is conducive to the asymptotic unconditional approach.

Suppose that the data have been stratified into J strata and consider the case of a dichotomous exposure variable. We suppose that in the jth stratum the development of disease in the exposed and unexposed cohorts is governed by binomial random variables A_{1j} and A_{2j} with parameters (π_{1j}, r_{1j}) and (π_{2j}, r_{2j}), respectively ($j = 1, 2, \ldots, J$). As in Section 4.1, we assume that subjects behave independently with respect to the development of disease. For the jth stratum, the data layout is given in Table 5.1 and the odds ratio is

$$OR_j = \frac{\pi_{1j}(1 - \pi_{2j})}{\pi_{2j}(1 - \pi_{1j})}.$$

TABLE 5.1   Observed Counts:
Closed Cohort Study

Disease	Exposure		
	yes	no	
yes	a_{1j}	a_{2j}	m_{1j}
no	b_{1j}	b_{2j}	m_{2j}
	r_{1j}	r_{2j}	r_j

Each of the J tables can be analyzed separately using the methods of Chapter 4. The stratum-specific estimate are

$$\hat{\pi}_{1j} = \frac{a_{1j}}{r_{1j}} \qquad \hat{\pi}_{2j} = \frac{a_{2j}}{r_{2j}}$$

$$\widehat{OR}_{uj} = \frac{\hat{\pi}_{1j}(1 - \hat{\pi}_{2j})}{\hat{\pi}_{2j}(1 - \hat{\pi}_{1j})} = \frac{a_{1j}b_{2j}}{a_{2j}b_{1j}}$$

and

$$\widehat{\text{var}}(\log \widehat{OR}_{uj}) = \frac{1}{a_{1j}} + \frac{1}{a_{2j}} + \frac{1}{b_{1j}} + \frac{1}{b_{2j}}.$$

It may be difficult to synthesize the results of such an analysis when there are many strata and the odds ratio estimates are heterogeneous. The situation is greatly simplified when there is homogeneity, in which case the common stratum-specific odds ratio will be denoted by OR. (In Chapter 2 we used the notation θ.) Much of this chapter is based on the homogeneity assumption. In order to avoid having to state this assumption repeatedly, we regard homogeneity as being present unless stated otherwise. In particular, reference to OR will automatically imply that homogeneity is being assumed and that $OR_j = OR$ for all j. Under homogeneity we have from (4.2) that

$$\pi_{1j} = \frac{OR\pi_{2j}}{OR\pi_{2j} + (1 - \pi_{2j})}. \tag{5.1}$$

Point Estimates and Fitted Counts
The unconditional maximum likelihood equations are

$$\sum_{j=1}^{J} a_{1j} = \sum_{j=1}^{J} \frac{\widehat{OR}_u \hat{\pi}_{2j} r_{1j}}{\widehat{OR}_u \hat{\pi}_{2j} + (1 - \hat{\pi}_{2j})} \tag{5.2}$$

and

$$m_{1j} = \frac{\widehat{OR}_u \hat{\pi}_{2j} r_{1j}}{\widehat{OR}_u \hat{\pi}_{2j} + (1 - \hat{\pi}_{2j})} + \hat{\pi}_{2j} r_{2j} \qquad (j = 1, 2, \ldots, J) \tag{5.3}$$

where $\widehat{OR}_u$ denotes the unconditional maximum likelihood estimate of OR. In the terminology of Section 2.5.4, $\widehat{OR}_u$ is a summary estimate of OR. Equations (5.2) and (5.3) are a system of $J + 1$ equations in the $J + 1$ unknowns $\widehat{OR}_u$ and $\hat{\pi}_{2j}$ ($j = 1, 2, \ldots, J$). When $J > 2$ it is not possible to solve for $\widehat{OR}_u$ and $\hat{\pi}_{2j}$ explicitly as was the case for $J = 2$. Below we describe two methods for solving these equations which are tailored to the odds ratio setting. More general numerical methods for solving multidimensional systems of equations are described in Appendix B. Once $\widehat{OR}_u$ and $\hat{\pi}_{2j}$ have been estimated, we have from (5.1) that

$$\hat{\pi}_{1j} = \frac{\widehat{OR}_u \hat{\pi}_{2j}}{\widehat{OR}_u \hat{\pi}_{2j} + (1 - \hat{\pi}_{2j})}. \tag{5.4}$$

The fitted counts for the jth stratum are defined to be

$$\hat{a}_{1j} = \hat{\pi}_{1j} r_{1j} \qquad \hat{a}_{2j} = \hat{\pi}_{2j} r_{2j}$$
$$\hat{b}_{1j} = (1 - \hat{\pi}_{1j}) r_{1j} \qquad \hat{b}_{2j} = (1 - \hat{\pi}_{2j}) r_{2j}. \tag{5.5}$$

Using (5.4) and (5.5) we can rewrite the maximum likelihood equations as

$$\sum_{j=1}^{J} a_{1j} = \sum_{j=1}^{J} \hat{a}_{1j} \tag{5.6}$$

and

$$m_{1j} = \hat{a}_{1j} + \hat{a}_{2j} \qquad (j = 1, 2, \ldots, J). \tag{5.7}$$

These equations exhibit what will be referred to as the "observed equals fitted" format. Clearly, $a_{1j} + b_{1j} = r_{1j} = \hat{a}_{1j} + \hat{b}_{1j}$ and $a_{2j} + b_{2j} = r_{2j} = \hat{a}_{2j} + \hat{b}_{2j}$, and so the observed and fitted column marginal totals agree. From (5.5) and (5.7), $a_{1j} + a_{2j} = m_{1j} = \hat{a}_{1j} + \hat{a}_{2j}$ and $b_{1j} + b_{2j} = m_{2j} = \hat{b}_{1j} + \hat{b}_{2j}$, and so the observed and fitted row marginal totals also agree. Therefore the table of fitted counts can be displayed as in Table 5.2. Note the similarity to Table 4.7 where, unlike the present analysis, we conditioned on the marginal totals.

A remarkable result is that (5.3) can be expressed as

$$\widehat{OR}_u = \frac{\hat{a}_{1j}(r_{2j} - m_{1j} + \hat{a}_{1j})}{(m_{1j} - \hat{a}_{1j})(r_{1j} - \hat{a}_{1j})} \tag{5.8}$$

TABLE 5.2   Fitted Counts: Closed Cohort Study

Disease	Exposure		
	yes	no	
yes	$\hat{a}_{1j}$	$m_{1j} - \hat{a}_{1j}$	m_{1j}
no	$r_{1j} - \hat{a}_{1j}$	$r_{2j} - m_{1j} + \hat{a}_{1j}$	m_{2j}
	r_{1j}	r_{2j}	r_j

(Gart, 1971, 1972; Breslow, 1976). Identity (5.8), which has an obvious similarity to (4.28), establishes a connection between the asymptotic unconditional methods of this section and the asymptotic conditional methods to be discussed in Section 5.2.

We now turn to the problem of solving the maximum likelihood equations. As in Section 4.4, we treat (5.8) as a second-degree polynomial in the unknown $\hat{a}_{1j}$ and use the quadratic formula to obtain

$$\hat{a}_{1j} = \frac{-y_j - \sqrt{y_j^2 - 4xz_j}}{2x} \tag{5.9}$$

where

$$x = \widehat{OR}_u - 1$$
$$y_j = -[(m_{1j} + r_{1j})\widehat{OR}_u - m_{1j} + r_{2j}]$$
$$z_j = \widehat{OR}_u m_{1j} r_{1j}.$$

Then (5.6) can be expressed as

$$\sum_{j=1}^J a_{1j} = \frac{-1}{2x} \sum_{j=1}^J \left(y_j + \sqrt{y_j^2 - 4xz_j} \right).$$

This is an equation in the single unknown $\widehat{OR}_u$ which can be solved by trial and error. Once $\widehat{OR}_u$ has been determined, x, y_j, and z_j can be calculated, which leads to the estimates $\hat{a}_{1j}$ and $\hat{a}_{2j} = m_{1j} - \hat{a}_{1j}$.

An alternative approach to solving the maximum likelihood equations is based on an ingenious idea of Clayton (1982) that will reappear in Section 10.3.1. Rewrite (5.2) as

$$0 = \sum_{j=1}^J \left[a_{1j} - \frac{\widehat{OR}_u \hat{\pi}_{2j} r_{1j}}{\widehat{OR}_u \hat{\pi}_{2j} + (1 - \hat{\pi}_{2j})} \right]$$

$$= \sum_{j=1}^J \left[\frac{a_{1j}(1 - \hat{\pi}_{2j}) - \widehat{OR}_u b_{1j} \hat{\pi}_{2j}}{\widehat{OR}_u \hat{\pi}_{2j} + (1 - \hat{\pi}_{2j})} \right]$$

$$= \sum_{j=1}^J \frac{a_{1j}(1 - \hat{\pi}_{2j})}{\widehat{OR}_u \hat{\pi}_{2j} + (1 - \hat{\pi}_{2j})} - \widehat{OR}_u \sum_{j=1}^J \frac{b_{1j} \hat{\pi}_{2j}}{\widehat{OR}_u \hat{\pi}_{2j} + (1 - \hat{\pi}_{2j})} \tag{5.10}$$

and solve for the $\widehat{OR}_u$ preceding the second summation to obtain

$$\widehat{OR}_u = \sum_{j=1}^J \frac{a_{1j}(1 - \hat{\pi}_{2j})}{\widehat{OR}_u \hat{\pi}_{2j} + (1 - \hat{\pi}_{2j})} \Bigg/ \sum_{j=1}^J \frac{b_{1j} \hat{\pi}_{2j}}{\widehat{OR}_u \hat{\pi}_{2j} + (1 - \hat{\pi}_{2j})}. \tag{5.11}$$

With $\hat{a}_{1j}$ given by (5.9), we have $\hat{\pi}_{2j} = (m_{1j} - \hat{a}_{1j})/r_{2j}$. This can be substituted in the right-hand side of (5.11), which results in an equation in the single unknown $\widehat{OR}_u$. The solution to this equation can be obtained using an iterative approach. The first step in the iteration is to select an initial value for $\widehat{OR}_u$, which we denote by $\widehat{OR}_u^{(1)}$. This is substituted in the right-hand side of (5.11), and the calculations are performed to get an updated value $\widehat{OR}_u^{(2)}$. Then $\widehat{OR}_u^{(2)}$ is substituted in the right-hand side of (5.11) to get the next updated value $\widehat{OR}_u^{(3)}$, and so on. This process is repeated until the desired accuracy is achieved. The initial value $\widehat{OR}_u^{(1)}$ is arbitrary, but the crude estimate of the odds ratio (4.6) is a reasonable choice.

Confidence Interval
Let

$$\hat{v}_j = \left(\frac{1}{\hat{a}_{1j}} + \frac{1}{\hat{a}_{2j}} + \frac{1}{\hat{b}_{1j}} + \frac{1}{\hat{b}_{2j}} \right)^{-1} \tag{5.12}$$

and let $\hat{V}_u = \sum_{j=1}^{J} \hat{v}_j$. An estimate of $\mathrm{var}(\log \widehat{OR}_u)$ is

$$\widehat{\mathrm{var}}(\log \widehat{OR}_u) = \frac{1}{\hat{V}_u} \tag{5.13}$$

and a $(1 - \alpha) \times 100\%$ confidence interval for OR is obtained by exponentiating

$$\left[\log \underline{OR}_u, \log \overline{OR}_u \right] = \log(\widehat{OR}_u) \pm \frac{z_{\alpha/2}}{\sqrt{\hat{V}_u}}$$

(Gart, 1962).

Wald and Likelihood Ratio Tests of Association
We say there is no association between exposure and disease if $\pi_{1j} = \pi_{2j}$ for all j. This is equivalent to $OR_j = 1$ for all j, and when homogeneity is present this can be succinctly expressed as $OR = 1$ or $\log(OR) = 0$. For the jth stratum, the expected counts are

$$\hat{e}_{1j} = \frac{r_{1j}m_{1j}}{r_j} \qquad \hat{e}_{2j} = \frac{r_{2j}m_{1j}}{r_j}$$

$$\hat{f}_{1j} = \frac{r_{1j}m_{2j}}{r_j} \qquad \hat{f}_{2j} = \frac{r_{2j}m_{2j}}{r_j}.$$

Let

$$\hat{v}_{0j} = \left(\frac{1}{\hat{e}_{1j}} + \frac{1}{\hat{e}_{2j}} + \frac{1}{\hat{f}_{1j}} + \frac{1}{\hat{f}_{2j}} \right)^{-1}$$

and let $\hat{V}_{0u} = \sum_{j=1}^{J} \hat{v}_{0j}$. Under the hypothesis of no association $H_0 : \log(OR) = 0$, an estimate of $\text{var}(\log \widehat{OR}_u)$ is

$$\widehat{\text{var}}_0(\log \widehat{OR}_u) = \frac{1}{\hat{V}_{0u}}.$$

The Wald and likelihood ratio tests of association are

$$X_w^2 = (\log \widehat{OR}_u)^2 \hat{V}_{0u} \qquad (df = 1) \tag{5.14}$$

and

$$X_{lr}^2 = 2\sum_{j=1}^{J}\left[a_{1j}\log\left(\frac{a_{1j}}{\hat{e}_{1j}}\right) + a_{2j}\log\left(\frac{a_{2j}}{\hat{e}_{2j}}\right) + b_{1j}\log\left(\frac{b_{1j}}{\hat{f}_{1j}}\right)\right.$$
$$\left. + b_{2j}\log\left(\frac{b_{2j}}{\hat{f}_{2j}}\right)\right] \qquad (df = 1) \tag{5.15}$$

respectively. An advantage of the likelihood ratio test over the Wald test is that (5.15) does not require an estimate of the odds ratio and thus avoids the necessity of having to solve the maximum likelihood equations.

Wald, Score, and Likelihood Ratio Tests of Homogeneity
All that has been said to this point is predicated on the homogeneity assumption, the validity of which should be assessed before proceeding with any of the above calculations. As was observed many years ago by Mantel and Haenszel (1959), in an epidemiologic study it is unrealistic to assume that stratum-specific odds ratios are ever going to be precisely equal. In practice, there are too many factors affecting the association between exposure and disease for homogeneity to be strictly true. From this perspective, the homogeneity assumption is merely a convenient fiction that is adopted in order to simplify the analysis and interpretation of data. A pragmatic approach to the assessment of homogeneity involves the following steps: Examine stratum-specific estimates and their confidence intervals in order to develop a sense of whether there are meaningful differences across strata (after accounting for random error); perform a formal test of homogeneity; and then synthesize this information along with substantive knowledge of the relationship between exposure and disease, taking into account the aims of the study. In particular, even if a formal statistical test provides evidence that heterogeneity is present, it may be decided in the interests of simplicity to proceed on the basis of homogeneity and summarize across strata rather than retain stratum-specific estimates.

The Wald, score, and likelihood ratio tests of homogeneity are

$$X_h^2 = \sum_{j=1}^{J} \hat{v}_j (\log \widehat{OR}_{uj} - \log \widehat{OR}_u)^2 \qquad (df = J - 1) \tag{5.16}$$

$$X_h^2 = \sum_{j=1}^{J} \frac{(a_{1j} - \hat{a}_{1j})^2}{\hat{v}_j} \qquad (df = J - 1) \qquad (5.17)$$

and

$$X_h^2 = 2 \sum_{j=1}^{J} \left[a_{1j} \log \left(\frac{a_{1j}}{\hat{a}_{1j}} \right) + a_{2j} \log \left(\frac{a_{2j}}{\hat{a}_{2j}} \right) + b_{1j} \log \left(\frac{b_{1j}}{\hat{b}_{1j}} \right) \right.$$

$$\left. + b_{2j} \log \left(\frac{b_{2j}}{\hat{b}_{2j}} \right) \right] \qquad (df = J - 1) \qquad (5.18)$$

respectively (Liang and Self, 1985; Rothman and Greenland, 1998, p. 275; Lachin, 2000, §4.6.2). Note that there are $J - 1$ degrees of freedom as opposed to the single degree of freedom for the tests of association. Arguing as in Section 4.1, it is readily demonstrated that $(a_{1j} - \hat{e}_{1j})^2 = (a_{2j} - \hat{e}_{2j})^2$, with similar identities for other rows and columns. It follows from (5.12) that (5.17) can be written as

$$X_h^2 = \sum_{j=1}^{J} \left[\frac{(a_{1j} - \hat{a}_{1j})^2}{\hat{a}_{1j}} + \frac{(a_{2j} - \hat{a}_{2j})^2}{\hat{a}_{2j}} + \frac{(b_{1j} - \hat{b}_{1j})^2}{\hat{b}_{1j}} + \frac{(b_{2j} - \hat{b}_{2j})^2}{\hat{b}_{2j}} \right].$$

The likelihood ratio tests, (5.15) and (5.18), have a similar appearance because in each case a comparison is being made between observed and fitted counts. In (5.15) the fitted (expected) counts are estimated under the hypothesis of no association, and in (5.18) they are estimated under the hypothesis of homogeneity. We can think of (5.15) and (5.18), along with the other tests of association and homogeneity, as tests of "goodness of fit" in which observed values are compared to fitted values, where the latter are based on a particular model. When the model fits the data well, the observed and fitted values will be close in value, the test of goodness of fit will result in a large p-value, and the model (hypothesis) will not be rejected. This type of reasoning is particularly important in the regression setting when a succession of increasingly complicated models are fitted to data and a decision needs to made as to which model fits the data best.

Test for Linear Trend

The test for linear trend in J (2 × 2) tables has many similarities to the test for linear trend in a single 2 × I table (4.38). Let s_j be the exposure level for the jth stratum with $s_1 < s_2 < \cdots < s_J$. Consider the scatter plot of $\log(\widehat{OR}_j)$ against s_j ($j = 1, 2, \ldots, J$) and let $\log(\widehat{OR}_j) = \hat{\alpha} + \hat{\beta} s_j$ be the best-fitting straight line, where α and β are constants. Note that as opposed to the 2 × I situation, where we were interested in testing for a linear trend in log-odds across exposure categories, here we are concerned with a linear trend in log-odds ratios across strata. Linear trend is said to be present if $\beta \neq 0$. The score test of $H_0: \beta = 0$, which will be referred to as the test for linear trend (in log-odds ratios), is

$$X_t^2 = \frac{\left[\sum_{j=1}^{J} s_j(a_{1j} - \hat{a}_{1j})\right]^2}{\sum_{j=1}^{J} s_j^2 \hat{v}_j - \left(\sum_{j=1}^{J} s_j \hat{v}_j\right)^2 \big/ \hat{v}_\bullet} \qquad \text{(df} = 1) \qquad (5.19)$$

where $\hat{v}_j$ is given by (5.12) (Breslow and Day, 1980, p. 142). Although X_t^2 has been presented in terms of log-odds ratios, it has an equivalent interpretation as a test for linear trend in odds ratios. Evidence for the presence of linear trend is also evidence that the stratum-specific odds ratios are unequal—that is, are heterogeneous. In this book we will usually interpret the test for linear trend (5.19) in this more limited sense—that is, as a test of homogeneity which has 1 degree of freedom.

Example 5.1 (Receptor Level–Breast Cancer)　Table 5.3 gives the breast cancer data discussed in Example 4.2 after stratifying by stage. The purpose of stratifying is twofold: to determine whether stage is a confounder of the association between receptor level and breast cancer survival, and to investigate whether it is an effect modifier of this association. We first consider the issue of confounding using the methods of Chapter 2, where we take E and F to be receptor level and stage, respectively. Stage is an overwhelmingly important predictor of survival in breast cancer patients, irrespective of receptor level, and so condition 1 of Section 2.5.3 is satisfied. There is more limited evidence in the oncology literature that stage is associated with receptor level, and so condition 2 may or may not be satisfied. For the moment assume that condition 2 does not hold. This means that stage fails one of the necessary requirements to be a confounder of the association between receptor level and breast cancer survival. From Example 4.2 the crude odds ratio estimate of the association between receptor level and breast cancer survival is $\widehat{OR}_u = 3.35$, and the 95% confidence interval for OR is [1.69, 6.70]. Following the discussion surrounding Table 2.6(b), we take $\widehat{OR}_u = 3.35$ to be an estimate of the overall odds ratio for the cohort.

Despite the above remarks, there are reasons to believe that stage is in fact a confounder. Consider Table 5.4, which gives the stratum-specific analysis of the breast cancer data according to stage of disease. The values of π_{2j} are quite different, a finding that is consistent with the remarks made above in connection with condition 1. However, based on these data it seems that $p_{1j} \neq p_{2j}$ for stages I and III, which means that condition 2 may be satisfied after all. According to this reasoning, stage

TABLE 5.3　Observed Counts: Receptor Level–Breast Cancer

Survival	Stage I Receptor level low	high		Stage II Receptor level low	high		Stage III Receptor level low	high	
dead	2	5	7	9	17	26	12	9	21
alive	10	50	60	13	57	70	2	6	8
	12	55	67	22	74	96	14	15	29

TABLE 5.4   Odds Ratio Estimates and 95% Confidence Intervals: Receptor Level–Breast Cancer

Stage	$\widehat{OR}_{uj}$	$\underline{OR}_{uj}$	$\overline{OR}_{uj}$	π_{2j}	p_{1j}	p_{2j}
I	2.00	.34	11.80	.09	.25	.38
II	2.32	.85	6.36	.23	.46	.51
III	4.00	.65	24.66	.60	.29	.10

satisfies the necessary conditions to be a confounder. From Tables 4.5(a) and 5.3, $cE = (31/144)48 = 10.33$ and $sE = (5/55)12 + (17/74)22 + (9/15)14 = 14.54$. cE is substantially smaller than sE, and so we have additional evidence that stage may be a confounder. For the sake of illustration we assume for the rest of the example that stage is indeed a confounder. Accordingly we take $s\widehat{OR} = [23(48 - 14.54)]/[14.54(48 - 23)] = 2.12$ to be an estimate of the overall odds ratio for the cohort. Note that this estimate makes no assumptions regarding homogeneity, an issue we now consider.

Before applying the stratified methods of this section, it is helpful to examine the strata separately using the techniques of Chapter 4. As can be seen from Table 5.4, there is little difference between the odds ratio estimates for stages I and II, but the estimate for stage III is noticeably larger, a finding that points to heterogeneity. In addition, there is something of an increasing trend in the odds ratio estimates, also suggesting heterogeneity. However, the 95% confidence intervals overlap to a considerable extent. In particular, each confidence interval contains the odds ratio estimates for the other two strata, a finding that is consistent with homogeneity. Overall, the evidence is mostly in favor of homogeneity. Assume for the moment that the increasing trend in odds ratio estimates is "real"—that is, not due to random error. The interpretation is that the odds ratio for the association between receptor level and breast cancer mortality increases as stage becomes more advanced. It is important not to make the mistake of interpreting this finding as an indication that the mortality risk from breast cancer increases with stage. Table 5.4 is concerned with odds ratios relating receptor level and breast cancer mortality, not with odds ratios relating stage and breast cancer mortality.

The maximum likelihood estimates are $\widehat{OR}_u = 2.51$, $\hat{\pi}_{21} = .086$, $\hat{\pi}_{22} = .226$, and $\hat{\pi}_{23} = .639$, and the fitted counts are given in Table 5.5. Note that (5.8) is

TABLE 5.5   Fitted Counts Under Homogeneity: Receptor Level–Breast Cancer

	Stage I			Stage II			Stage III		
	Receptor level			Receptor level			Receptor level		
Survival	low	high		low	high		low	high	
dead	2.28	4.72	7	9.29	16.71	26	11.42	9.58	21
alive	9.72	50.28	60	12.71	57.29	70	2.58	5.42	8
	12	55	67	22	74	96	14	15	29

TABLE 5.6   Expected Counts: Receptor Level–Breast Cancer

Survival	Stage I Receptor level low	high		Stage II Receptor level low	high		Stage III Receptor level low	high	
dead	1.25	5.75	7	5.96	20.04	26	10.14	10.86	21
alive	10.75	49.25	60	16.04	53.96	70	3.86	4.14	8
	12	55	67	22	74	96	14	15	29

satisfied in each stratum, for example, $(2.29 \times 50.29)/(4.72 \times 9.72) = 2.51$. From $\hat{V}_u = 1.29 + 3.79 + 1.31 = 6.40$ and $\widehat{\text{var}}(\log \widehat{OR}_u) = 1/6.40 = .156$, the 95% confidence interval for OR is $[1.16, 5.44]$, which we observe does not contain 1.

The expected counts are given in Table 5.6. From $\hat{V}_{0u} = .922 + 3.35 + 1.45 = 5.72$, the Wald and likelihood ratio tests of association are $X_w^2 = (\log 2.51)^2(5.72) = 4.83$ ($p = .03$) and $X_{lr}^2 = 5.64$ ($p = .02$). The interpretation is that, after adjusting for the confounding effects of stage, receptor level is associated with breast cancer survival.

On a cell-by-cell basis the observed and fitted counts are close in value and so the homogeneity model appears to fit the data well. The Wald, score, and likelihood ratio tests of homogeneity are

$$(5.16) = 1.29(.693 - .919)^2 + 3.79(.842 - .919)^2 + 1.31(1.39 - .919)^2$$

$$= .374 \; (p = .83)$$

$$(5.17) = \frac{(2 - 2.28)^2}{1.29} + \frac{(9 - 9.29)^2}{3.79} + \frac{(12 - 11.42)^2}{1.31} = .341 \; (p = .84)$$

and $(5.18) = .351$ ($p = .84$), each of which provides considerable evidence in favor of homogeneity. Setting $s_1 = 1$, $s_2 = 2$, and $s_3 = 3$, the test for linear trend is

$$X_t^2 = \frac{(.863)^2}{28.24 - (12.81)^2/6.40} = .286 \; (p = .59)$$

which is also consistent with homogeneity.

Based on the stratum-specific confidence intervals, the tests of homogeneity, and the test for linear trend, it is reasonable to conclude that there is a common stratum-specific odds ratio. From the overall confidence interval and the tests of association, we infer that this odds ratio is not equal to 1. So we take $\widehat{OR}_u = 2.51$ to be an estimate of the common stratum-specific odds ratio. In view of the discussion surrounding Table 2.6(b), the summary estimate ($\widehat{OR}_u = 2.51$) and the standardized estimate ($s\widehat{OR} = 2.12$) characterize different features of the cohort. In practice, only summary estimates are reported in the literature.

5.2 ASYMPTOTIC CONDITIONAL METHODS FOR J (2 × 2) TABLES

We now turn our attention to methods based on the asymptotic conditional approach. The techniques discussed in this section work well under the same large-strata conditions considered in the previous section. In addition, these methods are also valid when, within each stratum, the number of subjects in each exposure category is small, provided there are a large number of strata (Breslow, 1981). These will be referred to as the sparse-strata conditions. The presence of a large number of strata ensures that, even though stratum-specific sample sizes may be small, the overall sample size for the study is large. As a consequence of conditioning on the marginal totals, the stratum-specific nuisance parameters π_{2j} are eliminated and so, just as in the asymptotic unconditional case, there is a large amount of data available to estimate OR. For the remainder of this chapter the examples are based on data of the large-strata type. In Chapter 6 we consider a particular application in which sparse-strata, but not large-strata, conditions are satisfied. In that setting it is demonstrated that asymptotic unconditional methods may produce biased estimates.

Let A_{1j} denote the hypergeometric random variable for the jth stratum. From (4.13) the probability function is

$$P(A_{1j} = a_{1j}|OR) = \frac{1}{C_j}\binom{r_{1j}}{a_{1j}}\binom{r_{2j}}{m_{1j} - a_{1j}}OR^{a_{1j}} \tag{5.20}$$

where

$$C_j = \sum_{x=l_j}^{u_j}\binom{r_{1j}}{x}\binom{r_{2j}}{m_{1j} - x}OR^{x}$$

and where $l_j = \max(0, r_{1j} - m_{2j})$ and $u_j = \min(r_{1j}, m_{1j})$. From (4.14) and (4.15) the mean and variance of A_{1j} are

$$E(A_{1j}|OR) = \frac{1}{C_j}\sum_{x=l_j}^{u_j}x\binom{r_{1j}}{x}\binom{r_{2j}}{m_{1j} - x}OR^{x} \tag{5.21}$$

and

$$v_j = \text{var}(A_{1j}|OR) = \frac{1}{C_j}\sum_{x=l_j}^{u_j}[x - E(A_{1j}|OR)]^2\binom{r_{1j}}{x}\binom{r_{2j}}{m_{1j} - x}OR^{x}. \tag{5.22}$$

Point Estimates and Fitted Counts
The conditional maximum likelihood equation is

$$\sum_{j=1}^{J}a_{1j} = \sum_{j=1}^{J}E(A_{1j}|\widehat{OR}_{\text{c}}) \tag{5.23}$$

where

$$E(A_{1j}|\widehat{OR}_{\mathrm{c}}) = \frac{1}{\hat{C}_j} \sum_{x=l_j}^{u_j} x \binom{r_{1j}}{x} \binom{r_{2j}}{m_{1j}-x} (\widehat{OR}_{\mathrm{c}})^x$$

$$\hat{C}_j = \sum_{x=l_j}^{u_j} \binom{r_{1j}}{x} \binom{r_{2j}}{m_{1j}-x} (\widehat{OR}_{\mathrm{c}})^x$$

and where $\widehat{OR}_{\mathrm{c}}$ denotes the conditional maximum likelihood estimate of OR (Birch, 1964; Gart, 1970). Unlike the unconditional maximum likelihood equations, which involve $J + 1$ unknowns, (5.23) has the single unknown $\widehat{OR}_{\mathrm{c}}$, making it feasible to find a solution by trial and error. Once $\widehat{OR}_{\mathrm{c}}$ has been estimated, the fitted count for the index cell of the jth table is defined to be

$$\hat{a}_{1j} = E(A_{1j}|\widehat{OR}_{\mathrm{c}}). \tag{5.24}$$

The rest of the fitted counts are calculated along the lines of Table 4.7, thereby ensuring that the observed and fitted marginal totals agree. In view of (5.24) we can rewrite (5.23) as $\sum_{j=1}^{J} a_{1j} = \sum_{j=1}^{J} \hat{a}_{1j}$. This equation is formally the same as (5.6) except that the $\hat{a}_{1j}$ are based on (5.24) rather than (5.9).

Confidence Interval

Analogous to (4.20) and (4.21), an implicit $(1 - \alpha) \times 100\%$ confidence interval for OR is obtained by solving the equations

$$\frac{\sum_{j=1}^{J} a_{1j} - \sum_{j=1}^{J} E(A_{1j}|\underline{OR}_{\mathrm{c}})}{\sqrt{\sum_{j=1}^{J} \mathrm{var}(A_{1j}|\underline{OR}_{\mathrm{c}})}} = z_{\alpha/2}$$

and

$$\frac{\sum_{j=1}^{J} a_{1j} - \sum_{j=1}^{J} E(A_{1j}|\overline{OR}_{\mathrm{c}})}{\sqrt{\sum_{j=1}^{J} \mathrm{var}(A_{1j}|\overline{OR}_{\mathrm{c}})}} = -z_{\alpha/2}$$

for $\underline{OR}_{\mathrm{c}}$ and $\overline{OR}_{\mathrm{c}}$ (Mantel, 1977). Although these are complicated expressions, they are amenable to the trial and error approach. Given the estimate $\widehat{OR}_{\mathrm{c}}$, from (5.22) an estimate of $\mathrm{var}(A_{1j}|OR)$ is

$$\hat{v}_j = \frac{1}{\hat{C}_j} \sum_{x=l_j}^{u_j} (x - \hat{a}_{1j})^2 \binom{r_{1j}}{x} \binom{r_{2j}}{m_{1j}-x} (\widehat{OR}_{\mathrm{c}})^x \tag{5.25}$$

where we note from (5.24) that $\hat{a}_{1j} = E(A_{1j}|\widehat{OR}_{\mathrm{c}})$. Let $\hat{V}_{\mathrm{c}} = \sum_{j=1}^{J} \hat{v}_j$. As shown in Appendix C, an estimate of $\mathrm{var}(\log \widehat{OR}_{\mathrm{c}})$ is

$$\widehat{\mathrm{var}}(\log \widehat{OR}_{\mathrm{c}}) = \frac{1}{\hat{V}_{\mathrm{c}}} \tag{5.26}$$

and a $(1 - \alpha) \times 100\%$ confidence interval for OR is obtained by exponentiating

$$\left[\log \underline{OR}_c, \log \overline{OR}_c\right] = \log(\widehat{OR}_c) \pm \frac{z_{\alpha/2}}{\sqrt{\hat{V}_c}}$$

(Birch, 1964).

A first impression is that $\hat{V}_c$ and $\hat{V}_u$ are quite dissimilar due to the difference between (5.12) and (5.25). However, the Cornfield approximation to (5.25), given by the stratum-specific version of (4.31), provides a bridge between (5.12) and (5.25) and hence between $\hat{V}_u$ and $\hat{V}_c$. Also, (4.28) and (5.8) provide a connection between the fitted counts based on the asymptotic unconditional and asymptotic conditional methods.

Mantel–Haenszel Test of Association

When $OR = 1$ it follows from (4.17) and (4.18) that (5.21) and (5.22) simplify to

$$\hat{e}_{1j} = \frac{r_{1j}m_{1j}}{r_j} \tag{5.27}$$

and

$$\hat{v}_{0j} = \frac{r_{1j}r_{2j}m_{1j}m_{2j}}{r_j^2(r_j - 1)}. \tag{5.28}$$

The Mantel–Haenszel test of association is

$$X_{\text{mh}}^2 = \frac{\left(\sum_{j=1}^{J} a_{1j} - \sum_{j=1}^{J} \hat{e}_{1j}\right)^2}{\sum_{j=1}^{J} \hat{v}_{0j}} = \frac{(a_{1\bullet} - \hat{e}_{1\bullet})^2}{\hat{v}_{0\bullet}} \qquad (\text{df} = 1) \tag{5.29}$$

(Mantel and Haenszel, 1959). With l_j and u_j defined as above, let $R = \min(\hat{e}_{1\bullet} - l_\bullet, u_\bullet - \hat{e}_{1\bullet})$. Mantel and Fleiss (1980) show that the normal approximation underlying the Mantel–Haenszel test should be satisfactory provided $R \geq 5$. A more straightforward criterion given by Rothman and Greenland (1998, p. 275) requires that the summed observed counts, $\hat{a}_{1\bullet}$, $\hat{a}_{2\bullet}$, $\hat{b}_{1\bullet}$, and $\hat{b}_{2\bullet}$, and the summed expected counts, $\hat{e}_{1\bullet}$, $\hat{e}_{2\bullet}$, $\hat{f}_{1\bullet}$, and $\hat{f}_{2\bullet}$, should all be greater than or equal to 5. This shows that it is the overall counts, not stratum-specific counts, which determine the validity of the normal approximation.

Example 5.2 (Receptor Level–Breast Cancer) The asymptotic conditional estimate of the odds ratio is $\widehat{OR}_c = 2.47$. Table 5.7 gives the corresponding fitted counts. Unlike the asymptotic unconditional case, an identity of the form (5.8) is not necessarily satisfied. For example, $(2.29 \times 50.29)/(4.71 \times 9.71) = 2.52$, which does not equal $\widehat{OR}_c = 2.47$. Comparing Tables 5.5 and 5.7, the fitted counts based on the asymptotic unconditional and asymptotic conditional methods are nearly identical.

TABLE 5.7   Fitted Counts Under Homogeneity: Receptor Level–Breast Cancer

Survival	Stage I Receptor level low	high		Stage II Receptor level low	high		Stage III Receptor level low	high	
dead	2.29	4.71	7	9.27	16.73	26	11.44	9.56	21
alive	9.71	50.29	60	12.73	57.27	70	2.56	5.44	8
	12	55	67	22	74	96	14	15	29

The implicit 95% confidence interval for OR is $[1.15, 5.28]$. From $\hat{V}_c = 1.32 + 3.84 + 1.35 = 6.52$ and $\widehat{\mathrm{var}}(\log \widehat{OR}_c) = 1/6.52 = .153$, the explicit 95% confidence interval for OR is $[1.15, 5.32]$. The Mantel–Haenszel test is $X_{\mathrm{mh}}^2 = (23 - 17.35)^2/5.82 = 5.49$ ($p = .02$). Since $R = \min(17.35 - 6, 43 - 17.35) = 11.35$, the normal approximation is satisfactory. The interpretation of these results is virtually the same as for Example 5.1.

5.3   MANTEL–HAENSZEL ESTIMATE OF THE ODDS RATIO

Both the asymptotic unconditional and asymptotic conditional methods of estimating OR involve extensive calculations. We now discuss an alternative method of point estimation which is computationally straightforward and which produces excellent results under both large-strata and sparse-strata conditions. The celebrated Mantel–Haenszel estimate of OR is

$$\widehat{OR}_{\mathrm{mh}} = \frac{\sum_{j=1}^{J} R_j}{\sum_{j=1}^{J} S_j} = \frac{R_\bullet}{S_\bullet} \tag{5.30}$$

where

$$R_j = \frac{a_{1j} b_{2j}}{r_j}$$

and

$$S_j = \frac{a_{2j} b_{1j}}{r_j}$$

(Mantel and Haenszel, 1959). Rewriting (5.30) as $\widehat{OR}_{\mathrm{mh}} = (\sum_{j=1}^{J} S_j \widehat{OR}_{\mathrm{u}j})/S_\bullet$, we see that $\widehat{OR}_{\mathrm{mh}}$ is a weighted average of stratum-specific odds ratio estimates. It can be shown that $S_j = 1/\widehat{\mathrm{var}}_0(\widehat{OR}_{\mathrm{u}j})$ and so the weights entering into $\widehat{OR}_{\mathrm{mh}}$ are the reciprocals of stratum-specific variance estimates which are calculated under the hypothesis of no association. There is an interesting connection between X^2_{mh} and $\widehat{OR}_{mh}$. It is readily demonstrated that $a_{1j} - \hat{e}_{1j} = R_j - S_j$ and hence that $a_{1\bullet} - \hat{e}_{1\bullet} = R_\bullet - S_\bullet$. It follows that $X^2_{mh} = 0$ if and only if $\widehat{OR}_{mh} = 1$.

The Robins–Breslow–Greenland (RBG) estimate of $\text{var}(\log \widehat{OR}_{mh})$ is

$$\widehat{\text{var}}(\log \widehat{OR}_{mh}) = \frac{T_\bullet}{2(R_\bullet)^2} + \frac{U_\bullet + V_\bullet}{2(R_\bullet)(S_\bullet)} + \frac{W_\bullet}{2(S_\bullet)^2} \tag{5.31}$$

where

$$T_j = \frac{a_{1j}b_{2j}(a_{1j} + b_{2j})}{r_j^2}$$

$$U_j = \frac{a_{2j}b_{1j}(a_{1j} + b_{2j})}{r_j^2}$$

$$V_j = \frac{a_{1j}b_{2j}(a_{2j} + b_{1j})}{r_j^2}$$

$$W_j = \frac{a_{2j}b_{1j}(a_{2j} + b_{1j})}{r_j^2}$$

(Robins, Breslow, and Greenland, 1986; Robins, Greenland, and Breslow, 1986; Phillips and Holland, 1987). An important property of this estimate is that it is valid under both large-strata and sparse-strata conditions. A $(1 - \alpha) \times 100\%$ confidence interval for OR is obtained by exponentiating

$$\left[\log \underline{OR}_{mh}, \log \overline{OR}_{mh}\right] = \log(\widehat{OR}_{mh}) \pm z_{\alpha/2}\sqrt{\widehat{\text{var}}(\log \widehat{OR}_{mh})} \,.$$

When there is only one stratum, (5.31) simplifies to (4.7). Sato (1990) gives another estimate of $\text{var}(\log \widehat{OR}_{mh})$ which is applicable in both large-strata and sparse-strata settings.

Prior to (5.31) becoming available, the test-based method of estimating $\text{var}(\log \widehat{OR}_{mh})$ was commonly used. This approach lacks a sound theoretical basis but has the attraction of computational simplicity. The test-based method can be adapted to a variety of settings using arguments similar to what follows. It will have become apparent by now that tests of association developed using different theoretical approaches, such as X^2_w, X^2_{lr}, and X^2_{mh}, tend to have similar values. If we had an estimate of $\text{var}(\log \widehat{OR}_{mh})$ other than (5.31), we would expect the corresponding test of association to be close in value to X^2_{mh}. That is, we would have the approximate equality $X^2_{mh} = (\log \widehat{OR}_{mh})^2 / \widehat{\text{var}}_0(\log \widehat{OR}_{mh})$. The test-based approach "solves" this equation for $\widehat{\text{var}}_0(\log \widehat{OR}_{mh})$ and defines the estimate of $\widehat{\text{var}}(\log \widehat{OR}_{mh})$ to be

$$\widehat{\text{var}}_0(\log \widehat{OR}_{mh}) = \frac{(\log \widehat{OR}_{mh})^2}{X^2_{mh}}$$

(Miettinen, 1976). The subscript 0 is needed because the variance is being estimated under the null hypothesis $H_0 : OR = 1$. Strictly speaking, the test-based approach is valid only when $OR = 1$, but in practice this method produces satisfactory results for a broad range of odds ratios (Halperin, 1977; Miettinen, 1977). In what follows, the notation $\widehat{\text{var}}(\log \widehat{OR}_{\text{mh}})$ will be used only to denote the RBG estimate.

The Breslow–Day test of homogeneity is calculated by replacing $\widehat{OR}_{\text{u}}$ with $\widehat{OR}_{\text{mh}}$ in (5.9) and (5.12). The resulting estimates, denoted by $\hat{a}_{1j\,\text{mh}}$ and $\hat{v}_{j\,\text{mh}}$, are substituted in (5.17) to obtain

$$X_{\text{bd}}^2 = \sum_{j=1}^{J} \frac{(a_{1j} - \hat{a}_{1j\text{mh}})^2}{\hat{v}_{j\text{mh}}} - \frac{(a_{1\bullet} - \hat{a}_{1\bullet\text{mh}})^2}{\hat{v}_{\bullet\text{mh}}} \qquad (\text{df} = J - 1) \qquad (5.32)$$

(Breslow and Day, 1980, p. 142; Breslow, 1996). The second term in (5.32), which is due to Tarone (1985), corrects for using $\widehat{OR}_{\text{mh}}$ in place of the more efficient estimate $\widehat{OR}_{\text{u}}$. Since $\widehat{OR}_{\text{u}}$ is defined so as to satisfy $a_{1\bullet} = \hat{a}_{1\bullet}$ (5.6), it follows that the correction term will be small when $\widehat{OR}_{\text{mh}}$ is close to $\widehat{OR}_{\text{u}}$, as is often the case in practice. Liang and Self (1985) and Liang (1987) describe tests of homogeneity for the sparse-strata setting, but the formulas are complicated and will not be presented here.

Example 5.3 (Receptor Level–Breast Cancer)   The Mantel–Haenszel odds ratio estimate is $\widehat{OR}_{\text{mh}} = 9.32/3.67 = 2.54$, the RBG variance estimate is

$$\widehat{\text{var}}(\log \widehat{OR}_{\text{mh}}) = \frac{6.37}{2(9.32)^2} + \frac{2.55 + 2.95}{2(9.32)(3.67)} + \frac{1.12}{2(3.67)^2} = .159$$

and the 95% confidence interval for OR is [1.16, 5.55]. The Breslow–Day test of homogeneity is $X_{\text{bd}}^2 = .341$ ($p = .84$), which includes the correction term .001. The test-based estimate is $\widehat{\text{var}}_0(\log \widehat{OR}_{\text{mh}}) = (\log 2.54)^2/5.49 = .158$, which is almost identical to the RBG estimate.

5.4   WEIGHTED LEAST SQUARES METHODS FOR J (2 × 2) TABLES

Weighted least squares (WLS) methods for odds ratio analysis were introduced by Woolf (1955) and extended to the regression setting by Grizzle et al. (1969). Similar to the asymptotic unconditional methods, these techniques perform well under large-strata, but not sparse-strata, conditions. Following Section 1.2.2, define the weight for the jth stratum to be

$$\hat{w}_j = \frac{1}{\widehat{\text{var}}(\log \widehat{OR}_{\text{u}j})} = \left(\frac{1}{a_{1j}} + \frac{1}{a_{2j}} + \frac{1}{b_{1j}} + \frac{1}{b_{2j}} \right)^{-1}$$

and let $\hat{W}_{\text{ls}} = \sum_{j=1}^{J} \hat{w}_j$. The WLS estimate of $\log(OR)$ is defined to be the weighted average of the $\log(\widehat{OR}_{\text{u}j})$,

$$\log(\widehat{OR}_{ls}) = \frac{1}{\hat{W}_{ls}} \sum_{j=1}^{J} \hat{w}_j \log(\widehat{OR}_{uj}).$$ (5.33)

An estimate of OR is obtained by exponentiating (5.33),

$$\widehat{OR}_{ls} = \exp\left(\frac{1}{\hat{W}_{ls}} \sum_{j=1}^{J} \hat{w}_j \log(\widehat{OR}_{uj})\right).$$

From (1.25) an estimate of var($\log \widehat{OR}_{ls}$) is

$$\widehat{\text{var}}(\log \widehat{OR}_{ls}) = \frac{1}{\hat{W}_{ls}}$$ (5.34)

and a $(1 - \alpha) \times 100\%$ confidence interval for $\widehat{OR}_{ls}$ is obtained by exponentiating

$$\left[\log \underline{OR}_{ls}, \log \overline{OR}_{ls}\right] = \log(\widehat{OR}_{ls}) \pm \frac{z_{\alpha/2}}{\sqrt{\hat{W}_{ls}}}.$$ (5.35)

Let $\hat{W}_{0ls} = \hat{V}_{0u}$, where $\hat{V}_{0u}$ was defined in Section 5.1 in conjunction with the Wald test. The WLS test of association is

$$X_{ls}^2 = (\log \widehat{OR}_{ls})^2 \hat{W}_{0ls} \qquad (\text{df} = 1)$$ (5.36)

and the test of homogeneity is

$$X_h^2 = \sum_{j=1}^{J} \hat{w}_j (\log \widehat{OR}_{uj} - \log \widehat{OR}_{ls})^2 \qquad (\text{df} = J - 1).$$ (5.37)

Note the similarity of (5.34), (5.36), and (5.37) to the asymptotic unconditional formulas (5.13), (5.14), and (5.16), respectively. The difference is that the weighted least squares formulas are based on observed counts, whereas the asymptotic unconditional formulas use fitted counts.

Example 5.4 (Receptor Level–Breast Cancer)   From $\hat{W}_{ls} = 1.22 + 3.78 + 1.16 = 6.16$ and

$$\log(\widehat{OR}_{ls}) = \frac{(1.22 \times .693) + (3.78 \times .842) + (1.16 \times 1.39)}{6.16} = .915$$

the WLS estimate of the odds ratio is $\widehat{OR}_{ls} = \exp(.915) = 2.50$. From $\widehat{\text{var}}(\log \widehat{OR}_{ls}) = 1/6.16 = .162$, the 95% confidence interval for OR is $[1.13, 5.50]$. The test of association is $X_{ls}^2 = (\log 2.50)^2 (5.72) = 4.79$ ($p = .03$), where $\hat{W}_{0ls} = 5.72$ comes from Example 5.1. The test of homogeneity is

$$X_h^2 = 1.22(.693 - .915)^2 + 3.78(.842 - .915)^2 + 1.16(1.39 - .915)^2$$
$$= .338 \ (p = .84).$$

5.5   INTERPRETATION UNDER HETEROGENEITY

When homogeneity is present, one of the issues facing the data analyst is how to summarize stratum-specific estimates. The asymptotic unconditional, asymptotic conditional, Mantel–Haenszel, and weighted least squares methods provide four somewhat different answers to this question. The Mantel–Haenszel and weighted least squares estimates are weighted averages of stratum-specific odds ratio and log-odds ratio estimates, respectively, where the weights are reciprocals of estimated variances. As was pointed out in Section 1.2.1, this approach to weighting is highly efficient in the sense of ensuring that overall variance is kept to a minimum. The asymptotic unconditional and asymptotic conditional estimates are based on the maximum likelihood method and are therefore also optimal under asymptotic conditions.

When there is heterogeneity (interaction, effect modification) the situation is much different. In the first place, the fact that the stratum-specific odds ratios vary across strata says something about the relationship between exposure and disease which would be lost if the data were to be summarized. This provides a rationale for retaining the stratum-specific estimates and interpreting, to the extent possible, whatever patterns may be present. When there are many strata and no meaningful patterns are evident, it can be confusing as well as inconvenient to have to deal with many odds ratio estimates. In this situation it is useful to have a method of estimating an overall odds ratio.

If the stratifying variable is not a confounder, the crude odds ratio estimate serves this purpose. When confounding is present, the standardized estimate described in Section 2.5.4 can be used. An alternative is to estimate the overall odds ratio using a weighted average of stratum-specific estimates (Miettinen, 1972a). However, unlike the situation with $\widehat{OR}_{mh}$ and $\widehat{OR}_{ls}$, when there is heterogeneity the weights are chosen so as to reflect the distribution of the stratifying variable in the underlying population, as opposed to being defined in terms of inverse variances. When heterogeneity is present, summary estimates of the odds ratio, such as $\widehat{OR}_u$, $\widehat{OR}_c$, $\widehat{OR}_{mh}$, and $\widehat{OR}_{ls}$, do not estimate an epidemiologically meaningful parameter (Greenland, 1982). On the other hand, population weights can be used to form a weighted average when there is homogeneity, but this would not be optimal for variance estimation.

When heterogeneity is present, stratum-specific estimates of the odds ratio may be arrayed on both sides of 1, a phenomenon referred to as qualitative interaction (Peto, 1982; Gail and Simon, 1985). In this situation, exposure will appear to be detrimental in some strata and beneficial in others. As a consequence, the weighted average may be close to 1 even though stratum-specific estimates might be much larger or smaller. An appropriate interpretation of such a weighted average is that it represents a "net" measure of effect.

Similar issues arise when testing for association, as can be illustrated with the Mantel–Haenszel test X_{mh}^2. Rewriting the numerator of (5.29) as $\sum_{j=1}^{J}(a_{1j} - \hat{e}_{1j})$, consider

$$a_{1j} - \hat{e}_{1j} = \left(\frac{a_{2j}b_{1j}}{r_j}\right)(\widehat{OR}_{uj} - 1).$$

Under homogeneity, each of the $(\widehat{OR}_{uj} - 1)$ will tend to be near $(OR - 1)$. So the differences $(a_{1j} - \hat{e}_{1j})$ will tend to have the same sign: positive when $OR > 1$, and negative when $OR < 1$. When there is qualitative interaction, $\sum_{j=1}^{J}(a_{1j} - \hat{e}_{1j})$ will be a sum of positive and negative terms. We can interpret this quantity as the "net" difference between observed and expected counts. Even when some of the terms are quite large in absolute value, X_{mh}^2 may be small as a result of cancellation of positive and negative terms. In this case, X_{mh}^2 could have a large p-value and so the hypothesis of no association might not be rejected. Under these circumstances we can still interpret X_{mh}^2 as a test of association, provided we consider the null hypothesis to be one of no "net" association between exposure and disease.

Suppose that a test of association is performed and that it is not known whether homogeneity is present or not. If the p-value is small, we infer that exposure is associated with disease, in either absolute or net terms. However, if the p-value is large, there are two cases to consider: If homogeneity is present, then there is no association between exposure and disease; but if heterogeneity is present, all that can be said is that there is no net association. In the latter case, there may be important associations in certain of the strata which were not detected by the overall test. For this reason it is prudent to establish that there is homogeneity before performing a test of association. At a minimum, stratum-specific odds ratio estimates should be examined to determine whether the majority of them are on one side of 1 or the other.

5.6   SUMMARY OF 2 × 2 EXAMPLES AND RECOMMENDATIONS

Table 5.8 summarizes the results of the receptor level–breast cancer analyses based on the asymptotic unconditional (AU), asymptotic conditional (AC), Mantel–Haenszel (MH), and weighted least squares (WLS) methods. These findings are typical of data satisfying large-strata conditions, in that all four approaches give quite similar results. In terms of computational ease, the Mantel–Haenszel and weighted least squares methods are by far the most convenient.

TABLE 5.8   Summary of Receptor Level–Breast Cancer Results

Result	AU	AC	MH	WLS
$\widehat{OR}$	2.51	2.47	2.54	2.50
$[\underline{OR}, \overline{OR}]$	[1.16, 5.44]	[1.15, 5.32][a]	[1.16, 5.55][b]	[1.13, 5.50]
Association p-value	.02[c]	.02	—	.03
Homogeneity p-value	.84[d]	—	.84	.84
Trend p-value	.59	—	—	—

[a] Explicit
[b] RBG
[c] X_{lr}^2
[d] Likelihood ratio

There has been considerable theoretical research on the statistical properties of odds ratio methods, and this provides some guidance as to their applicability in different settings. Under large-strata conditions, $\widehat{OR}_u$, $\widehat{OR}_c$, $\widehat{OR}_{mh}$, and $\widehat{OR}_{ls}$ all have desirable asymptotic properties (Gart, 1962; Andersen, 1970; Hauck, 1979; Tarone et al., 1983) and perform well in finite samples, although $\widehat{OR}_{ls}$ can be biased (Hauck et al., 1982; Hauck, 1984; Donner and Hauck, 1986). Under sparse-strata conditions, $\widehat{OR}_c$ and $\widehat{OR}_{mh}$ continue to have attractive asymptotic and finite sample properties (Breslow, 1981; Hauck and Donner, 1988), but the same is no longer true of $\widehat{OR}_u$ and $O\hat{R}_{ls}$ (Lubin, 1981; Davis, 1985). $\widehat{OR}_{mh}$ is asymptotically efficient only when $OR = 1$ (Tarone et al., 1983), but nevertheless performs well for values of OR likely to be seen in practice. The preceding features of $\widehat{OR}_{mh}$, along with its ease of computation, make it a very desirable choice among available estimates. This is the conclusion of Hauck (1987, 1989) after an extensive review of the literature.

X^2_{mh} is optimal for detecting association when there is homogeneity (Gart and Tarone, 1983). The test for linear trend X^2_t is locally optimal against alternatives that can be expressed as smooth, monotonic functions of exposure (Tarone and Gart, 1980). The local property of X^2_t means that it is not sensitive to model misspecification of the exposure–disease relationship. Each of the tests of homogeneity considered above has low power to detect heterogeneity, especially under sparse-strata conditions (Greenland, 1983; Liang and Self, 1985; Jones et al., 1989; Paul and Donner, 1989; Paul and Donner, 1992).

The weighted least squares methods perform well under large-strata conditions and should be considered on grounds of computational ease. $\widehat{OR}_{mh}$, $\widehat{var}(\log \widehat{OR}_{mh})$, and X^2_{mh}, which we refer to subsequently as the MH–RBG methods, are also computationally straightforward and have the advantage of producing excellent results under both large-strata and sparse-strata conditions. For the asymptotic analysis of closed cohort data based on the odds ratio, the MH–RBG methods are recommended. When the overall sample size is small, these methods may not perform well, making it necessary to resort to exact calculations.

5.7   ASYMPTOTIC METHODS FOR J $(2 \times I)$ TABLES

In this section we consider methods for analyzing stratified data when the exposure variable is polychotomous. The data layout for the jth stratum is given in Table 5.9. We say there is no association between exposure and disease if $\pi_{1j} = \pi_{2j} = \cdots =$

TABLE 5.9   Observed Counts: Closed Cohort Study

Disease	Exposure category					
	1	2	$\cdots$	i	$\cdots$	I	
yes	a_{1j}	a_{2j}	$\cdots$	a_{ij}	$\cdots$	a_{Ij}	m_{1j}
no	b_{1j}	b_{2j}	$\cdots$	b_{ij}	$\cdots$	b_{Ij}	m_{2j}
	r_{1j}	r_{2j}	$\cdots$	r_{ij}	$\cdots$	r_{Ij}	r_j

π_{1j} for all j. The expected counts for the ith exposure category in the jth stratum are

$$\hat{e}_{ij} = \frac{r_{ij}m_{1j}}{r_j} \quad \text{and} \quad \hat{f}_{ij} = \frac{r_{ij}m_{2j}}{r_j}.$$

With $i = 1$ as the reference category, let $\widehat{OR}_{\mathrm{mh}i}$ denote the Mantel–Haenszel odds ratio estimate comparing the ith exposure category to the first category.

The Mantel–Haenszel test of association X_{mh}^2 has a generalization to the $J(2 \times I)$ setting, but the formula involves matrix algebra (Appendix E). Peto and Pike (1973) give a computationally convenient approximation to X_{mh}^2,

$$X_{\mathrm{pp}}^2 = \sum_{i=1}^{I} \frac{(a_{i\bullet} - \hat{e}_{i\bullet})^2}{\hat{g}_{i\bullet}} \qquad (\mathrm{df} = I - 1) \tag{5.38}$$

where

$$\hat{g}_{ij} = \left(\frac{m_{2j}}{r_j - 1}\right)\hat{e}_{ij}. \tag{5.39}$$

It can be shown that $X_{\mathrm{pp}}^2 \leq X_{\mathrm{mh}}^2$ and so X_{pp}^2 is conservative compared to X_{mh}^2 (Peto and Pike, 1973; Crowley and Breslow, 1975). Although X_{pp}^2 was not discussed in Section 5.2, it is still valid when $I = 2$. However, there is no need to rely on such an approximation since X_{mh}^2 is readily calculated. Let s_i be the exposure level for the ith category with $s_1 < s_2 < \cdots < s_I$. For each j, define

$$U_j = \sum_{j=1}^{J} s_i(a_{ij} - \hat{e}_{ij})$$

and

$$V_j = \left(\frac{m_{2j}}{r_j - 1}\right)\left[\sum_{i=1}^{I} s_i^2 \hat{e}_{ij} - \left(\sum_{i=1}^{I} s_i \hat{e}_{ij}\right)^2 \bigg/ \hat{e}_{\bullet j}\right].$$

U_j and V_j correspond to the numerator and denominator of (4.38), and so they can be used to test for linear trend in the jth stratum. An overall test for linear trend is

$$X_{\mathrm{t}}^2 = \frac{(U_{\bullet})^2}{V_{\bullet}} \qquad (\mathrm{df} = 1) \tag{5.40}$$

(Mantel, 1963; Birch, 1965; Breslow and Day, 1980, p. 149). As shown in Appendix E, a conservative approximation to (5.40) is

TABLE 5.10   Observed Counts: Stage–Breast Cancer

Survival	Low receptor level				High receptor level			
	Stage				Stage			
	I	II	III		I	II	III	
dead	2	9	12	23	5	17	9	31
alive	10	13	2	25	50	57	6	113
	12	22	14	48	55	74	15	144

$$X_t^2 = \frac{\left[\sum_{i=1}^I s_i(a_{i\bullet} - \hat{e}_{i\bullet})\right]^2}{\sum_{i=1}^I s_i^2 \hat{g}_{i\bullet} - \left(\sum_{i=1}^I s_i \hat{g}_{i\bullet}\right)^2 / \hat{g}_{\bullet\bullet}} \qquad (\text{df} = 1) \qquad (5.41)$$

that is, (5.41) ≤ (5.40) (Peto and Pike, 1973; Crowley and Breslow, 1975).

Example 5.5 (Stage–Breast Cancer)   Table 5.10 gives the observed counts corresponding to Table 4.16 after stratifying by receptor level. Arguing along the lines of Example 5.1, a rationale can be given for treating receptor level as a confounder of the association between stage of disease and breast cancer survival.

Table 5.11 gives the Mantel–Haenszel odds ratio estimates and the RBG 95% confidence intervals, with stage I as the reference category and with adjustment for receptor level. The adjusted results of Table 5.11 are close to the crude results of Table 4.17. According to the collapsibility approach to confounding discussed in

TABLE 5.11   Mantel–Haenszel Estimates and RBG 95% Confidence Intervals: Stage–Breast Cancer

Stage	$\widehat{OR}_{\text{mh}i}$	$\underline{OR}_{\text{mh}i}$	$\overline{OR}_{\text{mh}i}$
II	3.11	1.25	7.71
III	18.96	6.00	59.89

TABLE 5.12   Expected Counts: Stage–Breast Cancer

Survival	Low receptor level				High receptor level			
	Stage				Stage			
	I	II	III		I	II	III	
dead	5.75	10.54	6.71	23	11.84	15.93	3.23	31
alive	6.25	11.46	7.29	25	43.16	58.07	11.77	113
	12	22	14	48	55	74	15	144

Section 2.5.5, it appears that receptor level is unlikely to be an important confounder. There is a clear trend in odds ratio estimates in Table 5.11 (where $\widehat{OR}_{mh1} = 1$).

The expected counts are given in Table 5.12. The Mantel–Haenszel test is $X^2_{mh} = 30.82$ ($p < .001$) and the Peto–Pike approximation is

$$X^2_{pp} = \frac{(7 - 17.59)^2}{12.41} + \frac{(26 - 26.47)^2}{18.20} + \frac{(21 - 9.94)^2}{6.12} = 29.04 \ (p < .001).$$

Setting $s_1 = 1$, $s_2 = 2$, and $s_3 = 3$, the test for linear trend is (5.40) $= (21.65)^2/16.62 = 28.20$ ($p < .001$) and the approximation is

$$(5.41) = \frac{(21.65)^2}{140.28 - (67.17)^2/36.73} = 26.86 \ (p < .001).$$

As is often the case in practice, X^2_{pp} is only slightly less than X^2_{mh}, and (5.41) is only slightly less than (5.40).

CHAPTER 6

Risk Ratio Methods for Closed Cohort Data

Risk ratio methods for analyzing closed cohort data have many similarities to the odds ratio methods of Chapters 4 and 5. An important difference is that there is no conditional distribution that has the risk ratio as its parameter. However, risk ratio methods based on asymptotic unconditional, Mantel–Haenszel, and weighted least squares methods are available for the analysis of closed cohort data. As in the odds ratio setting, asymptotic unconditional and weighted least squares methods work well under large-strata, but not sparse-strata, conditions. In the absence of conditional techniques, and aside from exact methods which will not be discussed here, the Mantel–Haenszel methods are the only ones in wide use that are designed for the sparse-strata setting.

6.1 ASYMPTOTIC UNCONDITIONAL METHODS FOR A SINGLE 2 × 2 TABLE

The observed counts for the unstratified analysis are given in Table 4.1. Making the substitution $\pi_1 = RR\pi_2$, the joint probability function (4.1) can be reparameterized to obtain a likelihood that is a function of the parameters RR and π_2. This leads to the unconditional maximum likelihood equations,

$$a_1 = \widehat{RR}\hat{\pi}_2 r_1$$

and

$$\frac{a_1 - \widehat{RR}\hat{\pi}_2 r_1}{1 - \widehat{RR}\hat{\pi}_2} + \frac{a_2 - \hat{\pi}_2 r_2}{1 - \hat{\pi}_2} = 0$$

where $\widehat{RR}$ denotes the unconditional maximum likelihood estimate of RR. The solutions are

$$\widehat{RR} = \frac{\hat{\pi}_1}{\hat{\pi}_2} = \frac{a_1 r_2}{a_2 r_1} \tag{6.1}$$

143

and

$$\hat{\pi}_2 = \frac{a_2}{r_2}$$

where $\hat{\pi}_1 = a_1/r_1$. The unconditional maximum likelihood estimate of $\text{var}(\log \widehat{RR})$ is

$$
\begin{aligned}
\widehat{\text{var}}(\log \widehat{RR}) &= \frac{1 - \hat{\pi}_1}{\hat{\pi}_1 r_1} + \frac{1 - \hat{\pi}_2}{\hat{\pi}_2 r_2} \\
&= \frac{b_1}{a_1 r_1} + \frac{b_2}{a_2 r_2}
\end{aligned}
\tag{6.2}
$$

and a $(1 - \alpha) \times 100\%$ confidence interval for RR is obtained by exponentiating

$$\left[\log \underline{RR}, \log \overline{RR}\right] = \log(\widehat{RR}) \pm z_{\alpha/2} \sqrt{\frac{b_1}{a_1 r_1} + \frac{b_2}{a_2 r_2}} \,.$$

If either a_1 or a_2 equals 0, we replace (6.1) and (6.2) with

$$\widehat{RR} = \frac{(a_1 + .5)r_2}{(a_2 + .5)r_1}$$

and

$$\widehat{\text{var}}(\log \widehat{RR}) = \frac{b_1 + .5}{(a_1 + .5)r_1} + \frac{b_2 + .5}{(a_2 + .5)r_2}.$$

Since $\pi_1 = \pi_2$ is equivalent to $\log(RR) = 0$, the hypothesis of no association can be expressed as $H_0 : \log(RR) = 0$. Under H_0 an estimate of $\text{var}(\log \widehat{RR})$ is

$$\widehat{\text{var}}_0(\log \widehat{RR}) = \frac{\hat{f}_1}{\hat{e}_1 r_1} + \frac{\hat{f}_2}{\hat{e}_2 r_2} = \frac{r m_2}{r_1 r_2 m_1}.$$

The Wald test of association is

$$X_w^2 = \frac{(\log \widehat{RR})^2 r_1 r_2 m_1}{r m_2} \qquad (\text{df} = 1)$$

and the likelihood ratio test of association is precisely (4.12).

Example 6.1 (Receptor Level–Breast Cancer)   The data for this example are taken from Table 4.5(a). The estimate of the risk ratio is $\widehat{RR} = 2.23$, $\widehat{\text{var}}(\log \widehat{RR}) = .048$, and the 95% confidence interval for RR is [1.45, 3.42]. The Wald test of association is $X_w^2 = 9.02$ ($p = .003$) and, from Example 4.2, the likelihood ratio test of association is $X_{lr}^2 = 11.68$ ($p = .001$). The interpretation of these findings follows along the lines of Example 4.2.

6.2 ASYMPTOTIC UNCONDITIONAL METHODS FOR J (2 × 2) TABLES

We now consider asymptotic unconditional methods for stratified 2 × 2 tables. For the jth stratum, the data layout is given in Table 5.1 and the risk ratio is $RR_j = \pi_{1j}/\pi_{2j}$ ($j = 1, 2, \ldots, J$). Each of the J tables can be analyzed separately using the methods of the preceding section. The stratum-specific estimates are

$$\hat{\pi}_{1j} = \frac{a_{1j}}{r_{1j}} \qquad \hat{\pi}_{2j} = \frac{a_{2j}}{r_{2j}}$$

$$\widehat{RR}_j = \frac{\hat{\pi}_{1j}}{\hat{\pi}_{2j}} = \frac{a_{1j}r_{2j}}{a_{2j}r_{1j}}$$

and

$$\widehat{\text{var}}(\log \widehat{RR}_j) = \frac{b_{1j}}{a_{1j}r_{1j}} + \frac{b_{2j}}{a_{2j}r_{2j}}.$$

When there is homogeneity the common stratum-specific risk ratio is denoted by RR.

Point Estimates and Fitted Counts
The unconditional maximum likelihood equations are

$$\sum_{j=1}^{J} \frac{a_{1j} - \widehat{RR}\hat{\pi}_{2j}r_{1j}}{1 - \widehat{RR}\hat{\pi}_{2j}} = 0 \qquad (6.3)$$

and

$$\frac{a_{1j} - \widehat{RR}\hat{\pi}_{2j}r_{1j}}{1 - \widehat{RR}\hat{\pi}_{2j}} + \frac{a_{2j} - \hat{\pi}_{2j}r_{2j}}{1 - \hat{\pi}_{2j}} = 0 \qquad (j = 1, 2, \ldots, J). \qquad (6.4)$$

This is a system of $J + 1$ equations in the $J + 1$ unknowns $\widehat{RR}$ and $\hat{\pi}_{2j}$ ($j = 1, 2, \ldots, J$). A solution to these equations can be obtained using the general methods described in Appendix B. Unlike the situation for the odds ratio in Section 5.1, there is no guarantee that the $\hat{\pi}_{2j}$ which solve (6.3) and (6.4) will necessarily satisfy the constraints $0 \leq \hat{\pi}_{2j} \leq 1$. When the constraints are not satisfied, alternate methods must be used to maximize the likelihood. Once $\widehat{RR}$ and $\hat{\pi}_{2j}$ have been estimated, we have $\hat{\pi}_{1j} = \widehat{RR}\hat{\pi}_{2j}$. The fitted counts $\hat{a}_{1j}$, $\hat{a}_{2j}$, $\hat{b}_{1j}$, and $\hat{b}_{2j}$ are defined as in (5.5). We can rewrite the maximum likelihood equations as

$$\sum_{j=1}^{J} \frac{a_{1j} - \hat{a}_{1j}}{1 - \hat{\pi}_{1j}} = 0$$

and

$$\frac{a_{1j} - \hat{a}_{1j}}{1 - \hat{\pi}_{1j}} + \frac{a_{2j} - \hat{a}_{2j}}{1 - \hat{\pi}_{2j}} = 0 \qquad (j = 1, 2, \ldots, J)$$

which shows that they do not have the "observed equals fitted" format. By definition, $\hat{a}_{1j} + \hat{b}_{1j} = r_{1j}$ and $\hat{a}_{2j} + \hat{b}_{2j} = r_{2j}$, and so the observed and expected column marginal totals agree. Unlike the situation for the odds ratio in Section 5.1, the same cannot be said for the row marginal totals.

Confidence Interval
Let

$$\hat{v}_j = \left(\frac{\hat{b}_{1j}}{\hat{a}_{1j} r_{1j}} + \frac{\hat{b}_{2j}}{\hat{a}_{2j} r_{2j}} \right)^{-1}$$

and let $\hat{V} = \sum_{j=1}^{J} \hat{v}_j$. An estimate of $\mathrm{var}(\log \widehat{RR})$ is

$$\widehat{\mathrm{var}}(\log \widehat{RR}) = \frac{1}{\hat{V}}$$

and a $(1 - \alpha) \times 100\%$ confidence interval for RR is obtained by exponentiating

$$\left[\log \underline{RR}, \log \overline{RR} \right] = \log(\widehat{RR}) \pm \frac{z_{\alpha/2}}{\sqrt{\hat{V}}}$$

(Tarone et al., 1983; Gart, 1985).

Wald and Likelihood Ratio Tests of Association
Let

$$\hat{v}_{0j} = \left(\frac{\hat{f}_{1j}}{\hat{e}_{1j} r_{1j}} + \frac{\hat{f}_{2j}}{\hat{e}_{2j} r_{2j}} \right)^{-1} = \frac{r_{1j} r_{2j} m_{1j}}{r_j m_{2j}}$$

and let $\hat{V}_0 = \sum_{j=1}^{J} \hat{v}_{0j}$. Under the hypothesis of no association $H_0 : \log(RR) = 0$, an estimate of $\mathrm{var}(\log \widehat{RR})$ is

$$\widehat{\mathrm{var}}_0(\log \widehat{RR}) = \frac{1}{\hat{V}_0}.$$

The Wald test of association is

$$X_{\mathrm{w}}^2 = (\log \widehat{RR})^2 \hat{V}_0 \qquad (\mathrm{df} = 1)$$

and the likelihood ratio test of association is precisely (5.15).

TABLE 6.1   Risk Ratio Estimates and 95% Confidence
Intervals: Receptor Level–Breast Cancer

Stage	$\hat{RR}_j$	$\underline{RR}_j$	$\overline{RR}_j$
I	1.83	.40	8.35
II	1.78	.93	3.42
III	1.43	.90	2.27

Likelihood Ratio Test of Homogeneity
The likelihood ratio test of homogeneity is

$$X_h^2 = 2 \sum_{j=1}^{J} \left[a_{1j} \log \left(\frac{a_{1j}}{\hat{a}_{1j}} \right) + a_{2j} \log \left(\frac{a_{2j}}{\hat{a}_{2j}} \right) + b_{1j} \log \left(\frac{b_{1j}}{\hat{b}_{1j}} \right) + b_{2j} \log \left(\frac{b_{2j}}{\hat{b}_{2j}} \right) \right]$$

$$(\mathrm{df} = J - 1)$$

which is identical in form to (5.18) but which uses fitted counts based on the risk
ratio.

Example 6.2 (Receptor Level–Breast Cancer)   The data for this example are
taken from Table 5.3. Table 6.1 gives a stratum-specific analysis according to stage
of disease using the methods of the preceding section. The 95% confidence intervals
are fairly wide and each one contains the risk ratio estimates for the other two strata,
which suggests the presence of homogeneity. It is interesting that, unlike Table 5.4
where the odds ratio estimates have an increasing trend, the risk ratio estimates show
a decreasing trend.
The maximum likelihood estimates are $\hat{RR} = 1.56$, $\hat{\pi}_{21} = .095$, $\hat{\pi}_{22} = .242$,
and $\hat{\pi}_{23} = .558$, and the fitted counts are given in Table 6.2. Note that the observed
and fitted row marginal totals do not agree. Comparing the fitted counts in Tables
5.5 and 6.2 to the observed counts in Table 5.3, there seems to be little to choose
between the odds ratio and risk ratio models in terms of goodness of fit. From $\hat{V} =$
$1.54 + 8.49 + 15.78 = 25.81$ and $\hat{var}(\log \hat{RR}) = 1/25.81 = .039$, the 95% confidence
interval for RR is $[1.06, 2.29]$.

TABLE 6.2   Fitted Counts Under Homogeneity: Receptor Level–Breast Cancer

Survival	Stage I Receptor level low	high		Stage II Receptor level low	high		Stage III Receptor level low	high	
dead	1.78	5.23	7.01	8.28	17.87	26.15	12.18	8.38	20.56
alive	10.22	49.77	59.99	13.72	56.13	69.85	1.82	6.62	8.44
	12	55	67	22	74	96	14	15	29

With $\hat{V}_0 = 1.15 + 6.30 + 19.01 = 26.46$, the Wald test of association is $X_w^2 = $ $(\log 1.56)^2(26.46) = 5.20$ ($p = .02$). From Example 5.1, the likelihood ratio test of association is $X_{lr}^2 = 5.64$ ($p = .02$). The likelihood ratio test of homogeneity is $X_h^2 = .325$ ($p = .85$), and so there is considerable evidence in favor of homogeneity.

Recalling the results of Example 5.1, it seems that the data in Table 5.3 are homogeneous with respect to both the odds ratio and the risk ratio. This conclusion contradicts the observation made in Section 2.4.5 that at most one of these measures of effect can be homogeneous when the stratifying variable is a risk factor for the disease. As pointed out in Section 5.6, tests of homogeneity generally have low power and, as such, may fail to detect heterogeneity even when it is present. So an explanation for the preceding contradictory finding is that one or both of the odds ratio and risk ratio are in fact heterogeneous, but this was not detected by the tests of homogeneity.

6.3   MANTEL–HAENSZEL ESTIMATE OF THE RISK RATIO

Over the years the Mantel–Haenszel estimate of the odds ratio has proven to be so useful that analogous estimates of other measures of effect have been developed. These more recent estimates are also referred to as Mantel–Haenszel estimates. The Mantel–Haenszel estimate of the risk ratio is

$$\widehat{RR}_{mh} = \frac{\sum_{j=1}^{J} R_j}{\sum_{j=1}^{J} S_j} = \frac{R_\bullet}{S_\bullet} \tag{6.5}$$

where

$$R_j = \frac{a_{1j} r_{2j}}{r_j}$$

and

$$S_j = \frac{a_{2j} r_{1j}}{r_j}$$

(Rothman and Boice, 1979, p. 12; Nurminen, 1981; Tarone, 1981). Greenland and Robins (1985b) give an estimate of $\text{var}(\log \widehat{RR}_{mh})$ which is valid under both large-strata and sparse-strata conditions,

$$\widehat{\text{var}}(\log \widehat{RR}_{mh}) = \frac{T_\bullet}{R_\bullet S_\bullet} \tag{6.6}$$

where

$$T_j = \frac{r_{1j} r_{2j} m_{1j} - a_{1j} a_{2j} r_j}{r_j^2}.$$

A $(1 - \alpha) \times 100\%$ confidence interval for RR is obtained by exponentiating

$$\left[\log \underline{RR}_{\text{mh}}, \log \overline{RR}_{\text{mh}}\right] = \log(\widehat{RR}_{\text{mh}}) \pm z_{\alpha/2}\sqrt{\widehat{\text{var}}(\log \widehat{RR}_{\text{mh}})} \, .$$

When there is only one stratum, (6.5) and (6.6) simplify to (6.1) and (6.2).

Example 6.3 (Receptor Level–Breast Cancer)   Based on the above methods, $\widehat{RR}_{\text{mh}} = 14.79/9.14 = 1.62$, $\widehat{\text{var}}(\log \widehat{RR}_{\text{mh}}) = 5.40/(14.79 \times 9.14) = .040$, and a 95% confidence interval for RR is [1.09, 2.39].

6.4   WEIGHTED LEAST SQUARES METHODS FOR J (2×2) TABLES

For the weighted least squares methods, the weight for the jth stratum is defined to be

$$\hat{w}_j = \frac{1}{\widehat{\text{var}}(\log \widehat{RR}_j)} = \left(\frac{b_{1j}}{a_{1j}r_{1j}} + \frac{b_{2j}}{a_{2j}r_{2j}}\right)^{-1} .$$

The risk ratio formulas are the same as (5.33)–(5.37) except that $\hat{w}_j$ is defined as above and RR replaces OR.

Example 6.4 (Receptor Level–Breast Cancer)   From $\hat{W}_{\text{ls}} = 1.67 + 9.01 + 17.75 = 28.43$ and

$$\log(\widehat{RR}_{\text{ls}}) = \frac{(1.67 \times .606) + (9.01 \times .577) + (17.75 \times .357)}{28.43} = .441$$

the WLS estimate of the risk ratio is $\widehat{RR}_{\text{ls}} = \exp(.441) = 1.55$. From

$$\widehat{\text{var}}(\log \widehat{RR}_{\text{ls}}) = 1/28.43 = .035$$

the 95% confidence interval for RR is [1.08, 2.25]. The test of association is $X^2_{\text{ls}} = (\log 1.55)^2 (26.46) = 5.15$ ($p = .02$), where $\hat{W}_{0\text{ls}} = 26.46$ comes from Example 6.2. The test of homogeneity is

$$X^2_{\text{h}} = 1.67(.606 - .441)^2 + 9.01(.577 - .441)^2 + 17.75(.357 - .441)^2 = .337 \, (p = .84).$$

TABLE 6.3   Summary of Receptor Level–Breast Cancer Results

Result	AU	MH	WLS
RR	1.56	1.62	1.55
$[\underline{RR}, \overline{RR}]$	[1.06, 2.29]	[1.09, 2.39]	[1.08, 2.25]
Association p-value	.02[a]	—	.02
Homogeneity p-value	.85	—	.84

[a] Wald

6.5 SUMMARY OF EXAMPLES AND RECOMMENDATIONS

Table 6.3 summarizes the results of the stratified receptor level–breast cancer analyses based on the asymptotic unconditional (AU), Mantel–Haenszel (MH), and weighted least squares (WLS) methods. All three methods produce similar results. Less theoretical research has been done on the statistical properties of the risk ratio than on the odds ratio, but the evidence is that $\widehat{RR}$, $\widehat{RR}_{mh}$, and $\widehat{RR}_{ls}$ have properties that are broadly similar to their odds ratio counterparts (Tarone et al., 1983; Greenland and Robins, 1985b). When large-strata conditions are satisfied, $\widehat{RR}_{ls}$ has a clear advantage over $\widehat{RR}$ in terms of computational ease. $\widehat{RR}_{mh}$ can be used under large-strata conditions and it is the only one of the three estimates that is valid under sparse-strata conditions (Walker, 1985). However, $\widehat{RR}_{mh}$ can be inefficient and so its use should be restricted to the sparse-strata setting (Greenland and Robins, 1985b).

CHAPTER 7

Risk Difference Methods for Closed Cohort Data

Risk difference methods for analyzing closed cohort data are similar to those based on the risk ratio. In fact, the preceding chapter and the present one have so much in common that it is possible to use language here that is almost identical to that of Chapter 6.

7.1 ASYMPTOTIC UNCONDITIONAL METHODS FOR A SINGLE 2 × 2 TABLE

The observed counts for the unstratified analysis are given in Table 4.1. Making the substitution $\pi_1 = \pi_2 + RD$, the joint probability function (4.1) can be reparameterized to obtain a likelihood that is a function of the parameters RD and π_2. This leads to the unconditional maximum likelihood equations,

$$a_1 = (\hat{\pi}_2 + \widehat{RD})r_1$$

and

$$\frac{a_1 - (\hat{\pi}_2 + \widehat{RD})r_1}{(\hat{\pi}_2 + \widehat{RD})(1 - \hat{\pi}_2 - \widehat{RD})} + \frac{a_2 - r_2\hat{\pi}_2}{\hat{\pi}_2(1 - \hat{\pi}_2)} = 0$$

where $\widehat{RD}$ denotes the unconditional maximum likelihood estimate of RD. The solutions are

$$\widehat{RD} = \hat{\pi}_1 - \hat{\pi}_2 = \frac{a_1}{r_1} - \frac{a_2}{r_2} \qquad (7.1)$$

and

$$\hat{\pi}_2 = \frac{a_2}{r_2}$$

where $\hat{\pi}_1 = a_1/r_1$. The unconditional maximum likelihood estimate of $\mathrm{var}(\widehat{RD})$ is

$$\widehat{\mathrm{var}}(\widehat{RD}) = \frac{\hat{\pi}_1(1 - \hat{\pi}_1)}{r_1} + \frac{\hat{\pi}_2(1 - \hat{\pi}_2)}{r_2}$$
$$= \frac{a_1 b_1}{r_1^3} + \frac{a_2 b_2}{r_2^3} \qquad (7.2)$$

and a $(1 - \alpha) \times 100\%$ confidence interval for RD is

$$[\underline{RD}, \overline{RD}] = \widehat{RD} \pm z_{\alpha/2} \sqrt{\frac{a_1 b_1}{r_1^3} + \frac{a_2 b_2}{r_2^3}} .$$

Note that (7.2) is precisely the variance estimate that results from applying (1.9) to the random variable $\hat{\pi}_1 - \hat{\pi}_2$. Since $\pi_1 = \pi_2$ is equivalent to $RD = 0$, the hypothesis of no association can be expressed as $H_0 : RD = 0$. Under H_0 an estimate of $\mathrm{var}(\widehat{RD})$ is

$$\widehat{\mathrm{var}}_0(\widehat{RD}) = \frac{\hat{e}_1 \hat{f}_1}{r_1^3} + \frac{\hat{e}_2 \hat{f}_2}{r_2^3} = \frac{m_1 m_2}{r_1 r_2 r}.$$

The Wald test of association is

$$X_{\mathrm{w}}^2 = \frac{(\widehat{RD})^2 r_1 r_2 r}{m_1 m_2}$$
$$= \frac{(a_1 b_2 - a_2 b_1)^2 r}{r_1 r_2 m_1 m_2} \qquad (df = 1).$$

and the likelihood ratio test of association is precisely (4.12). Note that X_{w}^2 is identical to X_{p}^2 (4.10).

Example 7.1 (Receptor Level–Breast Cancer)   The data for this example are taken from Table 4.5(a). The estimate of the risk difference is $\widehat{RD} = .264$, $\widehat{\mathrm{var}}(\widehat{RD}) = (.0798)^2$, and the 95% confidence interval for RD is [.107, .420]. The Wald test of association is $X_{\mathrm{w}}^2 = 12.40$ ($p < .001$) and, from Example 4.2, the likelihood ratio test of association is $X_{\mathrm{lr}}^2 = 11.68$ ($p = .001$).

7.2 ASYMPTOTIC UNCONDITIONAL METHODS FOR J (2 × 2) TABLES

We now consider asymptotic unconditional methods for stratified 2×2 tables. For the jth stratum, the data layout is given in Table 5.1 and the risk difference is $RD_j = \pi_{1j} - \pi_{2j}$ ($j = 1, 2, \ldots, J$). Each of the J tables can be analyzed separately using the methods of the preceding section. The stratum-specific estimates are

$$\hat{\pi}_{1j} = \frac{a_{1j}}{r_{1j}} \qquad \hat{\pi}_{2j} = \frac{a_{2j}}{r_{2j}}$$

$$\widehat{RD}_j = \hat{\pi}_{1j} - \hat{\pi}_{2j} = \frac{a_{1j}}{r_{1j}} - \frac{a_{2j}}{r_{2j}}$$

and

$$\widehat{\mathrm{var}}(\widehat{RD}_j) = \frac{a_{1j}b_{1j}}{r_{1j}^3} + \frac{a_{2j}b_{2j}}{r_{2j}^3}.$$

When there is homogeneity the common stratum-specific value of the risk difference is denoted by RD.

Point Estimates and Fitted Counts

The unconditional maximum likelihood equations are

$$\sum_{j=1}^{J} \frac{a_{1j} - (\hat{\pi}_{2j} + \widehat{RD})r_{1j}}{(\hat{\pi}_{2j} + \widehat{RD})(1 - \hat{\pi}_{2j} - \widehat{RD})} = 0 \qquad (7.3)$$

and

$$\frac{a_{1j} - (\hat{\pi}_{2j} + \widehat{RD})r_{1j}}{(\hat{\pi}_{2j} + \widehat{RD})(1 - \hat{\pi}_{2j} - \widehat{RD})} + \frac{a_{2j} - \hat{\pi}_{2j}r_{2j}}{\hat{\pi}_{2j}(1 - \hat{\pi}_{2j})} = 0 \qquad (j = 1, 2, \ldots, J). \quad (7.4)$$

This is a system of $J + 1$ equations in the $J + 1$ unknowns $\widehat{RD}$ and $\hat{\pi}_{2j}$ ($j = 1, 2, \ldots, J$). A solution to these equations can be obtained using the general methods described in Appendix B.

As was the case for the risk ratio in Section 6.2, there is no guarantee that the $\hat{\pi}_{2j}$ which solve (7.3) and (7.4) will necessarily satisfy the constraints $0 \le \hat{\pi}_{2j} \le 1$. When the constraints are not satisfied, alternate methods must be used to maximize the likelihood. Once $\widehat{RD}$ and $\hat{\pi}_{2j}$ have been estimated, we have $\hat{\pi}_{1j} = \hat{\pi}_{2j} + \widehat{RD}$. The fitted counts $\hat{a}_{1j}$, $\hat{a}_{2j}$, $\hat{b}_{1j}$, and $\hat{b}_{2j}$ are defined as in (5.5). We can rewrite the maximum likelihood equations as

$$\sum_{j=1}^{J} \frac{a_{1j} - \hat{a}_{1j}}{\hat{\pi}_{1j}(1 - \hat{\pi}_{1j})} = 0$$

and

$$\frac{a_{1j} - \hat{a}_{1j}}{\hat{\pi}_{1j}(1 - \hat{\pi}_{1j})} + \frac{a_{2j} - \hat{a}_{2j}}{\hat{\pi}_{2j}(1 - \hat{\pi}_{2j})} = 0 \qquad (j = 1, 2, \ldots, J)$$

which shows that they do not have the "observed equals fitted" format.

Confidence Interval
Let

$$\hat{v}_j = \left(\frac{\hat{a}_{1j}\hat{b}_{1j}}{r_{1j}^3} + \frac{\hat{a}_{2j}\hat{b}_{2j}}{r_{2j}^3} \right)^{-1}$$

and let $\hat{V} = \sum_{j=1}^{J} \hat{v}_j$. An estimate of $\mathrm{var}(\widehat{RD})$ is

$$\widehat{\mathrm{var}}(\widehat{RD}) = \frac{1}{\hat{V}}$$

and a $(1 - \alpha) \times 100\%$ confidence interval for RD is

$$\left[\underline{RD}, \overline{RD} \right] = \widehat{RD} \pm \frac{z_{\alpha/2}}{\sqrt{\hat{V}}}.$$

Wald and Likelihood Ratio Tests of Association
Let

$$\hat{v}_{0j} = \left(\frac{\hat{e}_{1j}\hat{f}_{1j}}{r_{1j}^3} + \frac{\hat{e}_{2j}\hat{f}_{2j}}{r_{2j}^3} \right)^{-1} = \frac{r_{1j}r_{2j}r_j}{m_{1j}m_{2j}}$$

and let $\hat{V}_0 = \sum_{j=1}^{J} \hat{v}_{0j}$. Under the hypothesis of no association $H_0 : RD = 0$, an estimate of $\mathrm{var}(\widehat{RD})$ is

$$\widehat{\mathrm{var}}_0(\widehat{RD}) = \frac{1}{\hat{V}_0}.$$

The Wald test of association is

$$X_{\mathrm{w}}^2 = (\widehat{RD})^2 \hat{V}_0 \qquad (\mathrm{df} = 1)$$

and the likelihood ratio test of association is precisely (5.15).

Likelihood Ratio Test of Homogeneity
The likelihood ratio test of homogeneity is

$$X_{\mathrm{h}}^2 = 2 \sum_{j=1}^{J} \left[a_{1j} \log \left(\frac{a_{1j}}{\hat{a}_{1j}} \right) + a_{2j} \log \left(\frac{a_{2j}}{\hat{a}_{2j}} \right) + b_{1j} \log \left(\frac{b_{1j}}{\hat{b}_{1j}} \right) + b_{2j} \log \left(\frac{b_{2j}}{\hat{b}_{2j}} \right) \right]$$

$$(\mathrm{df} = J - 1)$$

which is identical in form to (5.18) but which uses fitted counts based on the risk difference.

TABLE 7.1 Risk Difference Estimates and 95%
Confidence Intervals: Receptor Level–Breast Cancer

Stage	$\widehat{RD}_j$	$\underline{RD}_j$	$\overline{RD}_j$
I	.076	−.148	.300
II	.179	−.047	.406
III	.257	−.051	.565

TABLE 7.2 Fitted Counts Under Homogeneity: Receptor Level–Breast Cancer

	Stage I			Stage II			Stage III		
	Receptor level			Receptor level			Receptor level		
Survival	low	high		low	high		low	high	
dead	2.94	4.61	7.55	8.68	17.24	25.92	11.44	9.84	21.28
alive	9.06	50.39	59.45	13.32	56.76	70.08	2.56	5.16	7.72
	12	55	67	22	74	96	14	15	29

Example 7.2 (Receptor Level–Breast Cancer)   The data for this example are
taken from Table 5.3. Table 7.1 gives a stratum-specific analysis according to stage
of disease using the methods of the preceding section. The 95% confidence intervals
are fairly wide and each one contains the risk difference estimates for the other two
strata, which suggests the presence of homogeneity.

The maximum likelihood estimates are $\widehat{RD} = .161$, $\hat{\pi}_{21} = .084$, $\hat{\pi}_{22} = .233$, and
$\hat{\pi}_{23} = .656$, and the fitted counts are given in Table 7.2. Note that the observed and
fitted row marginal totals do not agree. Comparing the fitted counts in Tables 5.5, 6.2,
and 7.2 to the observed counts in Table 5.3, there seems to be little to choose between
the odds ratio, risk ratio, and risk difference models in terms of goodness of fit. From
$\hat{V} = 59.47 + 75.35 + 38.90 = 173.72$ and $\widehat{\text{var}}(\widehat{RD}) = 1/173.72 = (.0759)^2$, the
95% confidence interval for RD is [.013, .310].

With $\hat{V}_0 = 105.29 + 85.87 + 36.25 = 227.41$, the Wald test of association is $X_w^2 =$
$(.161)^2(227.41) = 5.92$ ($p = .01$). From Example 5.1, the likelihood ratio test of
association is $X_{lr}^2 = 5.64$ ($p = .02$). The likelihood ratio test of homogeneity is
$X_h^2 = .856$ ($p = .65$), and so there is considerable evidence in favor of homogeneity.

7.3 MANTEL–HAENSZEL ESTIMATE OF THE RISK DIFFERENCE

The Mantel–Haenszel estimate of the risk difference is

$$\widehat{RD}_{\text{mh}} = \frac{\sum_{j=1}^{J} R_j - \sum_{j=1}^{J} S_j}{\sum_{j=1}^{J} T_j} = \frac{R_\bullet - S_\bullet}{T_\bullet} \tag{7.5}$$

where

$$R_j = \frac{a_{1j}r_{2j}}{r_j}$$

$$S_j = \frac{a_{2j}r_{1j}}{r_j}$$

and

$$T_j = \frac{r_{1j}r_{2j}}{r_j}$$

(Greenland and Robins, 1985b). It is easily shown that

$$\widehat{RD}_{\text{mh}} = \frac{1}{T_\bullet} \sum_{j=1}^{J} T_j \widehat{RD}_j$$

and so $\widehat{RD}_{\text{mh}}$ is a weighted average of stratum-specific risk difference estimates. Sato (1989) gives an estimate of $\text{var}(\widehat{RD}_{\text{mh}})$ which is valid under both large-strata and sparse-strata conditions,

$$\widehat{\text{var}}(\widehat{RD}_{\text{mh}}) = \frac{(\widehat{RD}_{\text{mh}}U_\bullet) + V_\bullet}{(T_\bullet)^2} \tag{7.6}$$

where

$$U_j = \frac{r_{1j}^2 a_{2j} - r_{2j}^2 a_{1j} + r_{1j}r_{2j}(r_{2j} - r_{1j})/2}{r_j^2}$$

and

$$V_j = \frac{a_{1j}b_{2j} + a_{2j}b_{1j}}{2r_j}.$$

A $(1 - \alpha) \times 100\%$ confidence interval for RD is

$$\left[\underline{RD}_{\text{mh}}, \overline{RD}_{\text{mh}}\right] = \widehat{RD}_{\text{mh}} \pm z_{\alpha/2}\sqrt{\widehat{\text{var}}(\widehat{RD}_{\text{mh}})}.$$

When there is only one stratum, (7.5) and (7.6) simplify to (7.1) and (7.2).

Example 7.3 (Receptor Level–Breast Cancer)   Based on the above methods,

$$\widehat{RD}_{\text{mh}} = (14.79 - 9.14)/34.05 = .166$$

$$\widehat{\text{var}}(\widehat{RD}_{\text{mh}}) = [.166(1.12) + 6.49]/(34.05)^2 = (.0759)^2$$

and a 95% confidence interval for RD is [.0171, .315].

7.4 WEIGHTED LEAST SQUARES METHODS FOR J (2 × 2) TABLES

For the weighted least squares methods, the weight for the jth stratum is defined to be

$$\hat{w}_j = \frac{1}{\widehat{\text{var}}(\widehat{RD}_j)} = \left(\frac{a_{1j}b_{1j}}{r_{1j}^3} + \frac{a_{2j}b_{2j}}{r_{2j}^3} \right)^{-1}.$$

The risk difference formulas are the same as (5.33)–(5.37) except that $\hat{w}_j$ is defined as above and RD replaces $\log(OR)$.

Example 7.4 (Receptor Level–Breast Cancer) From $\hat{W}_{ls} = 76.47 + 74.74 + 40.41 = 191.62$, the WLS estimate of the risk difference is

$$\widehat{RD}_{ls} = \frac{(76.47 \times .076) + (74.74 \times .179) + (40.41 \times .257)}{191.62} = .154.$$

From $\widehat{\text{var}}(\widehat{RD}_{ls}) = 1/191.62 = (.0722)^2$, the 95% confidence interval for RD is [.013, .296]. The test of association is $X_{ls}^2 = (.154)^2(227.41) = 5.42$ ($p = .02$), where $\hat{W}_{0ls} = 227.41$ comes from Example 7.2. The test of homogeneity is

$$X_h^2 = 76.47(.076 - .154)^2 + 74.74(.179 - .154)^2$$
$$+ 40.41(.257 - .154)^2 = .946 \ (p = .62).$$

7.5 SUMMARY OF EXAMPLES AND RECOMMENDATIONS

Table 7.3 summarizes the results of the stratified receptor level–breast cancer analyses based on the asymptotic unconditional (AU), Mantel–Haenszel (MH), and weighted least squares (WLS) methods. All three methods produce similar results. The properties of $\widehat{RD}$, $\widehat{RD}_{mh}$, and $\widehat{RD}_{ls}$ are similar to those of $\widehat{RR}$, $\widehat{RR}_{mh}$, and $\widehat{RR}_{ls}$ as described in Section 6.5 (Greenland and Robins, 1985b), and so the corresponding recommendations are made.

TABLE 7.3 Summary of Receptor Level–Breast Cancer Results

Result	AU	MH	WLS
RD	.161	.166	.154
$[\underline{RD}, \overline{RD}]$	[.013, .310]	[.017, .315]	[.013, .296]
Association p-value	.01[a]	—	.02
Homogeneity p-value	.65	—	.62

[a] Wald

CHAPTER 8

Survival Analysis

In Chapters 3–7, methods were presented for analyzing data from closed cohort studies. As described in Section 2.2.1, the key features of the closed cohort design are that all subjects have the same maximum observation time and no subjects become unobservable, for example, due to being lost to follow-up or withdrawing from the study. These assumptions are very restrictive and rarely satisfied in practice. More general methods for analyzing cohort data are available, which are referred to collectively as survival analysis. In this chapter we discuss some of the fundamental ideas in survival analysis such as censoring, survival functions, hazard functions, the proportional hazards assumption, and competing risks. A counterfactual definition of confounding in open cohort studies is given in Appendix G. There are many books that can be consulted for additional material on survival analysis, including Kalbfleisch and Prentice (1980), Lawless (1982), Cox and Oakes (1984), Lee (1992), Collett (1994), Marubini and Valsecchi (1995), Parmar and Machin (1995), Kleinbaum (1996), Klein and Moeschberger (1997), and Hosmer and Lemeshow (1999).

8.1 OPEN COHORT STUDIES AND CENSORING

Cohort studies are usually designed with a specific endpoint in mind. For the sake of concreteness we take the endpoint to be a particular disease. During the course of follow-up a subject either develops the disease or not. If the disease occurs, follow-up ceases for that individual as far as the cohort study is concerned. For that subject the length of follow-up is defined to be the time from the beginning of follow-up until the onset of disease, regardless of what happens subsequently. If the disease does not develop, follow-up continues until the subject becomes unobservable or reaches the termination date (end) of the study, whichever comes first. In this case, length of follow-up is defined to be the time from the beginning of follow-up until either of the preceding two events. We refer to a cohort study in which subjects have different maximum observation times as an open cohort study.

Consider an open cohort study conducted during a given (calendar) time period $[\tau_0, \tau_1]$, where τ_1 is the termination date of the study. Let τ' be a fixed time such that $\tau_0 < \tau' \leq \tau_1$. Suppose that subjects are recruited into the study on an ongoing basis throughout $[\tau_0, \tau']$ and that follow-up begins immediately after recruitment. This method of accrual is referred to as staggered entry because not all members of the cohort are placed under observation at the same time. As a result of staggered entry, subjects inevitably have different maximum observation times. For example, someone recruited at time τ_0 will have a maximum observation time of $\tau_1 - \tau_0$, while an individual recruited at time τ' will have a maximum observation time of $\tau_1 - \tau'$. Even if no subjects become unobservable, staggered entry and varying maximum observation times will result in subjects having different lengths of follow-up.

For historical reasons it is usual in the survival analysis literature to refer to the study endpoint as "death" and to the length of follow-up for a given subject as the "survival" time. These and related conventions are adopted irrespective of whether the study has a mortality endpoint or not. So, for example, when we speak of a subject surviving to the end of the study we mean that, for this individual, the endpoint of interest did not occur. For a given subject, let t denote the survival time and define an indicator variable as follows: $\delta = 1$ if the subject dies, and $\delta = 0$ otherwise. When $\delta = 0$ we say that t is a censored survival time, and when $\delta = 1$ that t is uncensored. In this way, the outcome for each subject is made dichotomous— that is, censored or not. Survival data on each subject can be compactly written in vector form (t, δ), which we refer to as an observation. We say that an observation is censored or uncensored according to whether $\delta = 0$ or $\delta = 1$, respectively.

Figure 8.1(a) depicts an open cohort study involving six subjects in which the maximum observation time is 10 years. The horizontal axis is calendar time and, in the above notation, $\tau_0 = 0$, $\tau' = 5$, and $\tau_1 = 10$. The line for each subject, which we refer to as a follow-up line, stretches between the calendar time points that the individual was under observation. A solid dot indicates that the subject died, and a circle means that the subject was censored. So, subject 1 entered at the beginning of recruitment, was followed for 10 years, and exited the study alive. Subject 2 also

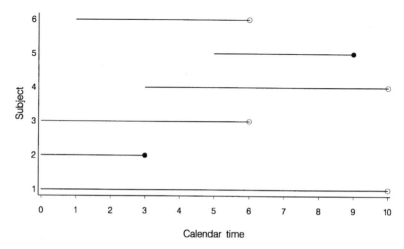

FIGURE 8.1(a)   Follow-up times for censored survival data.

entered at the beginning of recruitment but died after 3 years of follow-up. Subject 6 was enrolled at the 1-year point, was followed for 5 years, and exited the study alive. Figure 8.1(a) involves two types of "time": calendar time on the horizontal axis and survival time as depicted by the follow-up lines. If it can be assumed that such factors as recruitment and outcome are independent of calendar time, it is appropriate to "collapse" over the calendar time dimension. This results in Figure 8.1(b) in which all follow-up lines have been given the same starting point. Note that now the horizontal axis is labeled survival time.

Cohort data may contain information on several endpoints of interest. For example, as part of an ongoing follow-up of a group of patients with coronary artery disease, information might be collected on such endpoints as nonfatal myocardial infarction (heart attack), whether revascularization surgery was performed, and fatal myocardial infarction. The same individual could generate the observation (2.5, 1) when nonfatal myocardial infarction is the endpoint, (4.0, 1) when revascularization is the endpoint, and (6.0, 0) when fatal myocardial infarction is the endpoint. The interpretation is that this person had a nonfatal myocardial infarction 2.5 years into follow-up, underwent revascularization surgery 1.5 years later, and exited the database alive 2 years after that. The important point is that each choice of endpoint leads to a different definition of survival time and, by virtue of that, to a different cohort study.

According to the above definition of censoring, all subjects who do not develop the disease are lumped together as censored observations. However, the causes of censoring, in particular the reasons for becoming unobservable, may differ among subjects in ways that are important to the interpretation of study findings. For example, consider a cohort of patients with a particular type of cancer who have been treated with an innovative therapy and who are now being followed for death due to that disease. A subject who is censored as a result of being struck dead by lightning

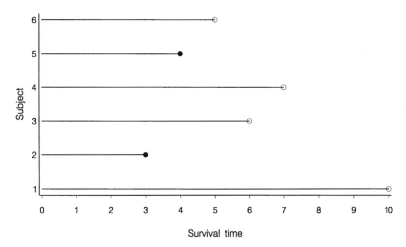

FIGURE 8.1(b)   Survival times for censored survival data.

presumably had a mortality risk from cancer that was no different from any other randomly selected member of the cohort. This type of censoring is said to be uninformative because a knowledge of the censoring mechanism does not tell us anything about the risk of experiencing the endpoint of interest. When censoring is uninformative, individuals censored at a given point during follow-up are a random sample of the members of the cohort surviving to that time point (Clayton and Hills, 1993, §7.5).

Now consider a subject who is censored as a result of being lost to follow-up after moving out of the study area. Suppose the reason this person decided to move was a dramatic remission of disease. Had this person remained in the study, there is a less than average chance that death from cancer would have occurred during follow-up. This type of censoring is said to be informative because a knowledge of the censoring mechanism tells us something about the risk of experiencing the endpoint of interest. When censoring is informative, individuals censored at a given point during follow-up are a nonrandom sample of the members of the cohort surviving to that time point, and this can lead to biased risk estimates. In the present example, the type of censoring described would result in the mortality risk being overestimated by the study. Consider a comparative study in which informative censoring takes place in both the exposed and unexposed cohorts. In most situations it is reasonable to assume that the risk estimates for both cohorts will be biased in the same direction. Consequently, when the risk estimates are combined into a measure of effect, the biases will tend to cancel each other out, to a greater or lesser extent. This means that informative censoring is usually of greater concern when the data are being analyzed in absolute rather than relative terms.

In a particular study, the endpoint might be quite narrowly defined—for example, death from a specific cause, onset of a certain illness, or recovery following a particular type of treatment. In each instance, only the specified endpoint is of interest and all other exits from the cohort are treated as censored observations. For example, consider a cohort study of breast cancer patients where the endpoint is death from this disease. In this setting, any reason for a subject becoming unobservable—in particular, death from a cause other than breast cancer—results in a censored observation. In a sense, the survival analysis is conducted as if death from breast cancer is the only possible cause of death and that, if followed long enough, all subjects would eventually die of this disease. Although such an assumption is usually unrealistic, it offers certain conceptual advantages. In particular, when cohorts are being compared in the same study or across studies, observed mortality differences will be specific to the endpoint of interest and not obscured (confounded) by extraneous factors related to censoring.

The methods of analyzing censored survival data presented in this book are all based on the assumption that censoring is uninformative, an assumption that may not be satisfied in practice. When censoring is informative, this must be considered at some point in the survival analysis. One approach is to model the censoring mechanism as part of the survival analysis in an effort to account for informative censoring. This requires information on the reasons for censoring and usually this degree of detail is unavailable. A practical alternative is to perform the survival analysis under the assumption that censoring is uninformative and then use qualitative arguments based on what may be known or suspected about the censoring mechanism to decide whether a parameter estimate is significantly biased.

8.2 SURVIVAL FUNCTIONS AND HAZARD FUNCTIONS

In the statistical theory of survival analysis, survival time is regarded as a continuous random variable that we denote by T. Accordingly, the survival time t discussed in the preceding section is a realization of T. As with any continuous random variable, T has an associated probability function $f(t)$. In survival analysis it is generally more convenient to characterize T in terms of two other functions, namely, the survival function $S(t)$ and the hazard function $h(t)$. The survival function is defined to be $S(t) = P(T \geq t)$; that is, $S(t)$ equals the probability of surviving until (at least) time t. Suppose that the sample space of T is $[0, \tau]$, where τ is the maximum survival time possible according to the study design. For example, $\tau = 5$ in a cohort study of cancer patients in which the maximum length of observation is set at 5 years. By definition, $S(0) = 1$, which means that the entire cohort is alive at $t = 0$. As intuition suggests, $S(t)$ is a nonincreasing function of t, so that $t_1 < t_2$ implies $S(t_1) \geq S(t_2)$. The graph of $S(t)$ provides a convenient method of depicting the survival experience of the cohort over the course of follow-up.

Let t be an arbitrary but fixed time and let ε be a small positive number. The interval $[t, t + \varepsilon)$ is the set of survival times greater than or equal to t and strictly less than $t + \varepsilon$. The probability of dying in $[t, t + \varepsilon)$ is $S(t) - S(t + \varepsilon)$, a quantity that approaches 0 as ε approaches 0. Now consider

$$\frac{S(t) - S(t + \varepsilon)}{\varepsilon} \tag{8.1}$$

which is the probability "per unit time" of dying in $[t, t + \varepsilon)$. As shown in Appendix F, as ε approaches 0, (8.1) has a limiting value equal to $f(t)$. This shows that, for a given time t, $f(t)$ has the rather unusual units of "per unit time." For instance, if survival time is measured in years, the units are "per year," which is sometimes written as year^{-1}. With t and ε as before, the conditional probability of dying in $[t, t + \varepsilon)$, given survival to t, is

$$Q_\varepsilon(t) = \frac{S(t) - S(t + \varepsilon)}{S(t)}$$

and the conditional probability per unit time of dying in $[t, t + \varepsilon)$, given survival to t, is

$$\frac{Q_\varepsilon(t)}{\varepsilon} = \frac{S(t) - S(t + \varepsilon)}{S(t)\varepsilon}. \tag{8.2}$$

As discussed in Appendix F, as ε approaches 0, (8.2) has a limiting value that we denote by $h(t)$. For a given time t we refer to $h(t)$ as the hazard at time t. When considered in its entirety, we refer to $h(t)$ as the hazard function. In life table theory the hazard function is sometimes termed the force of mortality, and in the epidemiologic literature it is often referred to as the incidence density. Like $f(t)$, $h(t)$ also has the units "per unit time." It follows from the definition of $h(t)$ that the product $h(t)\varepsilon$ is approximately equal to $Q_\varepsilon(t)$, with the approximation improving as ε gets smaller. We will see in Chapter 9 that the hazard function is closely related to the death rate, a measure of mortality risk that is widely used in demography and epidemiology.

Since T is a continuous random variable, the probability of dying at any given time is 0. It is only when we consider the probability of dying in an interval of time that a nonzero probability is obtained. For this reason we sometimes refer to $f(t)$ and $h(t)$ as "instantaneous" probabilities. We can characterize $S(t), f(t)$, and $h(t)$ as follows: $S(t)$ is the probability that an individual alive at time 0 will survive to time t; $f(t)$ is the instantaneous probability per unit time that an individual alive at time 0 will die at time t; and $h(t)$ is the instantaneous probability per unit time that an individual alive at time t will die "in the next instant." In Appendix F we show that $S(t)$, $f(t)$, and $h(t)$ are mathematically equivalent in the sense that each of them can be expressed in terms of the others. A useful identity relating the three functions is

$$h(t) = \frac{f(t)}{S(t)}. \tag{8.3}$$

Consider an open cohort study involving r subjects and denote the observations by $(t_1, \delta_1), \ldots, (t_i, \delta_i), \ldots, (t_r, \delta_r)$. Suppose that the observations are a sample from the distribution of T. Consequently each subject has the survival function $S(t)$, the probability function $f(t)$, and the hazard function $h(t)$. We seek an expression for the unconditional likelihood of the observations under the assumption that censoring is uninformative. If subject i survives to t_i, the corresponding term in the likelihood is $S(t_i)$, while if subject i dies at t_i the term is $f(t_i)$. Making use of δ_i, we can write the contribution of subject i to the likelihood as

$$S(t_i)^{1-\delta_i} f(t_i)^{\delta_i} = S(t_i)\left[\frac{f(t_i)}{S(t_i)}\right]^{\delta_i} = S(t_i)h(t_i)^{\delta_i} \tag{8.4}$$

where the last equality follows from (8.3). Therefore the likelihood is

$$L = \prod_{i=1}^{r} S(t_i)h(t_i)^{\delta_i}. \tag{8.5}$$

Example 8.1 (Canadian Females, 1990–1992) In this example we consider the ordinary life table cohort for Canadian females in 1990–1992. As explained in Chapter 15, this is a hypothetical cohort based on cross-sectional data in which a group of newborns is followed until the last subject dies. For the moment we treat the example as if it represents findings from an actual cohort study in which a large group of newborns has been followed until death. Figures 8.2(a)–8.2(c) are smooth curves that were created by interpolating published Statistics Canada ordinary life table functions. The data used by Statistics Canada to create the ordinary life table consist of census counts for mid-1991, numbers of births during 1990–1992, and numbers of deaths from all causes during 1990–1992.

Note that Figures 8.2(a) and 8.2(b) have been truncated at age 100, and Figure 8.2(c) at age 70. In Figure 8.2(a), the survival function is nonincreasing and ultimately decreases to 0 at the upper limit of life length. Although somewhat difficult to appreciate from the graph, there is a sharp drop in the survival curve in the first year of life due to perinatal and other causes of death following the newborn period. After that there is a very gradual decline until late middle age, and then a precipi-

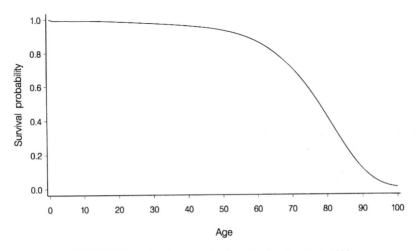

FIGURE 8.2(a)   Survival curve for Canadian females, 1990–1992

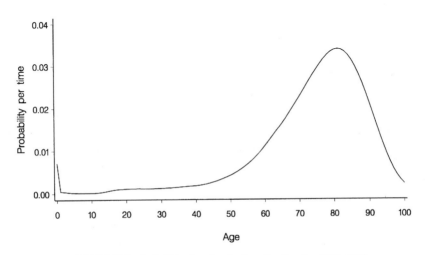

FIGURE 8.2(b)   Probability function for Canadian females, 1990–1992

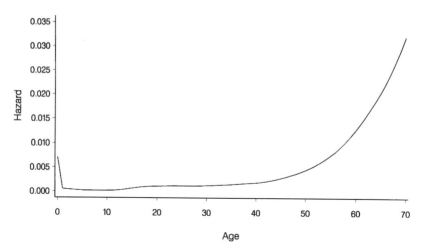

FIGURE 8.2(c)   Hazard function for Canadian females, 1990–1992

tous drop as old age approaches. In Figure 8.2(b), the probability function reflects these same phenomena in that there is a steep slope in the first year of life, a gentle increase throughout late middle age, and then a large peak in the curve in the senior years. The area under the curve between any two ages equals the (unconditional) probability of dying in that age interval. This explains why so much of the area is concentrated at older ages. The curve declines rapidly as 100 is approached because very few members of the cohort survive long enough to die at such an old age. In Figure 8.2(c), the hazard function demonstrates the same patterns observed in the survival function and probability function. In particular, we note the rapid increase in the hazard function as extreme old age approaches. This shows that for someone who has lived to be very old, the risk of dying in the next instant gets progressively greater as time passes.

8.3   HAZARD RATIO

Despite the mathematical equivalence of the probability function, survival function, and hazard function, the conditional nature of the hazard function makes it convenient for comparing mortality risk across cohorts. This is because a comparison at a given time point based on the hazard function involves only those individuals who have survived to that point. By contrast, the probability function and the survival function are unconditional and thus reflect the mortality experience of subjects who died prior to the time of interest, along with those who have survived. For this reason, survival models are often defined in terms of hazard functions.

Consider two cohorts, one of which is exposed and the other unexposed. Denote the corresponding survival functions by $S_1(t)$ and $S_2(t)$, and the hazard functions by $h_1(t)$ and $h_2(t)$, respectively. The ratio of hazard functions $h_1(t)/h_2(t)$ is central to modern survival analysis. Of particular importance is the situation where $h_1(t)/h_2(t)$ is constant—that is, independent of t. In this case we write

$$HR = \frac{h_1(t)}{h_2(t)} \tag{8.6}$$

and refer to HR as the hazard ratio. In the epidemiologic literature the hazard ratio is sometimes referred to as the incidence density ratio. When (8.6) holds we say that $h_1(t)$ and $h_2(t)$ satisfy the proportional hazards assumption, or that they are proportional. When the proportional hazards assumption is satisfied, the parameter HR is a convenient measure of effect for an open cohort study in much the same way that RD, RR, and OR serve in this capacity for a closed cohort study. It is clear from (8.6) that, similar to the risk ratio and odds ratio, the hazard ratio is a multiplicative measure of effect. For example, suppose that $HR = 3$. Then, at every time point, a subject in the exposed cohort has a hazard that is three times as great as the hazard facing a member of the unexposed cohort. It is important to appreciate that the proportional hazards assumption specifies that the ratio of the hazard functions is constant, not the individual hazard functions. In fact, let $h_2(t) \geq 0$ have an arbitrary functional form and, for a given constant $\psi > 0$, define $h_1(t) = \psi h_2(t)$. Then $h_1(t)$ and $h_2(t)$ are proportional and ψ is the hazard ratio. In Appendix F it is demonstrated that the proportional hazards assumption is equivalent to

$$S_1(t) = [S_2(t)]^{HR}. \tag{8.7}$$

Most of the methods for analyzing censored survival data that are presented in this book are based on the proportional hazards assumption. As a result, much of the discussion focuses on the hazard ratio. However, it should be remembered that the hazard ratio is a relative measure of effect and, by virtue of that, tells us nothing about absolute risk. Therefore, as part of a survival analysis it is important to examine survival curves in their entirety in order to gain a more complete appreciation of mortality risks.

8.4 COMPETING RISKS

When vital statistics mortality data are being coded, it is usual to identify a single entity as "the" cause of death. Suppose that all causes of death have been grouped into K mutually exclusive "causes" ($k = 1, 2, \ldots, K$). Consider a cohort study of mortality where, along with survival time T, the cause of death of each subject is recorded. We can imagine that, under given study conditions, each subject has a set of "potential" survival times $(T_1, \ldots, T_k, \ldots, T_K)$, one for each cause (Gail, 1975). If death is due to cause k, the observed survival time is T_k. In this case no physical meaning is attached to the remaining potential survival times (Prentice et al., 1978). Associated with each cause k is a hazard function, denoted by $h^k(t)$, which has the following definition: $h^k(t)$ is the instantaneous probability per unit time that an individual alive at time t will die in the next instant of cause k, in the presence of other causes of death.

The phrase "in the presence of other causes of death" is included in the definition because the risk of dying of cause k may be related to other causes. For example, suppose that two of the causes of death are myocardial infarction and stroke. These two conditions have a number of risk factors in common, and so the risk of dying of one of them may be related to the risk of dying of the other. Since an individual can die of only one cause, the K causes are said to be competing risks. In the statistics literature, $h^k(t)$ is referred to as the crude hazard function for cause k (Chiang, 1968, Chapter 11; Elandt-Johnson and Johnson, 1980, Chapter 9; Tsiatis, 1998). This use of the term crude differs from our previous usage and refers only to the fact that competing risks are present. Since causes of death are mutually exclusive and exhaustive, the cause-specific hazard functions satisfy the fundamental identity

$$h(t) = \sum_{k=1}^{K} h^k(t). \tag{8.8}$$

In Section 8.1 it was remarked that the usual approach to survival analysis is unrealistic because only one endpoint is permitted. The competing risks model offers an alternative approach to analyzing survival data in that several endpoints (risks) can be accommodated simultaneously. In the breast cancer example, consider the following five endpoints: death from breast cancer, death from any other cancer, death from any noncancer cause, withdrawal from the study, and any other reason for becoming unobservable. In this way, competing risks analysis is able to take explicit account of causes of death other than breast cancer, thereby providing a more realistic model of the survival experience of the cohort. Unfortunately, procedures for competing risks analysis are not included as part of standard statistical packages.

The censoring and competing risks approaches to survival analysis can be reconciled when the risks are "independent." Deciding whether the independence assumption is satisfied requires substantive knowledge of the competing risks and is not an issue that can be resolved using statistical methods (Prentice et al., 1978; Prentice and Kalbfleisch, 1988). For example, consider the following three causes of death: motor vehicle accidents, myocardial infarction, and stroke. Although a myocardial infarction or a stroke might cause a driver to have a motor vehicle accident, and a motor vehicle accident might precipitate a myocardial infarction or a stroke, for the most part, deaths due to motor vehicle accidents and these two circulatory conditions can be viewed as independent mortality risks. On the other hand, myocardial infarction and stroke share a number of risk factors and are therefore not independent.

When the independence assumption is satisfied, $h^k(t)$ depends only on cause k. In this case we drop the phrase "in the presence of other causes of death" from the earlier interpretation and refer to $h^k(t)$ as the net hazard function for cause k (Chiang, 1968, Chapter 11; Elandt-Johnson and Johnson, 1980, Chapter 9; Tsiatis, 1998). Consider a cohort study in which subjects either reach the endpoint of interest or are censored. This type of study can be given a competing risks interpretation by defining two risks: Risk 1 is the endpoint of interest and risk 2 is censoring. According to this model, censoring is uninformative precisely when the risks are independent. When there is independence, survival and censoring in the cohort are governed by the net hazard functions $h^1(t)$ and $h^2(t)$, respectively.

CHAPTER 9

Kaplan–Meier and Actuarial Methods for Censored Survival Data

In this chapter we describe the Kaplan–Meier and actuarial methods of estimating a survival function from censored survival data. An important feature of these methods is that, aside from uninformative censoring, they are based on relatively few assumptions. In the case of the Kaplan–Meier method, nothing is assumed about the functional form of either the survival curve or the hazard function. An approach to comparing survival curves is presented which is based on the stratified odds ratio techniques of Chapter 5. The MH–RBG methods are shown to be especially useful in this regard. References for this chapter are those given at the beginning of Chapter 8.

9.1 KAPLAN–MEIER SURVIVAL CURVE

Consider an open cohort study involving r subjects and let the observations be $(t_1, \delta_1), \ldots, (t_i, \delta_i), \ldots, (t_r, \delta_r)$. We assume that $t_i > 0$ for all i, which ensures that each member of the cohort is followed for at least a small amount of time. In this chapter, uncensored survival times—that is, those at which a death occurs—will be referred to as death times. Suppose that among the r survival times there are J death times: $\tau_1 < \cdots < \tau_j < \cdots < \tau_J$. Let $\tau_0 = 0$ and denote the maximum survival time by τ_{J+1}, that is, $\tau_{J+1} = \max(t_1, t_2, \ldots, t_r)$. The Kaplan–Meier approach to censored survival data begins by partitioning the period of follow-up into $J + 1$ intervals using the death times as cutpoints: $[\tau_0, \tau_1), [\tau_1, \tau_2), \ldots, [\tau_j, \tau_{j+1}), \ldots, [\tau_{J-1}, \tau_J)$, $[\tau_J, \tau_{J+1}]$, where we note that the last interval contains τ_{J+1}. We refer to $[\tau_j, \tau_{j+1})$ as the jth interval. In many applications there will be considerable censoring at τ_{J+1} due to subjects surviving to the end of the study. If τ_{J+1} is a death time, then $\tau_J = \tau_{J+1}$ and the last interval shrinks to the single point τ_J. Let a_j be the number of deaths at τ_j and let c_j be the number of censored observations in the jth interval $(j = 0, 1, \ldots, J)$. By definition, $a_0 = 0$.

The group of subjects "at risk" at τ_j, also referred to as the jth risk set, consists of those individuals with a survival time greater than or equal to τ_j $(j = 0, 1, \ldots, J)$. So the jth risk set consists of three types of individuals: those who survive beyond τ_j, those who are censored at τ_j, and those who die at τ_j. Defining the first two groups of subjects to be at risk at τ_j is reasonable since, if death occurs, it will happen some time after τ_j. However, including subjects who die at τ_j in the risk set is not intuitive. This convention has its origins in the theoretical development of the Kaplan–Meier method, which can be viewed as a limiting case of the actuarial method discussed below. Indeed, another name for the Kaplan–Meier approach to censored survival data is the product-limit method. Loosely speaking, the jth risk set consists of all subjects who are alive "just prior to" τ_j. Let r_j denote the number of subjects in the jth risk set ($j = 0, 1, \ldots, J$), and denote by r_{J+1} the number of subjects who survive to τ_{J+1}. Note that c_J includes the r_{J+1} subjects who survive to the end of the study. We need to separate this group of individuals from those who are censored for other reasons. Define c'_j as follows: $c'_j = c_j$ for $j < J$, and $c'_J = c_J - r_{J+1}$. Since subjects exit the cohort only by death or censoring, it follows that

$$r_{j+1} = r_j - a_j - c'_j \tag{9.1}$$

($j = 0, 1, \ldots, J$). Therefore,

$$r_0 - r_{J+1} = \sum_{j=0}^{J}(r_j - r_{j+1}) = \sum_{j=0}^{J} a_j + \sum_{j=0}^{J} c'_j = a_\bullet + c'_\bullet$$

and so $r_{J+1} = r_0 - a_\bullet - c'_\bullet$. This identity says that the number of subjects surviving to the end of the study equals the number who started minus those who died or were censored prior to τ_{J+1}.

We now derive an estimate of $S(t)$ based on the above partition and certain conditional probabilities. For brevity we denote $S(\tau_j)$ by S_j so that, in particular, $S_0 = 1$. For $j > 0$, consider the interval $[\tau_j - \varepsilon, \tau_j + \varepsilon)$, where ε is a positive number that is small enough to ensure that the interval does not contain any death times other than τ_j or any censoring times. For $j = 0$, we use the interval $[\tau_0, \tau_0 + \varepsilon)$ and make some obvious modifications to the following arguments. Let p_j denote the conditional probability of surviving to $\tau_j + \varepsilon$, given survival to $\tau_j - \varepsilon$ ($j = 0, 1, \ldots, J$). That is, p_j is the conditional probability of surviving to just after τ_j, given survival to just prior to τ_j. Let $q_j = 1 - p_j$ be the corresponding conditional probability of dying. Since there are r_j subjects at risk at $\tau_j - \varepsilon$, and a_j of them die prior to $\tau_j + \varepsilon$, the binomial estimates are

$$\hat{q}_j = \frac{a_j}{r_j} \tag{9.2}$$

and

$$\hat{p}_j = 1 - \hat{q}_j = \frac{r_j - a_j}{r_j} \tag{9.3}$$

($j = 0, 1, \ldots, J$). These estimates are valid regardless of how small we take ε. In the limit, as ε goes to 0, $[\tau_j - \varepsilon, \tau_j + \varepsilon)$ shrinks to the single point τ_j, in keeping with the product-limit approach. Since $a_0 = 0$, it follows that $\hat{q}_0 = 0$ and $\hat{p}_0 = 1$.

The probability of surviving from τ_0 to τ_1 is p_1. For those who survive to τ_1, the conditional probability of surviving to τ_2 is p_2, and so the probability of surviving from τ_0 to τ_2 is $S_2 = p_1 p_2$. Likewise, the probability of surviving from τ_0 to τ_3 is $S_3 = p_1 p_2 p_3$. Proceeding in this way we obtain the probability $S_j = p_1 p_2 \cdots p_j$ of surviving from τ_0 to τ_j ($j = 1, 2, \ldots, J$). The Kaplan–Meier estimate of S_j is

$$\hat{S}_j = \hat{p}_1 \hat{p}_2 \cdots \hat{p}_j \tag{9.4}$$

($j = 1, 2, \ldots, J$). We define $\hat{S}_0 = 1$ and $\hat{S}_{J+1} = \hat{S}_J$. Since there are no deaths in the jth interval other than at τ_j, $\hat{S}(t)$ equals $\hat{S}_j$ for all t in the interval, and so the graph of $\hat{S}(t)$ is a step function.

When the Kaplan–Meier estimate of the survival function is graphed as a function of time, it will be referred to as the Kaplan–Meier survival curve. A schematic representation of a Kaplan–Meier survival curve is shown in Figure 9.1. Note that each of the line segments making up the steps includes the left endpoint (indicated by a solid dot) but not the right endpoint (indicated by a circle), except for the final line segment that includes both endpoints. This is consistent with the way intervals were defined. Most software packages join the steps with vertical lines to enhance visual appearance. If τ_{J+1} is a death time, the final line segment shrinks to a single point.

An estimate of $\text{var}(\hat{S}_j)$ is given by "Greenwood's formula,"

$$\widehat{\text{var}}(\hat{S}_j) = (\hat{S}_j)^2 \sum_{i=1}^{j} \frac{\hat{q}_i}{\hat{p}_i r_i} \tag{9.5}$$

and a $(1 - \alpha) \times 100\%$ confidence interval for $\hat{S}_j$ is

$$[\underline{S}_j, \overline{S}_j] = \hat{S}_j \pm z_{\alpha/2} \sqrt{\widehat{\text{var}}(\hat{S}_j)}$$

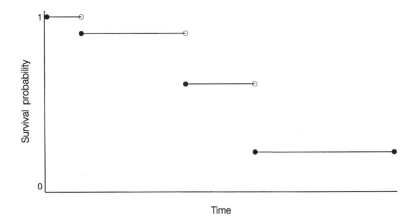

FIGURE 9.1 Schematic Kaplan–Meier survival curve

($j = 1, 2, \ldots, J$) (Greenwood, 1926). The normal approximation can be improved by using the log–minus-log transformation $\log(-\log \hat{S}_j)$. This leads to the Kalbfleisch–Prentice estimate,

$$\widehat{\text{var}}[\log(-\log \hat{S}_j)] = \frac{1}{(\log \hat{S}_j)^2} \sum_{i=1}^{j} \frac{\hat{q}_i}{\hat{p}_i r_i}$$

($j = 1, 2, \ldots, J$) (Kalbfleisch and Prentice, 1980, p. 15). A $(1 - \alpha) \times 100\%$ confidence interval for $\hat{S}_j$ is obtained from

$$[\log(-\log \underline{S}_j), \log(-\log \overline{S}_j)] = \log(-\log \hat{S}_j) \mp z_{\alpha/2}\sqrt{\widehat{\text{var}}[\log(-\log \hat{S}_j)]} \quad (9.6)$$

by inverting the log–minus-log transformation, as illustrated in Example 9.1. Note the $\mp$ sign in (9.6) rather than the usual $\pm$ sign.

It is of interest to examine the estimates $\hat{S}_j$ and $\widehat{\text{var}}(\hat{S}_j)$ under the assumption that there is no censoring except at τ_{J+1}—that is, assuming $c'_j = 0$ ($j = 0, 1, \ldots, J$). With this assumption, (9.1) simplifies to $r_{j+1} = r_j - a_j$ ($j = 0, 1, \ldots, J$). From this identity, as well as (9.3)–(9.5), $\hat{p}_0 = 1$, and $\hat{q}_0 = 0$, it follows that

$$\begin{aligned}
\hat{S}_j &= \hat{p}_1 \hat{p}_2 \cdots \hat{p}_j = \hat{p}_0 \hat{p}_1 \cdots \hat{p}_j \\
&= \left(\frac{r_0 - a_0}{r_0}\right)\left(\frac{r_1 - a_1}{r_1}\right) \cdots \left(\frac{r_j - a_j}{r_j}\right) \\
&= \left(\frac{r_0 - a_0}{r_0}\right)\left(\frac{r_1 - a_1}{r_0 - a_0}\right) \cdots \left(\frac{r_{j+1}}{r_{j-1} - a_{j-1}}\right) \\
&= \frac{r_{j+1}}{r_0}
\end{aligned} \quad (9.7)$$

and

$$\begin{aligned}
\widehat{\text{var}}(\hat{S}_j) &= (\hat{S}_j)^2 \sum_{i=1}^{j} \frac{\hat{q}_i}{\hat{p}_i r_i} = (\hat{S}_j)^2 \sum_{i=0}^{j} \frac{\hat{q}_i}{\hat{p}_i r_i} \\
&= (\hat{S}_j)^2 \sum_{i=0}^{j} \left(\frac{1}{r_{i+1}} - \frac{1}{r_i}\right) = (\hat{S}_j)^2 \left(\frac{1}{r_{j+1}} - \frac{1}{r_0}\right) \\
&= \frac{\hat{S}_j(1 - \hat{S}_j)}{r_0}.
\end{aligned} \quad (9.8)$$

Identities (9.7) and (9.8) show that when there is no censoring except at τ_{J+1}, the Kaplan–Meier and Greenwood estimates simplify to estimates based on the binomial distribution. The numerator of $\hat{S}_j$ in (9.7) is r_{j+1} because, in the absence of censoring, the risk set at τ_{j+1} is the same as the group of survivors to τ_j.

TABLE 9.1 Survival Times: Stage–Receptor Level–Breast Cancer

Stage	Receptor level	r_0	Survival times
I	Low	12	50* 51 51* 53*(2) 54*(2) 55* 56 56* 57* 60*
I	High	57	10 34 34* 47(2) 49*(2) 50*(7) 51*(6) 52*(5) 53*(6) 54*(5) 55*(2) 56*(2) 57*(5) 58*(5) 59*(4) 60*(3)
II	Low	23	4* 9 13 21 29(2) 40 46 49*(2) 52*(2) 53* 54* 55*(2) 56* 57 57* 58*(2) 59* 60*
II	High	75	11 16 21 23(2) 24 33(2) 36(2) 36* 37 45 46 49*(2) 50*(6) 51*(4) 52*(5) 53*(5) 54*(4) 55*(4) 56*(6) 57*(4) 58(2) 58*(8) 59*(5) 60*(6)
III	Low	15	9 12 14 15 15* 17 21 22 23(2) 31 34 35 53* 60*
III	High	17	7* 9 17 21* 22(2) 34(2) 41 49* 52* 55 56* 58*(2) 59*(2)

Example 9.1 (Receptor Level–Stage–Breast Cancer) The data in Table 9.1 are based on the cohort of 199 breast cancer patients described in Example 4.2. Recall that these individuals are a random sample of patients registered on the Northern Alberta Breast Cancer Registry during 1985. For the present example this cohort was followed to the end of 1989. Therefore the maximum observation times range between 4 and 5 years depending on the date of registration. Survival time was defined to be the length of time from registration until death from breast cancer or censoring. Survival times were first calculated to the nearest day and then rounded up to the nearest month to ensure that at least a few subjects had the same survival time. So the maximum survival time is 60 months. As discussed in Example 4.2, Registry patients receive regular checkups and their vital status is monitored on an ongoing basis. It is therefore reasonable to assume that members of the cohort were alive at the end of 1989 in the absence of information to the contrary. Therefore the reasons for censoring in this cohort are death from a cause other than breast cancer and exiting the study alive at the end of 1989.

For present purposes, we interpret the survival times as continuous, so that, for example, $t = 50$ is to be read as $t = 50.0$. The asterisks in Table 9.1 denote censored survival times and so 50* means that the subject was censored at $t = 50$, while 51 indicates that the subject died (of breast cancer) at $t = 51$. Strictly speaking, we should refer to the entries in Table 9.1 as observations since, for example, 50* and 51 are actually shorthand for (50, 0) and (51, 1). In Table 9.1, numbers in parentheses denote the multiplicity of survival times, so that 53*(2) represents two survival times of 53*. Note that when death and censoring take place at the same time, the censoring time has been recorded to the right of the death time. This convention is adopted since, when there is a tie, the (unobserved) death time for the censored individual (when it occurs) will be larger than the (observed) censoring time.

TABLE 9.2   Kaplan–Meier Estimates and Kalbfleisch–Prentice 95% Confidence Intervals: Breast Cancer

j	τ_j	a_j	r_j	$\hat{p}_j$	$\hat{S}_j$	$\underline{S}_j$	$\overline{S}_j$
0	0	0	199	1.0	1.0	—	—
1	9	3	197	.985	.985	.954	.995
2	10	1	194	.995	.980	.947	.992
3	11	1	193	.995	.975	.940	.989
4	12	1	192	.995	.970	.933	.986
5	13	1	191	.995	.964	.927	.983
6	14	1	190	.995	.959	.920	.979
7	15	1	189	.995	.954	.914	.976
8	16	1	187	.995	.949	.908	.972
9	17	2	186	.989	.939	.895	.965
10	21	3	184	.984	.924	.877	.953
11	22	3	180	.983	.908	.858	.941
12	23	4	177	.977	.888	.835	.925
13	24	1	173	.994	.883	.829	.920
14	29	2	172	.988	.872	.817	.912
15	31	1	170	.994	.867	.811	.908
16	33	2	169	.988	.857	.800	.899
17	34	4	167	.976	.836	.777	.881
18	35	1	162	.994	.831	.771	.877
19	36	2	161	.988	.821	.760	.868
20	37	1	158	.994	.816	.754	.863
21	40	1	157	.994	.811	.748	.859
22	41	1	156	.994	.805	.743	.854
23	45	1	155	.994	.800	.737	.850
24	46	2	154	.987	.790	.726	.841
25	47	2	152	.987	.779	.714	.831
26	51	1	129	.992	.773	.708	.826
27	55	1	77	.987	.763	.695	.818
28	56	1	67	.985	.752	.680	.810
29	57	1	55	.982	.738	.662	.800
30	58	2	43	.953	.704	.615	.776

Table 9.2 gives the Kaplan–Meier estimates of the survival probabilities as well as the Kalbfleisch–Prentice 95% confidence intervals. The first step in creating Table 9.2 was to list the death times in increasing order and then count the number of deaths at each death time. The first death time is $\tau_1 = 9$ and the number of deaths is $a_1 = 3$. In total, there were 30 distinct death times and 49 deaths in the cohort. The next step was to determine the number of subjects in each risk set using (9.1). For this purpose, the survival times were listed in increasing order and (9.1) was applied in a recursive fashion, starting with $j = 0$. To illustrate, for $j = 0$ the interval is $[0, 9.0)$, $r_0 = 199$, $a_0 = 0$, and $c'_0 = 2$. So $r_1 = 199 - 0 - 2 = 197$. Observe that the two subjects with censoring times prior to the first death time make

no contribution to the Kaplan–Meier calculations, and so the effective size of the cohort is $r_1 = 197$.

To calculate the 95% confidence interval for $\hat{S}_j$ using the Kalbfleisch–Prentice method it is necessary to invert the log–minus-log transformation. This is illustrated for $\bar{S}_2$. From $\hat{S}_2 = .980$ and

$$\sum_{i=1}^{2} \frac{\hat{q}_i}{\hat{p}_i r_i} = \frac{.015}{.985(197)} + \frac{.005}{.995(194)} = (.010)^2$$

we have $\widehat{\mathrm{var}}[\log(-\log \hat{S}_2)] = (.010)^2/[\log(.980)]^2 = (.495)^2$ and $\log(-\log \bar{S}_2) = \log[-\log(.980)] - 1.96(.495) = -4.87$. To invert the log–minus-log transformation, the exp–minus-exp transformation is applied to obtain $\bar{S}_2 = \exp[-\exp(-4.87)] = .992$. Based on Greenwood's formula, the upper 95% confidence bound for $\hat{S}_1$ is $\bar{S}_1 = 1.002$. The Kalbfleisch–Prentice approach has the attractive property that the upper and lower bounds are always between 0 and 1.

Figure 9.2 shows the Kaplan–Meier survival curve and Kalbfleisch–Prentice 95% confidence intervals for the breast cancer cohort. Strictly speaking, the endpoints of the confidence intervals should be plotted only for the death times rather than joined as they have been here. An appropriate alternative is to estimate what is referred to as a confidence band (Marubini and Valsecchi, 1995, §3.4.2; Hosmer and Lemeshow, 1999, §2.3). A confidence band places simultaneous upper and lower bounds on the entire survival curve, but there is the drawback that the computations are somewhat involved. Confidence bands tend to be wider than joined confidence intervals.

There were seven deaths in the cohort from causes other than breast cancer, and these were treated as censored observations. The question as to whether these deaths might somehow be related to breast cancer needs to be decided on substantive grounds. This decision has implications for whether censoring is deemed to be

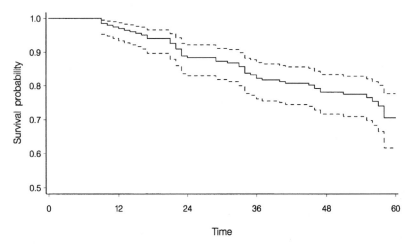

FIGURE 9.2   Kaplan–Meier survival curve and Kalbfleisch–Prentice 95% confidence intervals: Breast cancer cohort

informative or uninformative. Even if we assume that censoring is uninformative, it would be incorrect to interpret $\hat{S}_j$ as an estimate of the probability of not dying of breast cancer before τ_j. In light of remarks made at the end of Section 8.1, the correct interpretation adds the caveat that breast cancer is assumed to be the only cause of death. This is a convenient fiction when cohorts are being compared, but in reality, competing risks are virtually always present. As a result, when evaluating the findings of a survival analysis it is crucial to consider the possible effects of informative censoring and competing risks.

9.2   ODDS RATIO METHODS FOR CENSORED SURVIVAL DATA

Usually one of the aims of a survival analysis is to determine whether a given exposure is related to survival. One approach is to stratify the cohort according to exposure categories and compare the resulting Kaplan–Meier survival curves. For simplicity, suppose that exposure is dichotomous and that, at every death time, the survival curve for the exposed cohort lies below that for the unexposed cohort. A finding such as this suggests that exposure is associated with a decrease in survival. The question then arises as to how the observed difference in survival should be measured. One possibility is to pick a particular follow-up time and use methods based on the binomial distribution. For example, in the oncology literature 5-year survival probabilities are often used to compare outcomes following treatment. This approach has the attraction of simplicity but suffers from the drawback that, except for a single time point, the information in the survival curves is largely wasted. In this section we describe an alternative approach which uses the entire survival curve. The key idea is that death times are used to "stratify" the data, which are then analyzed using the odds ratio methods of Chapter 5 (Breslow, 1975, 1979).

9.2.1   Methods for *J* (2 × 2) Tables

Suppose that exposure is dichotomous. Let $S_1(t)$ and $h_1(t)$ be the survival function and hazard function for the exposed cohort, and let $S_2(t)$ and $h_2(t)$ be the corresponding functions for the unexposed cohort. Let the death times for the exposed and unexposed cohorts taken together be $\tau_1 < \cdots < \tau_j < \cdots < \tau_J$. For each τ_j we form a 2 × 2 table as depicted in Table 9.3. For the exposed cohort, a_{1j} is the number of deaths at τ_j and r_{1j} is the number of subjects at risk. We define a_{2j} and r_{2j} in an analogous manner for the unexposed cohort and derive the remaining table entries by addition or subtraction. As before, we refer to the r_j subjects at risk at τ_j as the jth risk set.

In Appendix F it is demonstrated that the odds ratio and risk ratio associated with Table 9.3 are approximately equal to $h_1(\tau_j)/h_2(\tau_j)$. If we treat the set of death times as a stratifying variable, we can adapt the methods of Chapters 5 and 6 to the analysis of censored survival data. In practice, the number of deaths at each death time may be relatively small. Often only one of the cohorts has a death at τ_j, in which case either a_{1j} or a_{2j} is 0. On the other hand, when there are many deaths, r_{1j} and r_{2j}

TABLE 9.3 Observed Counts: Open Cohort Study

Survival	Exposure		
	yes	no	
dead	a_{1j}	a_{2j}	m_{1j}
alive	$r_{1j} - a_{1j}$	$r_{2j} - a_{2j}$	m_{2j}
	r_{1j}	r_{2j}	r_j

may be small toward the end of follow-up. For these reasons we adopt methods that are suited to sparse-strata conditions, namely, the asymptotic conditional and MH–RBG methods. Since asymptotic conditional methods are not available for the risk ratio, we do not consider this measure of effect in what follows.

We now make the crucial assumption that the hazard functions, $h_1(t)$ and $h_2(t)$, are proportional. As a result, $h_1(\tau_j)/h_2(\tau_j) = HR$ for all j. Since we are treating the set of death times as a stratifying variable, the proportional hazards assumption is equivalent to the hazard ratio being homogeneous over "time." This means that we can apply the odds ratio methods developed in Chapter 5 under the assumption of homogeneity. In what follows we use odds ratio notation and terminology to frame the discussion, allowing the corresponding hazard ratio interpretation to be understood. In particular, $\widehat{OR}_c$ and $\widehat{OR}_{mh}$ will be viewed as estimates of HR, and X^2_{mh} will be regarded as a test of $H_0 : HR = 1$ (Mantel, 1966). In the survival analysis setting, the Mantel–Haenszel test is usually referred to as the logrank test (Peto, 1972; Peto and Peto, 1972). We adopt this terminology but will continue to use the notation X^2_{mh}.

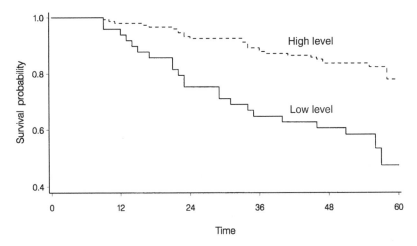

FIGURE 9.3(a) Kaplan–Meier survival curves: Breast cancer cohort stratified by receptor level

TABLE 9.4   Odds Ratio Analysis: Receptor Level–Breast Cancer

j	τ_j	a_{1j}	a_{2j}	r_{1j}	r_{2j}	$\hat{e}_{1j}$	$\hat{v}_{0j}$
1	9	2	1	49	148	.746	.555
2	10	0	1	47	147	.242	.184
3	11	0	1	47	146	.244	.184
4	12	1	0	47	145	.245	.185
5	13	1	0	46	145	.241	.183
6	14	1	0	45	145	.237	.181
7	15	1	0	44	145	.233	.179
8	16	0	1	42	145	.225	.174
9	17	1	1	42	144	.452	.348
10	21	2	1	41	143	.668	.514
11	22	1	2	39	141	.650	.503
12	23	2	2	38	139	.859	.663
13	24	0	1	36	137	.208	.165
14	29	2	0	36	136	.419	.329
15	31	1	0	34	136	.200	.160
16	33	0	2	33	136	.391	.312
17	34	1	3	33	134	.790	.623
18	35	1	0	32	130	.198	.159
19	36	0	2	31	130	.385	.309
20	37	0	1	31	127	.196	.158
21	40	1	0	31	126	.197	.158
22	41	0	1	30	126	.192	.155
23	45	0	1	30	125	.194	.156
24	46	1	1	30	124	.390	.312
25	47	0	2	29	123	.382	.307
26	51	1	0	26	103	.202	.161
27	55	0	1	15	62	.195	.157
28	56	1	0	12	55	.179	.147
29	57	1	0	9	46	.164	.137
30	58	0	2	6	37	.279	.234
Total	—	22	27	—	—	10.20	7.99

Example 9.2 (Receptor Level–Breast Cancer)   Figure 9.3(a) shows the Kaplan–Meier survival curves for the breast cancer cohort after stratifying by receptor level. It appears that subjects with low receptor level are at greater mortality risk than those with high receptor level. Table 9.4 gives the elements needed to calculate the logrank test. Each row of Table 9.4 corresponds to a 2×2 table of the form of Table 9.3. To illustrate, for $j = 1$ the table is

Survival	Receptor level		
	low	high	
dead	2	1	3
alive	47	147	194
	49	148	197

Note that due to stratification, $a_{1j} + a_{2j}$ and $r_{1j} + r_{2j}$ from Table 9.4 equal a_j and r_j from Table 9.2. The logrank test is $X^2_{mh} = (22 - 10.20)^2/7.99 = 17.43$ ($p < .001$), which provides considerable evidence that survival differs according to receptor level status.

Table 9.5 gives point estimates and 95% confidence intervals for OR based on the asymptotic conditional, MH–RBG, and asymptotic unconditional approaches. Even though there are only a few deaths at each death time, the risk sets are relatively large and thus large-strata conditions are satisfied. For this reason the asymptotic unconditional estimates have been included. As can be seen, the three methods produce virtually identical results.

9.2.2   Assessment of the Proportional Hazards Assumption

Graphical Method
We now turn to the problem of determining whether the proportional hazards assumption is satisfied. Following Section 9.1, we use the notation $S_{1j} = S_1(\tau_j)$ and $S_{2j} = S_2(\tau_j)$. From (8.7) the proportional hazards assumption is equivalent to $S_1(t) = [S_2(t)]^{HR}$, which in turn is equivalent to

$$\log[-\log S_1(t)] - \log[-\log S_2(t)] = \log(HR). \qquad (9.9)$$

Accordingly we can assess the proportional hazards assumption by graphing $\log(-\log \hat{S}_{1j})$ and $\log(-\log \hat{S}_{2j})$ together as functions of time and determining whether the curves are separated by a more or less constant vertical distance. If so, this is an indication that (9.9) is satisfied. Since the logarithmic function is undefined at 0, we only consider values of τ_j such that $\log(\hat{S}_{1j})$ and $\log(\hat{S}_{2j})$ are nonzero. Although somewhat subjective, the graphical method is computationally straightforward and tends to be quite revealing.

TABLE 9.5   Odds Ratio Estimates and 95% Confidence Intervals: Receptor Level–Breast Cancer

Method	$\widehat{OR}$	$\underline{OR}$	$\overline{OR}$
Asymptotic conditional	3.15	1.79	5.56
MH–RBG	3.19	1.80	5.65
Asymptotic unconditional	3.18	1.80	5.64

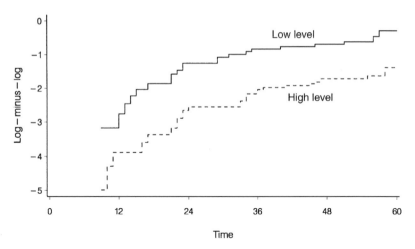

FIGURE 9.3(b)   Log–minus-log curves: Breast cancer cohort stratified by receptor level

Example 9.3 (Receptor Level–Breast Cancer)   Figure 9.3(b) is obtained directly from Figure 9.3(a) by applying the log–minus-log transformation. The distance between the curves is essentially constant throughout the period of follow-up, which supports the validity of the proportional hazards assumption.

Test for Linear Trend

As remarked above, the proportional hazards assumption is equivalent to the hazard ratio being homogeneous over time. In theory, this assumption could be assessed using, for example, the Breslow–Day test of homogeneity (5.32). It was pointed out in Section 5.6 that tests of homogeneity generally have low power, especially under sparse-strata conditions. An alternative is to evaluate homogeneity using a test for linear trend. This approach is best suited to the situation where there is a progressive increase or decrease in the log-odds ratios, as manifested by a corresponding widening or narrowing of the distance between the log–minus-log curves. The test for linear trend (5.19) can be adapted to the asymptotic conditional setting with time taken to be the stratifying variable (Breslow, 1984b). According to this approach, $\widehat{OR}_c$ is estimated using (5.23) and then $\hat{a}_{1j}$ and $\hat{v}_j$ are estimated using (5.24) and (5.25). The exposure level for the jth stratum is defined to be $s_j = j$.

Example 9.4 (Receptor Level–Breast Cancer)   With $\widehat{OR}_c = 3.15$, the test for linear trend is

$$X_t^2 = \frac{(-21.31)^2}{3390.2 - (176.49)^2/11.92} = .585 \; (p = .44)$$

which provides virtually no evidence against the proportional hazards assumption. This finding is consistent with our empirical assessment of Figure 9.3(b).

TABLE 9.6   Survival Times: Histologic Grade–Ovarian Cancer

Grade	r_0	Survival time
Low	15	28 89 175 195 309 377* 393* 421* 447* 462 709* 744* 770* 1106* 1206*
High	20	34 88 137 199 280 291 299* 300* 309 351 358 369(2) 370 375 382 392 429* 451 1119*

Example 9.5 (Histologic Grade–Ovarian Cancer)   Table 9.6 gives data from a cohort study of women with stage II or stage IIIA ovarian cancer, where the endpoint is progression of disease (Fleming et al., 1980). Survival time is measured in days and histologic grade is an indicator of the malignant potential of the tumor. These data have been analyzed by Breslow (1984b). The last death in the cohort is at day 462, after which the Kaplan–Meier survival curves remain horizontal until the end of follow-up—day 1206 (low grade) and day 1119 (high grade). In Figure 9.4(a) the Kaplan–Meier survival curves have been truncated at day 500. Until day 350 the two cohorts have almost identical survival, but then subjects with high-grade tumors experience much faster progression of disease. Table 9.7 gives point estimates and 95% confidence intervals for OR based on the asymptotic conditional, MH–RBG, and asymptotic unconditional approaches. The different methods produce similar results. The logrank test is $X_{mh}^2 = (16 - 10.67)^2/5.11 = 5.57$ ($p = .02$), and so there is evidence of a mortality difference between the two cohorts.

From the log–minus-log plots in Figure 9.4(b) it is clear that the proportional hazards assumption is not satisfied. With $\widehat{OR}_c = 3.09$, the test for linear trend is

$$X_t^2 = \frac{(26.92)^2}{665.3 - (46.56)^2/4.02} = 5.73 \qquad (p = .02)$$

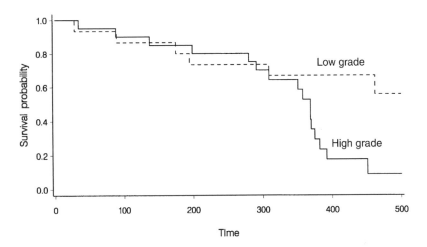

FIGURE 9.4(a)   Kaplan–Meier survival curves: Ovarian cancer cohort stratified by grade

TABLE 9.7   Odds Ratio Estimates and 95% Confidence
Intervals: Histologic Grade–Ovarian Cancer

Method	$\widehat{OR}$	$\underline{OR}$	$\overline{OR}$
Asymptotic conditional	3.09	1.16	8.21
MH–RBG	2.83	1.11	7.25
Asymptotic unconditional	3.32	1.20	9.16

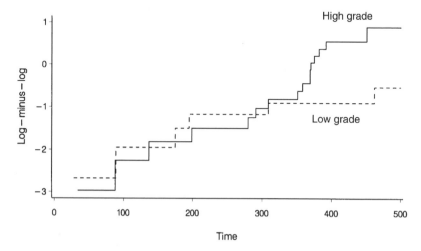

FIGURE 9.4(b)   Log–minus-log curves: Ovarian cancer cohort stratified by grade

which is consistent with the graphical assessment. The Breslow–Day test of homogeneity is $X_{bd}^2 = 23.80$ ($p = .20$, df $= 19$), which includes the correction term of .094. Despite the large magnitude of X_{bd}^2, with 19 degrees of freedom the p-value is also large. This illustrates that a test for linear trend may be able to detect heterogeneity across exposure categories that would be missed by a test of homogeneity. Since we have established that heterogeneity is present, the summary odds ratio estimates in Table 9.7 no longer have a meaningful epidemiologic interpretation (Section 5.5).

9.2.3   Methods for J ($2 \times I$) Tables

We now consider the analysis of censored survival data when the exposure variable is polychotomous. As in the dichotomous case, we begin by identifying the death times τ_j for all exposure categories combined. Let a_{ij} be the number of deaths in the ith exposure category at τ_j and let r_{ij} be the corresponding number of subjects at risk ($i = 1, 2, \ldots, I$; $j = 1, 2, \ldots, J$). The data layout is given in Table 9.8.

It was pointed out in Section 5.7 that the logrank test X_{mh}^2 for the J ($2 \times I$) setting satisfies the inequality $X_{pp}^2 \leq X_{mh}^2$, where X_{pp}^2 is given by (5.38). When censoring patterns do not differ greatly across exposure categories (subcohorts), X_{pp}^2 provides

TABLE 9.8 Observed Counts: Open Cohort Study

Survival	Exposure category							
	1	2	$\ldots$	i	$\ldots$	I		
dead	a_{1j}	a_{2j}	$\ldots$	a_{ij}	$\ldots$	a_{Ij}	m_{1j}
alive	$r_{1j} - a_{1j}$	$r_{2j} - a_{2j}$	$\ldots$	$r_{ij} - a_{ij}$	$\ldots$	$r_{Ij} - a_{Ij}$	m_{2j}
	r_{1j}	r_{2j}	$\ldots$	r_{ij}	$\ldots$	r_{Ij}	r_j

a good approximation to X_{mh}^2 (Crowley and Breslow, 1975). Peto and Peto (1972) consider

$$X_{\text{oe}}^2 = \sum_{i=1}^{I} \frac{(a_{i\bullet} - \hat{e}_{i\bullet})^2}{\hat{e}_{i\bullet}} \qquad (\text{df} = I - 1) \qquad (9.10)$$

for the analysis of censored survival data. Confusingly, (9.10) is sometimes referred to as the logrank test. Since there must be at least one death per stratum, $m_{1j} \geq 1$ and hence $\hat{g}_{ij} \leq \hat{e}_{ij}$, where $\hat{g}_{ij}$ is given by (5.39). It follows that $\hat{g}_{i\bullet} \leq \hat{e}_{i\bullet}$ and consequently that $X_{\text{oe}}^2 \leq X_{\text{pp}}^2$. Evidently, X_{oe}^2 will be close in value to X_{pp}^2 when the m_{1j} are small—that is, when there are few deaths at each death time. In summary, we have the inequalities $X_{\text{oe}}^2 \leq X_{\text{pp}}^2 \leq X_{\text{mh}}^2$.

Consider

$$X_{\text{t}}^2 = \frac{\left[\sum_{i=1}^{I} s_i (a_{i\bullet} - \hat{e}_{i\bullet})\right]^2}{\sum_{i=1}^{I} s_i^2 \hat{e}_{i\bullet} - \left(\sum_{i=1}^{I} s_i \hat{e}_{i\bullet}\right)^2 \Big/ \hat{e}_{\bullet\bullet}} \qquad (\text{df} = 1). \qquad (9.11)$$

As shown in Appendix E, (9.11) $\leq$ (5.41). It was pointed out in Section 5.7 that (5.41) $\leq$ (5.40). So we have the inequalities (9.11) $\leq$ (5.41) $\leq$ (5.40). As illustrated in the following example, for censored survival data, X_{oe}^2 and (9.11) are usually sufficiently accurate approximations to X_{mh}^2 and (5.40) for practical purposes.

Example 9.6 (Stage–Breast Cancer) Figure 9.5(a) shows the Kaplan–Meier survival curves for the breast cancer cohort stratified by stage. There is a clear pattern of increasing mortality for women with more advanced disease. The log–minus-log plots, shown in Figure 9.5(b), are generally supportive of the proportional hazards assumption, although proportionality for stage I is perhaps questionable. Table 9.9 gives the Mantel–Haenszel odds ratio estimates and RBG 95% confidence intervals with stage I taken as the reference category (where $\widehat{OR}_{\text{mh1}} = 1$). Note that the confidence intervals do not contain 1 and that there is only a small degree of overlap. The tests of association are $X_{\text{oe}}^2 = 52.05$, $X_{\text{pp}}^2 = 52.44$, and $X_{\text{mh}}^2 = 52.97$ ($p < .001$), and the tests for trend are (9.11) $= 38.03$, (5.41) $= 38.31$, and (5.40) $= 38.62$ ($p < .001$).

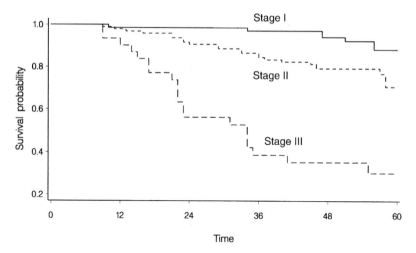

FIGURE 9.5(a)   Kaplan–Meier survival curves: Breast cancer cohort stratified by stage

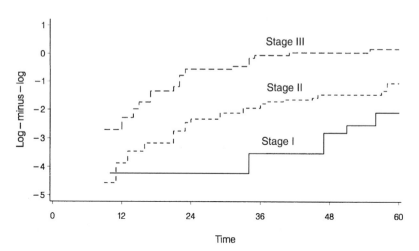

FIGURE 9.5(b)   Log–minus-log curves: Breast cancer cohort stratified by stage

TABLE 9.9   Mantel–Haenszel Odds Ratio Estimates and RBG 95% Confidence Intervals: Stage–Breast Cancer

Stage	$\widehat{OR}_{\mathrm{mh}i}$	$\underline{OR}_{\mathrm{mh}i}$	$\overline{OR}_{\mathrm{mh}i}$
II	2.89	1.17	7.14
III	14.53	5.42	38.95

9.2.4 Adjustment for Confounding

To this point, receptor level and stage have been analyzed separately as risk factors for breast cancer mortality. In Section 5.1 we gave arguments both for and against treating stage as a confounder of the risk relationship between receptor level and breast cancer survival. Similar reasoning applies here, with certain modifications in keeping with the definition of confounding in an open cohort study as outlined in Appendix G. If stage is not a confounder, any of the crude estimates in Table 9.5 can be used as an estimate of the overall hazard ratio. Here we use the term "crude" in the sense of not having been adjusted for stage. The fact that we have "adjusted" for time is implicit in our remarks.

If we regard stage as a confounder, we are led to consider Table 9.10, which gives stage-specific and stage-adjusted odds ratio estimates. The stage-specific estimates were calculated by stratifying by time for each stage separately, and the stage-adjusted estimates were obtained by stratifying jointly by these variables. It is noteworthy that the asymptotic unconditional estimates continue to be close to the asymptotic conditional and Mantel–Haenszel estimates despite the fact that sample sizes are sometimes relatively small. In particular, for stage III, there are only 15 and 17 individuals in the low and high receptor level categories, respectively. With stage taken to be a confounder, and assuming that there is homogeneity across stage, any of the adjusted estimates in Table 9.10 can be used as an estimate of the overall hazard ratio. We note that the adjusted estimates are smaller than the crude estimates in Table 9.5, suggesting that stage may be an important confounder.

In Table 9.10 the estimated odds ratios for stage III are substantially larger than the estimates for stages I and II. This suggests that the above homogeneity assumption may not be valid and that stage is an effect modifier of the association between receptor level and breast cancer mortality. This observation should be assessed formally using a test of homogeneity. In general it would be desirable to have a range of techniques that can be applied when the data have been stratified by two or more variables in addition to time. Such methods are available, but the formulas are cumbersome and will not be presented here. In Chapter 10, methods are described for analyzing censored survival data using the Poisson distribution. The Poisson formulas are much less complicated than those based on the odds ratio approach, and stratification by two confounders is readily handled.

In the preceding analysis we treated stage as a confounder of the risk relationship between receptor level and breast cancer survival. A corresponding analysis consid-

TABLE 9.10 Odds Ratio Estimates Stratified by Stage: Receptor Level–Breast Cancer

Method	Stage			Adjusted
	I	II	III	
Asymptotic conditional	2.26	2.09	3.02	2.44
Mantel–Haenszel	2.27	2.10	3.20	2.53
Asymptotic unconditional	2.31	2.11	3.26	2.52

TABLE 9.11   Mantel–Haenszel Odds Ratio Estimates
Stratified by Receptor Level: Stage–Breast Cancer

	Receptor level		
Stage	Low	High	Adjusted
II	2.51	3.00	2.83
III	20.13	10.77	14.24

ers receptor level to be a confounder of the risk relationship between stage and breast cancer survival. Table 9.11 gives receptor level-specific and receptor level-adjusted Mantel–Haenszel odds ratio estimates with stage I as the reference category. There is evidence of heterogeneity across receptor level categories. The adjusted estimates of Table 9.11 are close in value to the crude estimates of Table 9.9, suggesting that receptor level may not be a confounder.

In addition to stratifying separately by stage and receptor level, it is of interest to stratify by these variables jointly, which results in six receptor level–stage categories. Table 9.12 gives the resulting Mantel–Haenszel odds ratio estimates where the high receptor level–stage I category has been chosen as the reference category. For stages I and II there is roughly a doubling of the estimates as we move from high to low receptor level, but for stage III the estimate approximately triples. This suggests that there may be an interaction between receptor level and stage. Figure 9.6 shows the six Kaplan–Meier survival curves corresponding to this stratification, where the curves are labeled as follows: (stage, receptor level). The appearance is a bit confusing due to crossing-over of curves, especially for stage I subjects. Overall the pattern is broadly consistent with the estimates in Table 9.12.

9.2.5   Recommendations

In addition to the research cited in Section 5.6 in connection with closed cohort studies, there is further research on the use of odds ratio methods for analyzing censored survival data (Peto, 1972; Peto and Peto, 1972; Lininger et al., 1979; Bernstein et al., 1981; Crowley et al., 1982; Robins et al., 1986). Following Chapter 5, we recommend the MH–RBG methods for the analysis of censored survival data. The validity of the odds ratio approach rests on the proportional hazards assumption, which can

TABLE 9.12   Mantel–Haenszel Odds Ratio Estimates:
High Receptor Level–Stage I as Reference Category

	Receptor level	
Stage	High	Low
I	1.0	2.27
II	3.00	6.35
III	10.77	32.55

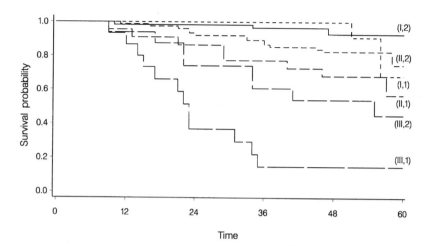

FIGURE 9.6 Kaplan–Meier survival curves: Breast cancer cohort stratified by stage and receptor level

be assessed using either the test for linear trend or the less objective, but possibly more informative, graphical method.

9.3 ACTUARIAL METHOD

In some cohort studies, exact death times and censoring times are not available. This is often the case with large surveillance systems such as cancer registries, where patient visits are scheduled on a routine basis. For those individuals who die or are censored between appointments, all that may be known is that they survived to the last follow-up time. In this case we say that the survival times are interval-censored and that the data are grouped. The actuarial method is a classical approach to the analysis of interval-censored survival data which has its roots in life table analysis.

The actuarial method differs from the Kaplan–Meier method in that intervals are determined by the investigator rather than based on observed death times. Let $\tau_0 = 0$, let τ_{J+1} be the maximum observation time, and let $\tau_1 < \tau_2 < \cdots < \tau_J$ be J intermediate time points. The actuarial approach begins by partitioning the period of follow-up into $J + 1$ intervals: $[\tau_0, \tau_1), [\tau_1, \tau_2), \ldots, [\tau_j, \tau_{j+1}), \ldots, [\tau_{J-1}, \tau_J), [\tau_J, \tau_{J+1}]$. As before, we refer to $[\tau_j, \tau_{j+1})$ as the jth interval. Let a_j and c_j be the numbers of deaths and censored observations in the jth interval, respectively $(j = 0, 1, \ldots, J)$. With interval-censored data we have no knowledge of the precise death times or censoring times, but this does not affect the counts a_j and c_j. Although the definitions of a_j and c_j are formally the same as those used in the Kaplan–Meier setting, a difference here is that deaths in the jth interval are permitted to occur throughout the interval rather than only at τ_j. A further difference is that a_0 is not necessarily equal to 0. The jth risk set is defined to be the group of subjects surviving to at least τ_j $(j = 0, 1, \ldots, J)$. We adopt the convention that subjects who die at τ_j

are included in the risk set. Let r_j denote the number of subjects in the jth risk set ($j = 0, 1, \ldots, J$), and denote by r_{J+1} the number of subjects who survive to τ_{J+1}. As before, we define $c'_j = c_j$ for $j < J$ and $c'_J = c_J - r_{J+1}$.

In order to estimate the survival curve it is necessary to make certain assumptions about its functional form and the distribution of censoring times. Specifically, we assume that $S(t)$ is a continuous function that is linear on each of the intervals. In other words, the graph of $S(t)$ is a series of line segments that meet at values corresponding to the endpoints of intervals. We also assume that censoring for reasons other than survival to τ_{J+1} takes place uniformly throughout each interval. Consequently, all censoring, except that due to survival to τ_{J+1}, occurs on average at the midpoint of each interval. Let p_j denote the conditional probability of surviving to τ_{j+1}, given survival to τ_j, and let $q_j = 1 - p_j$ be the corresponding conditional probability of dying ($j = 0, 1, \ldots, J$).

The actuarial approach to estimating the survival function proceeds along the lines of the Kaplan–Meier method. The denominator of $\hat{q}_j$ is r_j, and the numerator is defined to be the total number of deaths in the jth interval. The latter quantity is the sum of the a_j observed deaths plus the number of unobserved deaths among the c'_j censored subjects. With the preceding assumptions about the survival curve and censoring patterns, the number of unobserved deaths is estimated to be $(\hat{q}_j/2)c'_j$. So an estimate of q_j is $\hat{q}_j = [a_j + (\hat{q}_j/2)c'_j]/r_j$, which can be solved for $\hat{q}_j$ to give

$$\hat{q}_j = \frac{a_j}{r_j - (c'_j/2)} \tag{9.12}$$

($j = 0, 1, \ldots, J$). The denominator $r_j - (c'_j/2)$ will be denoted by r'_j and referred to as the "effective" sample size. This terminology is appropriate since r'_j can be thought of as the number of subjects who would need to be at risk in the absence of censoring in order to give the estimate (9.12). Note that r'_j may not be an integer. With $\hat{p}_j = 1 - \hat{q}_j$, we have the estimates

$$\hat{S}_j = \hat{p}_0 \hat{p}_1 \cdots \hat{p}_{j-1} \tag{9.13}$$

$$\widehat{\mathrm{var}}(\hat{S}_j) = (\hat{S}_j)^2 \sum_{i=0}^{j-1} \frac{\hat{q}_i}{\hat{p}_i r'_i}$$

and

$$\widehat{\mathrm{var}}[\log(-\log \hat{S}_j)] = \frac{1}{(\log \hat{S}_j)^2} \sum_{i=0}^{j-1} \frac{\hat{q}_i}{\hat{p}_i r'_i}$$

($j = 1, 2 \ldots, J + 1$). A graph of the actuarial survival curve is obtained by plotting the $\hat{S}_j$ and then joining these points by straight line segments.

Example 9.7 (Receptor Level–Breast Cancer)   Table 9.13 gives the actuarial analysis of the breast cancer data after stratifying by receptor level. The period of

TABLE 9.13 Actuarial Analysis: Receptor Level–Breast Cancer

j	τ_j	a_j	r_j	c'_j	r'_j	$\hat{p}_j$	$\hat{S}_j$	$\underline{S}_j$	$\overline{S}_j$
0	0	5	199	2	198.0	.975	1.0	—	—
1	12	17	192	2	191.0	.911	.975	.940	.989
2	24	11	173	1	172.5	.936	.888	.835	.925
3	36	10	161	1	160.5	.938	.831	.771	.877
4	48	6	150	132	84.0	.929	.780	.715	.832
5	60	—	12	—	—	—	.724	.648	.786

follow-up has been divided into 12-month blocks, and the 95% confidence intervals were estimated using the Kalbfleisch–Prentice method. Figure 9.7 shows the graph of the actuarial survival curve and the 95% confidence intervals. Not surprisingly, Figures 9.7 and 9.2 are similar.

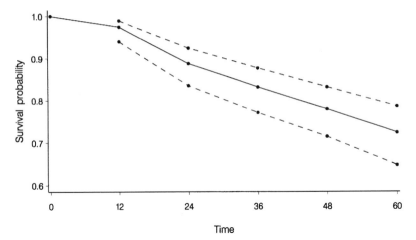

FIGURE 9.7 Actuarial survival curve and Kalbfleisch–Prentice 95% confidence intervals: Breast cancer cohort

CHAPTER 10

Poisson Methods for Censored Survival Data

The Kaplan–Meier method is based on relatively few assumptions; in particular, nothing is specified regarding the functional form of either the survival function or the hazard function. Censoring is assumed to be uninformative, but this is a feature of virtually all of the commonly used methods of survival analysis. Since so little structure is imposed, it is appropriate to view a Kaplan–Meier survival curve as a type of scatter plot of censored survival data. The appearance of a Kaplan–Meier curve can be used to form ideas about the nature of the underlying survival function and hazard function, in much the same way as a scatter plot is used as a visual aid in linear regression.

Despite these advantages, there are difficulties with the Kaplan–Meier approach. Kaplan–Meier curves are not designed to "smooth" the data while accounting for random variation in the way that a linear regression line is fitted to points in a scatter plot. As a result, Kaplan–Meier survival curves can be erratic in appearance and sensitive to small changes in survival times and censoring patterns, especially when the number of deaths is small. The Kaplan–Meier survival curves for the six receptor level–stage strata shown in Figure 9.6 are relatively well-behaved, but it is easy to imagine how complicated such a graph might otherwise be.

In this chapter we describe parametric methods of survival analysis based on the Weibull, exponential, and Poisson distributions. The computations required by the exponential and Poisson models are relatively straightforward, and the results are readily interpreted. However, this convenience is gained at the expense of having to make strong assumptions about the functional form of the hazard function, a decision that needs to be justified in any application.

10.1 POISSON METHODS FOR SINGLE SAMPLE SURVIVAL DATA

In theory, a hazard function can have almost any functional form. The estimated hazard function for Canadian females in 1990–1992 shown in Figure 8.2(c) has quite a complicated shape. This is to be expected because the cohort was followed over the entire life cycle and, as is well known, mortality risk is highly dependent on age. There may be a degree of systematic or random error in Figure 8.2(c), but Statistics Canada vital statistics data are very reliable and the sample size is so large that the complicated appearance must be accepted as a realistic depiction of the underlying hazard function. In practice, most cohort studies have a relatively small sample size and a fairly short period of follow-up. This means that the period of observation will usually be too short for the hazard function to exhibit much variation over time, and the sample size will be too small for it to be possible to discern subtle changes in the hazard function, even if they should be present. As a consequence, it is usually appropriate in epidemiologic studies to model the hazard function using relatively uncomplicated functional forms. Two of the most widely used are the Weibull and exponential distributions (Kalbfleisch and Prentice, 1980; Lawless, 1982; Cox and Oakes, 1984; Lee, 1992; Collett, 1994; Klein and Moeschberger, 1997).

10.1.1 Weibull and Exponential Distributions

The Weibull distribution has the survival function $S(t) = \exp[-(\lambda t)^\alpha]$ and hazard function $h(t) = \alpha\lambda(\lambda t)^{\alpha-1}$. Here λ and α are parameters satisfying the conditions $\lambda > 0$ and $\alpha > 0$. We refer to λ as the rate parameter and to α as the shape parameter. Figure 10.1(a) shows graphs of the hazard function for $\lambda = 1$ and $\alpha = .5, 1, 1.5$, and 3. Setting $\lambda = 1$ reflects the choice of time units but does not influence the basic shapes of the curves. When $\alpha = 1$, $h(t)$ is constant; when $\alpha < 1$, $h(t)$ is a decreasing function of time; and when $\alpha > 1$, $h(t)$ is increasing. The corresponding survival curves are shown in Figure 10.1(b). The Weibull distribution is applicable to a range of situations commonly encountered in epidemiology. For example, consider a cohort of surgical patients who are being monitored after having just undergone

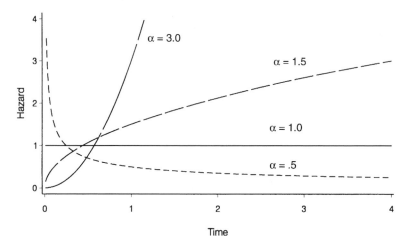

FIGURE 10.1(a)   Weibull hazard functions for selected values of α, with $\lambda = 1$

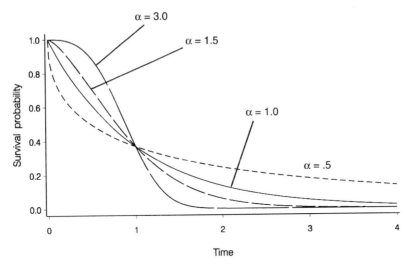

FIGURE 10.1(b) Weibull survival functions for selected values of α, with $\lambda = 1$

major surgery. Suppose that for the first few days after surgery the mortality risk is high, but after that it gradually declines. In this case a Weibull distribution with $\alpha < 1$ would be appropriate. As another example, consider a cohort of cancer patients who are undergoing long-term follow-up after entering remission. Suppose that for the first few years the risk of relapse is relatively low, but as time progresses more and more patients have a recurrence. In this case, a Weibull distribution with $\alpha > 1$ would be a reasonable choice. The ovarian cancer cohort with high-grade disease in Figure 9.4(a) exhibits the latter type of survival experience.

Consider an open cohort study of r subjects and, in the notation of Section 8.1, let (t_i, δ_i) be the observation for the ith subject ($i = 1, 2, \ldots, r$). Maximum likelihood methods can be used to estimate λ and α from these data but, except when $\alpha = 1$, closed-form expressions are not available. When $\alpha = 1$ the Weibull distribution simplifies to the exponential distribution, in which case $S(t) = e^{-\lambda t}$ and $h(t) = \lambda$. The exponential distribution rests on the assumption that the hazard function is constant over the entire period of follow-up. This assumption is evidently a very strong one and will often be unrealistic. However, when the sample size is small and the period of follow-up is relatively short, the exponential distribution provides a useful approach to analyzing censored survival data. The attraction of the exponential distribution is that the parameter λ is easily estimated, as shown below. Since the exponential hazard function has the same value at any point during follow-up, the exponential distribution is said to be "memoryless."

Let d denote the number of deaths in the cohort. This represents a change of notation from Chapter 9 where we used the symbol a. We adopt this convention as a way of distinguishing the formulas based on the exponential and Poisson distributions from those based on the binomial approach. It follows immediately from the definition of δ_i that $d = \sum_{i=1}^{r} \delta_i$. By definition, the ith subject was under observation for t_i time units. Therefore the total amount of time that the entire cohort was under observation is $n = \sum_{i=1}^{r} t_i$, which we refer to as the amount of person-time. For example, when time is measured in years or months, n is said to be the number of person-years or person-months, respectively. Observe that because n is defined to be a sum across all cohort members, the contributions of individual subjects are effectively lost. Consequently, 1 person followed for n years, and n individuals followed for 1 year, will both result in n person-years of observation. This is related to the memoryless property mentioned above.

Consider the exponential distribution with parameter λ. For subject i, $S(t_i) = e^{-\lambda t_i}$ and $h(t_i) = \lambda$, and so from (8.5) the unconditional likelihood is

$$L(\lambda) = \prod_{i=1}^{r} e^{-\lambda t_i} \lambda^{\delta_i} = e^{-\lambda n} \lambda^d. \tag{10.1}$$

From (10.1), the maximum likelihood estimates of λ, $\text{var}(\hat{\lambda})$ and $S(t) = e^{-\lambda t}$ are

$$\hat{\lambda} = \frac{d}{n}$$

$$\widehat{\text{var}}(\hat{\lambda}) = \frac{d}{n^2} = \frac{\hat{\lambda}}{n} \tag{10.2}$$

and

$$\hat{S}(t) = \exp(-\hat{\lambda}t).$$

For example, based on Figure 8.1(b), $\hat{\lambda} = 2/35 = .057$, $\widehat{\text{var}}(\hat{\lambda}) = 2/(35)^2 = (.040)^2$, and $\hat{S}(t) = \exp(-.057t)$. In Chapter 12 it is pointed out that the term "rate" is used throughout epidemiology to denote a variety of different types of parameters. To the extent that established conventions permit, we will restrict the use of this term to parameters that have an interpretation as follows: number of events of a given type, divided by the corresponding amount of person-time. The rate parameter λ satisfies this condition and so, in the exponential context, λ will be referred to as a hazard rate.

It is important not to confuse rates with probabilities. One major difference between these two quantities is that a rate has the units "per unit time" whereas a probability does not have any units. The absolute magnitude of a rate depends on the particular units of time chosen. Suppose that $d = 5$ (persons) and $n = 1$ person-year, in which case $\hat{\lambda} = 5$ "per year." This would more often be expressed as 5 "deaths per person-year" or 5 "deaths per person per year." Since 1 person-year is the same as .01 person-centuries, it is equally true that $\hat{\lambda} = 500$ "per century." So a rate can be made arbitrarily large or small in absolute terms by a suitable choice of time units, while a probability is always between 0 and 1.

10.1.2   Assessment of the Exponential Assumption

Graphical Assessment
The validity of the exponential assumption can be assessed graphically by plotting the estimated exponential survival curve $\hat{S}(t) = \exp(-\hat{\lambda}t)$ and the Kaplan–Meier survival curve and deciding subjectively whether the latter appears to be exponential in appearance. In a sense we are using the Kaplan–Meier curve as the "observed" survival curve and determining whether the "fitted" survival curve from the exponential model is adequate.

Cox–Oakes Test of Exponentiality
The graphical method is usually quite revealing but can be criticized for lacking objectivity. Cox and Oakes (1984, p. 43) describe a test of exponentiality based on the Weibull distribution. The concept is similar to that used to develop the test for linear trend in Section 4.6. Under $H_0 : \alpha = 1$, the Weibull distribution simplifies to the exponential distribution, and in this case the estimate of λ is $\hat{\lambda}_0 = d/n$. We can think of the ith subject as being equivalent to a cohort with a sample size of 1. From this perspective, δ_i is the number of observed deaths, t_i is the amount of person-time, and $\hat{e}_i = \hat{\lambda}_0 t_i$ is the expected number of deaths under the null hypothesis. Note that $\hat{e}_\bullet = \sum_{i=1}^{r} \hat{\lambda}_0 t_i = \hat{\lambda}_0 n = d$. Let $\hat{S}(t) = \exp[-(\hat{\lambda}t)^{\hat{\alpha}}]$ be the "best-fitting" Weibull survival curve for the observations (t_i, δ_i) and let $s_i = \log(\hat{\lambda}_0 t_i)$. The score test of $H_0 : \alpha = 1$, which will be referred as the Cox–Oakes test of exponentiality, is

$$X_{\text{co}}^2 = \frac{\left[d + \sum_{i=1}^{r} s_i(\delta_i - \hat{e}_i) \right]^2}{d + \sum_{i=1}^{r} s_i^2 \hat{e}_i - \left(\sum_{i=1}^{r} s_i \hat{e}_i \right)^2 \Big/ \hat{e}_\bullet} \qquad (\text{df} = 1).$$

Large values of X_{co}^2 provide evidence against the exponential assumption. It is important to appreciate that not rejecting H_0 is not the same as saying that survival is exponential. The correct interpretation is as follows: Given that we have decided to fit the data using a Weibull model, not rejecting H_0 means there is no reason not to choose the exponential model (which is a particular type of Weibull model). This means that there should be grounds for considering a Weibull model in the first place, an issue that can be addressed by examining the Kaplan–Meier curve and making a subjective judgment.

Example 10.1 (Breast Cancer)   The data for this example are taken from Table 9.1. For the cohort of breast cancer patients, $d = 49$ and $n = 9471$. Based on the exponential model, $\hat{\lambda} = 49/9471 = 5.17 \times 10^{-3}$ (deaths per person-month); and for the Wiebull model, $\hat{\lambda} = 8.01 \times 10^{-3}$ and $\hat{\alpha} = 1.49$. Figure 10.2(a) shows the exponential and Kaplan–Meier survival curves for these data. There are few deaths in the first 12 months of follow-up, and this causes the Kaplan–Meier curve to plateau before beginning a gradual decline. Other than this, the exponential model provides a reasonably good fit to the Kaplan–Meier survival curve. Figure 10.2(b) shows the Weibull and Kaplan–Meier survival curves. The Weibull model fits the data slightly better than the exponential model, especially during the first 12 months.

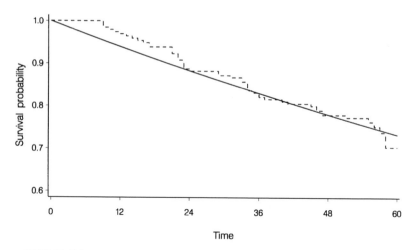

FIGURE 10.2(a)   Exponential and Kaplan–Meier survival curves: Breast cancer cohort

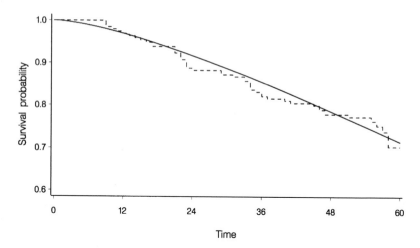

FIGURE 10.2(b)   Weibull and Kaplan–Meier survival curves: Breast cancer cohort

The Cox–Oakes test of exponentiality is $X^2_{co} = 5.84$ ($p = .02$), which provides moderate evidence that the exponential assumption may not be satisfied. Despite this finding, it might be argued that the exponential model provides a fit that is "good enough" for practical purposes. This is the difference between "statistical significance" and what is referred to in the medical literature as "clinical significance" (Sackett et al., 1985). In the present case it needs to be decided on substantive grounds whether the low mortality risk during the first 12 months is a meaningful finding (clinically significant) or can be ignored. Since it is reasonable that mortality risk might be low around the time of registration—for example, as a result of re-

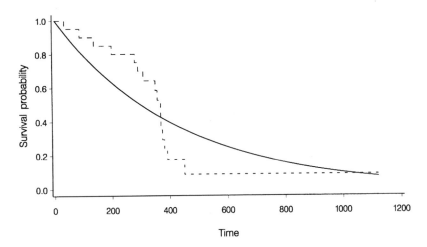

FIGURE 10.3(a)   Exponential and Kaplan–Meier survival curves: Ovarian cancer cohort, high grade

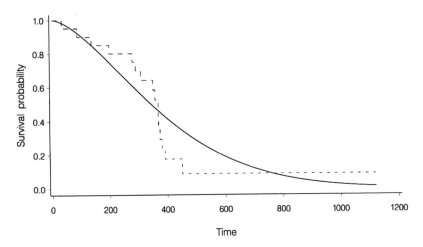

FIGURE 10.3(b)   Weibull and Kaplan–Meier survival curves: Ovarian cancer cohort, high grade

cent treatment for breast cancer—an argument can be made for adopting the Weibull model.

Example 10.2 (Ovarian Cancer: High Grade)   The data for this example are taken from Table 9.6, where we restrict the analysis to high-grade tumors. In Figure 9.4(a), there is relatively low mortality until about day 350, after which there is a sharp drop in survival. Based on the exponential model, $\hat{\lambda} = 16/6902 = 2.32 \times 10^{-3}$ (deaths per person-day); and for the Weibull model, $\hat{\lambda} = 2.30 \times 10^{-3}$ and $\hat{\alpha} = 1.58$. Figures 10.3(a) and 10.3(b) compare the exponential and Weibull survival curves,

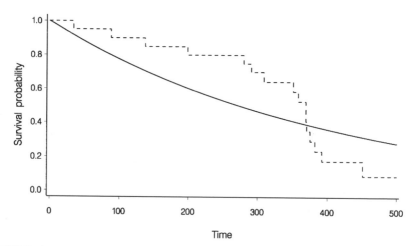

FIGURE 10.4(a)   Exponential and Kaplan–Meier survival curves: Ovarian cancer cohort, high grade, recoded data

respectively, to the Kaplan–Meier survival curve. Note that, as opposed to Figure 9.4(a), the horizontal axis is not truncated at day 500. As can be seen, the exponential model does not fit the data at all well, while the Weibull model provides only a slight improvement. Perhaps surprisingly, the Cox–Oakes test of exponentiality is $X^2_{co} = 3.65$ ($p = .06$), and so there is little evidence that the exponential assumption is not satisfied. The explanation for this finding is that the family of Weibull models provides such a poor fit in general that the exponential model cannot be rejected as a possibility.

The last death in this cohort occurred at day 451, yet follow-up continued for one individual until day 1196, thus creating a long tail on the right. For illustrative purposes the data were reanalyzed under the assumption that follow-up ended at day 500. So the survival time $t = 1196$ was recoded to $t = 500$. With this revision to the data, for the exponential model, $\hat{\lambda} = 16/6283 = 2.55 \times 10^{-3}$; and for the Wiebull model, $\hat{\lambda} = 2.67 \times 10^{-3}$ and $\hat{\alpha} = 2.55$. The resulting Kaplan–Meier, exponential, and Weibull survival curves are shown in Figures 10.4(a) and 10.4(b). For these hypothetical data the Weibull model provides a fairly good fit. The Cox–Oakes test of exponentiality is $X^2_{co} = 7.20$ ($p = .01$), which suggests that the much larger p-value in the earlier analysis was due to the poor fit of the best-fitting Weibull model.

Consider a cohort with hazard function $h(t)$ and survival function $S(t)$, and consider an exposure variable with K categories. Suppose that the cohort is stratified according to exposure category at the start of follow-up and that $h_k(t)$ is the hazard function for the kth subcohort ($k = 1, 2, \ldots, K$). Denote by p_k the proportion of the overall cohort in the kth subcohort at the start of follow-up. As demonstrated in Appendix G,

$$h(t) = \frac{\sum_{k=1}^{K} p_k S_k(t) h_k(t)}{\sum_{k=1}^{K} p_k S_k(t)}. \tag{10.3}$$

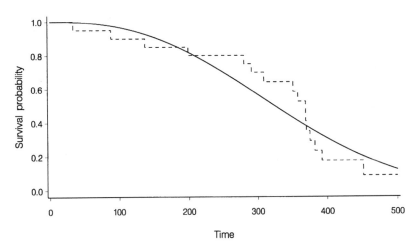

FIGURE 10.4(b) Weibull and Kaplan–Meier survival curves: Ovarian cancer cohort, high grade, recoded data

Therefore, at each follow-up time the overall hazard is a weighted average of stratum-specific hazards, where the weights are functions of time. As observed by Vaupel and Yashin (1985), the fact that the weights are time-dependent can lead to surprising consequences. This is illustrated for the case of two strata. Assume that the stratum-specific survival functions are exponential with hazard rates λ_1 and λ_2. Since $p_1 + p_2 = 1$, (10.3) becomes

$$h(t) = \frac{p_1 e^{-\lambda_1 t} \lambda_1 + (1 - p_1) e^{-\lambda_2 t} \lambda_2}{p_1 e^{-\lambda_1 t} + (1 - p_1) e^{-\lambda_2 t}}. \tag{10.4}$$

This shows that even though the stratum-specific hazard functions are exponential, the overall hazard function is not. However, when λ_1 and λ_2 are sufficiently small (death is a rare event), $e^{-\lambda_1 t}$ and $e^{-\lambda_2 t}$ will be close to 1 and so $h(t)$ will be approximately equal to $p_1 \lambda_1 + (1 - p_1) \lambda_2$, a constant. We illustrate these observations using a graphical approach. For this discussion we assume, without loss of generality, that the entire period of follow-up is a single time unit so that $0 \leq t \leq 1$. It can be shown that (10.4) is a strictly decreasing function of time, and so $h(t)$ has a maximum value when $t = 0$; that is, $h(0) = p_1 \lambda_1 + (1 - p_1) \lambda_2$. Since we are primarily interested in the shape of (10.4), it is sufficient to consider $h(t)/h(0)$. We further specialize by setting $p_1 = .5$ and $\lambda_1 = 2\lambda_2$ so that (10.4) has the single parameter λ_2. Figure 10.5 shows graphs of $h(t)/h(0)$ for $\lambda_2 = .1$, 1, and 10. As can be seen, for $\lambda_2 = .1$ the curve is virtually a horizontal line (constant), but this is not true for $\lambda_2 = 1$ and 10.

10.1.3 Poisson Distribution

Consider a cohort in which survival time follows an exponential distribution with hazard rate λ. As before, let (t_i, δ_i) be the observation for the ith subject ($i =$

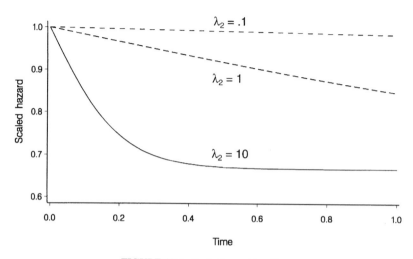

FIGURE 10.5   Scaled hazard functions

$1, 2, \ldots, r$). In most applications, both $d = \sum_{i=1}^{r} \delta_i$ and $n = \sum_{i=1}^{r} t_i$ are random variables. This is because, at the start of follow-up, it is usually not known how many deaths there will be in the cohort and how much person-time will be experienced. Let t_i' be the maximum observation time for the ith subject as determined by the study design. For example, in a study with staggered entry as described in Section 8.1, t_i' is the time from the beginning of follow-up of the ith subject until the end of the study. Since each t_i' is a known constant, so is $n' = \sum_{i=1}^{r} t_i'$. From $t_i \leq t_i'$ it follows that $n \leq n'$. We now make two crucial assumptions: Death is a rare event, and there is little censoring except possibly due to survival to the end of the study. Under these conditions, n is approximately equal to n' and so n can be treated as a constant. To illustrate with a simple example, consider a closed cohort study in which 1000 subjects are followed for up to 10 years with death from any cause as the endpoint. Suppose that there are only five deaths in the cohort, in which case n satisfies $9950 \leq n \leq 10,000$. Even if all deaths occur just after the start of follow-up, n will still be close to $n' = 10,000$. With n assumed to be constant, it can be shown that d is a Poisson random variable with parameter $\nu = \lambda n$ (Chiang, 1980, §8.2; Grimmett and Stirzaker, 1982, §6.8). The fact that d is Poisson is not unreasonable because, as noted in Section 1.1.2, the Poisson distribution is used to model counts of rare events. Berry (1983) and Breslow and Day (1987, §4.2) provide more detailed arguments which justify treating d as a Poisson random variable.

Recall from Section 1.1.2 that the Poisson probability function is

$$P(D = d | \nu) = \frac{e^{-\nu} \nu^d}{d!} \tag{10.5}$$

and that $E(D) = \text{var}(D) = \nu$. In view of the above remarks we now reparameterize (10.5) by setting $\nu = \lambda n$ to obtain

$$P(D = d|\lambda) = \frac{e^{-\lambda n}(\lambda n)^d}{d!}. \tag{10.6}$$

With this parameterization we say that D is a Poisson random variable with parameters (λ, n).

As pointed out in Example 1.8, when specifying the likelihood it is appropriate to ignore terms that do not involve the parameter of interest. Accordingly, the likelihood based on (10.6) is

$$L(\lambda) = e^{-\lambda n}\lambda^d.$$

This is the same as (10.1), the likelihood for the exponential distribution (Holford, 1980). It follows that the maximum likelihood estimates of λ and $\text{var}(\hat{\lambda})$ based on (10.6) are the same as those derived using the exponential approach; that is, $\hat{\lambda} = d/n$ and $\widehat{\text{var}}(\hat{\lambda}) = d/n^2$. Accordingly, it does not matter whether we treat λ as an exponential or a Poisson parameter. For the remainder of this chapter we focus on the Poisson interpretation. As will be seen, the resulting formulas exhibit a striking resemblance to those based on the binomial and hypergeometric distributions presented in Chapters 3–5 (Breslow and Day, 1980, 1987). Consequently, many of the remarks that are relevant to the Poisson approach have essentially been covered in earlier discussions. This makes it possible to describe what follows more briefly than would otherwise be the case.

10.1.4 Exact Methods for a Single Sample

Hypothesis Test
To perform an exact test of the hypothesis $H_0 : \lambda = \lambda_0$ we define lower and upper tail probabilities as follows:

$$P(D \le d|\lambda_0) = \exp(-\lambda_0 n) \sum_{x=0}^{d} \frac{(\lambda_0 n)^x}{x!} \tag{10.7}$$

and

$$P(D \ge d|\lambda_0) = 1 - \exp(-\lambda_0 n) \sum_{x=0}^{d-1} \frac{(\lambda_0 n)^x}{x!}. \tag{10.8}$$

The two-sided p-value is calculated using either the cumulative or doubling method as described in Section 3.1 for the binomial distribution.

Example 10.3 Let $d = 2$ and $n = 10$, and consider $H_0 : \lambda_0 = .4$. The Poisson distribution with parameters $(.4, 10)$ is shown in Table 10.1 for $d \le 12$. Based on the doubling method the p-value is $p = 2(.238) = .476$.

TABLE 10.1 Probability Function (%) for the Poisson Distribution with Parameters (.4, 10).

d	$P(D = d \mid .4)$	$P(D \leq d \mid .4)$	$P(D \geq d \mid .4)$
0	1.83	1.83	100
1	7.33	9.16	98.17
2	14.65	23.81	90.84
3	19.54	43.35	76.19
4	19.54	62.88	56.65
5	15.63	78.51	37.12
6	10.42	88.93	21.49
7	5.95	94.89	11.07
8	2.98	97.86	5.11
9	1.32	99.19	2.14
10	.53	99.72	.81
11	.19	99.91	.28
12	.06	99.97	.09
$\vdots$	$\vdots$	$\vdots$	$\vdots$

Confidence Interval

A $(1 - \alpha) \times 100\%$ confidence interval for λ is obtained by solving the equations

$$\frac{\alpha}{2} = P(D \geq d \mid \underline{\lambda}) = 1 - \exp(-\underline{\lambda}n) \sum_{x=0}^{d-1} \frac{(\lambda n)^x}{x!} \tag{10.9}$$

and

$$\frac{\alpha}{2} = P(D \leq d \mid \overline{\lambda}) = \exp(-\overline{\lambda}n) \sum_{x=0}^{d} \frac{(\overline{\lambda} n)^x}{x!} \tag{10.10}$$

for $\underline{\lambda}$ and $\overline{\lambda}$.

Example 10.4 Let $d = 2$ and $n = 10$. From

$$.025 = 1 - \exp(-10\underline{\lambda}) \sum_{x=0}^{1} \frac{(10\underline{\lambda})^x}{x!}$$

$$= 1 - \exp(-10\underline{\lambda})(1 + 10\underline{\lambda})$$

and

$$.025 = \exp(-10\overline{\lambda}) \sum_{x=0}^{2} \frac{(10\overline{\lambda})^x}{x!}$$

$$= \exp(-10\overline{\lambda})(1 + 10\overline{\lambda} + 50\overline{\lambda}^2)$$

a 95% confidence interval for λ is [.024, .723].

10.1.5   Asymptotic Methods for a Single Sample

Confidence Interval

Applying arguments used in the binomial case to (10.9) and (10.10), an implicit $(1 - \alpha) \times 100\%$ confidence interval for λ is obtained by solving the equation

$$\frac{(d - \lambda n)^2}{\lambda n} = (z_{\alpha/2})^2$$

using the quadratic formula. The result is

$$[\underline{\lambda}, \overline{\lambda}] = \frac{-b \pm \sqrt{b^2 - 4ac}}{2a}$$

where

$$a = n^2$$
$$b = -n \left[2d + (z_{\alpha/2})^2 \right]$$
$$c = d^2.$$

An explicit $(1 - \alpha) \times 100\%$ confidence interval for λ is

$$[\underline{\lambda}, \overline{\lambda}] = \frac{d}{n} \pm \frac{z_{\alpha/2}\sqrt{d}}{n} = \frac{d}{n} \left(1 \pm \frac{z_{\alpha/2}}{\sqrt{d}} \right). \tag{10.11}$$

Hypothesis Test

Under the null hypothesis $H_0 : \lambda = \lambda_0$, the maximum likelihood estimates of the mean and variance of $\hat{\lambda}$ are $E_0(\hat{\lambda}) = \lambda_0$ and $\text{var}_0(\hat{\lambda}) = \lambda_0/n$. A test of H_0 is

$$X^2 = \frac{(\hat{\lambda} - \lambda_0)^2}{\lambda_0/n} = \frac{(d - \lambda_0 n)^2}{\lambda_0 n} \qquad (\text{df} = 1). \tag{10.12}$$

Example 10.5   Table 10.2 gives 95% confidence intervals for λ where, in each case, $\hat{\lambda} = .2$. The performance of the methods is similar to what was observed in Table 3.2. The implicit method produces results that are reasonably close to the exact

TABLE 10.2   95% Confidence Intervals (%) for λ

Method	$d = 2$ $n = 10$		$d = 5$ $n = 25$		$d = 10$ $n = 50$	
	$\underline{\lambda}$	$\overline{\lambda}$	$\underline{\lambda}$	$\overline{\lambda}$	$\underline{\lambda}$	$\overline{\lambda}$
Exact	2.42	72.25	6.49	46.68	9.59	36.78
Implicit	5.48	72.93	8.54	46.82	10.86	36.82
Explicit	−7.72	47.72	2.47	37.53	7.60	32.40

TABLE 10.3 p-Values for Hypothesis Tests of $H_0 : \lambda = .4$

Method	$d = 2$ $n = 10$	$d = 5$ $n = 25$	$d = 10$ $n = 50$
Exact[a]	.453	.116	.019
Asymptotic	.317	.114	.025

[a]Cumulative

method for $d = 5$ and $d = 10$, while the explicit method leaves something to be desired, especially for $d = 2$.

Example 10.6   Table 10.3 gives p-values for hypothesis tests of $H_0 : \lambda = .4$ where, in each case, $\hat{\lambda} = .2$. The asymptotic and exact p-values are reasonably close in value.

Example 10.7 (Breast Cancer)   From Example 10.1, the estimated death rate for the entire breast cancer cohort is $\hat{\lambda} = 49/9471 = 5.17 \times 10^{-3}$ (deaths per person-month). Based on the implicit method, the 95% confidence interval for λ is $[3.91, 6.84] \times 10^{-3}$.

10.2   POISSON METHODS FOR UNSTRATIFIED SURVIVAL DATA

In this section we present methods for comparing cohorts across two or more categories of exposure. The techniques to be described correspond closely to the odds ratio methods of Chapter 4, and so it is possible to omit certain details that were covered as part of that discussion.

10.2.1   Asymptotic (Unconditional) Methods for a Single 1 × 2 Table

Consider Table 4.1, which gives the crude 2×2 table for a closed cohort study. Since $b_1 = r_1 - a_1$ and $b_2 = r_2 - a_2$, we might have used Table 10.4 as an alternative method of presenting the data.

When Poisson methods are used to analyze data from an open cohort study, the data can be presented as in Table 10.5. The correspondence between Table 10.4 and Table 10.5 is evident and continues the theme of drawing a parallel between the binomial and Poisson distributions. We will refer to Table 10.5 as a 1 × 2 table.

Suppose that survival in the exposed and unexposed cohorts is governed by Poisson random variables D_1 and D_2 with parameters (λ_1, n_1) and (λ_2, n_2), respectively. The random variables D_1 and D_2 are assumed to be independent, and so their joint probability function is the product of the individual probability functions,

$$P(D_1 = d_1, D_2 = d_2 | \lambda_1, \lambda_2) = \frac{\exp(-\lambda_1 n_1)(\lambda_1 n_1)^{d_1}}{d_1!} \times \frac{\exp(-\lambda_2 n_2)(\lambda_2 n_2)^{d_2}}{d_2!}.$$
$$(10.13)$$

TABLE 10.4   Observed Counts: Closed Cohort Study

	Exposure		
	yes	no	
deaths	a_1	a_2	m_1
	r_1	r_2	r

TABLE 10.5   Observed Counts and Person-Time: Censored Survival Data

	Exposure		
	yes	no	
deaths	d_1	d_2	m
person-time	n_1	n_2	n

Since the hazard functions, λ_1 and λ_2, are both constants, the proportional hazards assumption is satisfied. Denote the hazard ratio by $HR = \lambda_1/\lambda_2$. In order to make the role of HR explicit, we substitute $\lambda_1 = HR\lambda_2$ in (10.13), which reparameterizes the joint probability function in terms of HR and λ_2:

$$P(D_1 = d_1, D_2 = d_2 | HR, \lambda_2) = \frac{\exp(-HR\lambda_2 n_1)(HR\lambda_2 n_1)^{d_1}}{d_1!}$$

$$\times \frac{\exp(-\lambda_2 n_2)(\lambda_2 n_2)^{d_2}}{d_2!}. \qquad (10.14)$$

We view (10.14) as a likelihood which is a function of the parameters HR and λ_2.

Point Estimate
The unconditional maximum likelihood equations are

$$d_1 = \widehat{HR}\hat{\lambda}_2 n_1$$

and

$$m = \widehat{HR}\hat{\lambda}_2 n_1 + \hat{\lambda}_2 n_2.$$

An important result demonstrated below is that the unconditional and conditional maximum likelihood estimates of HR are identical. So there is no need to add a subscript u to the notation for the hazard ratio estimate. This also explains the use of parentheses in the title of this and subsequent sections dealing with asymptotic methods. Solving the above equations gives

$$\widehat{HR} = \frac{\hat{\lambda}_1}{\hat{\lambda}_2} = \frac{d_1 n_2}{d_2 n_1} \tag{10.15}$$

and

$$\hat{\lambda}_2 = \frac{d_2}{n_2}$$

where $\hat{\lambda}_1 = d_1/n_1$. If either d_1 or d_2 equals 0, we replace (10.15) with

$$\widehat{HR} = \frac{(d_1 + .5)n_2}{(d_2 + .5)n_1}.$$

Log-Hazard Ratio Transformation

The log-hazard ratio $\log(HR)$ is the counterpart in open cohort studies to the log-odds ratio in closed cohort studies. The maximum likelihood estimate of $\log(HR)$ is $\log(\widehat{HR})$. In view of the results of Section 4.1 on the odds ratio, it will come as no surprise that the distribution of $\widehat{HR}$ can be quite skewed, while the distribution of $\log(\widehat{HR})$ is generally relatively symmetric. Since $\widehat{HR} = (d_1/d_2)(n_2/n_1)$ and $\log(\widehat{HR}) = \log(d_1/d_2) + \log(n_2/n_1)$, the basic shapes of the distributions of $\widehat{HR}$ and $\log(\widehat{HR})$ do not depend on the constants n_1 and n_2. Accordingly, the following illustration is presented in terms of Poisson random variables rather than hazard rate estimates. Let D_1 and D_2 be Poisson random variables with parameters $\nu_1 = 2$ and $\nu_2 = 4$, respectively. The random variable D_1/D_2 has a range stretching from .020 to 49. The distribution is highly skewed, with outcomes less than or equal to 5 representing 99.3% of the probability. Figure 10.6(a), which was constructed in a manner similar to Figure 4.1(a), shows the graph of D_1/D_2 after truncation on the right at 5. Even though truncation has removed an extremely long tail, the graph is

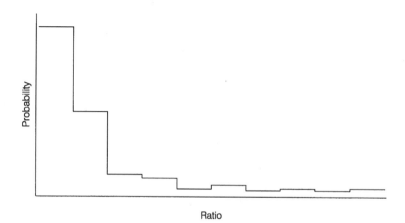

FIGURE 10.6(a)   Distribution of ratio of Poisson random variables with parameters 2 and 4

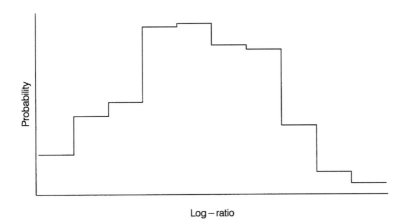

FIGURE 10.6(b) Distribution of log-ratio of Poisson random variables with parameters 2 and 4

still very skewed. Figure 10.6(b) shows the graph of $\log(D_1/D_2)$, which is seen to be relatively symmetric. This is a general finding and supports the use of $\log(\widehat{HR})$ in preference to $\widehat{HR}$ when calculations are based on a normal approximation.

Confidence Interval
The maximum likelihood estimate of $\text{var}(\log \widehat{HR})$ is

$$\widehat{\text{var}}(\log \widehat{HR}) = \frac{1}{d_1} + \frac{1}{d_2} \tag{10.16}$$

and a $(1 - \alpha) \times 100\%$ confidence interval for HR is obtained by exponentiating

$$\left[\log \underline{HR}, \log \overline{HR}\right] = \log(\widehat{HR}) \pm z_{\alpha/2}\sqrt{\frac{1}{d_1} + \frac{1}{d_2}}.$$

If either d_1 or d_2 equals 0, we replace (10.16) with

$$\widehat{\text{var}}(\log \widehat{HR}) = \frac{1}{d_1 + .5} + \frac{1}{d_2 + .5}.$$

Wald and Likelihood Ratio Tests of Association
We say there is no association between exposure and survival if $\lambda_1 = \lambda_2$. Under the hypothesis of no association $H_0 : \lambda_1 = \lambda_2$, the expected counts are

$$\hat{e}_1 = \frac{n_1 m}{n} \quad \text{and} \quad \hat{e}_2 = \frac{n_2 m}{n}$$

where we note that $\hat{e}_1 + \hat{e}_2 = m$. Since $\lambda_1 = \lambda_2$ is equivalent to $\log(HR) = 0$, the hypothesis of no association can be written as $H_0 : \log(HR) = 0$. Under H_0 an estimate of $\text{var}(\log \widehat{HR})$ is

$$\widehat{\text{var}}_0(\log \widehat{HR}) = \frac{1}{\hat{e}_1} + \frac{1}{\hat{e}_2} = \frac{n^2}{n_1 n_2 m}.$$

The Wald and likelihood ratio tests of association are

$$X_w^2 = \frac{(\log \widehat{HR})^2 n_1 n_2 m}{n^2} \qquad (\text{df} = 1)$$

and

$$X_{lr}^2 = 2\left[d_1 \log\left(\frac{d_1}{\hat{e}_1}\right) + d_2 \log\left(\frac{d_2}{\hat{e}_2}\right)\right] \qquad (\text{df} = 1)$$

respectively.

Example 10.8 (Receptor Level–Breast Cancer)   The data for this example are taken from Table 9.1. Table 10.6 gives observed and expected counts and person-months for the breast cancer cohort according to receptor level. The graphs of the corresponding Kaplan–Meier and exponential survival curves are shown in Figure 10.7. The exponential model provides a reasonable fit to the data during the latter part of follow-up, but does not perform quite as well early on, especially for the low receptor level cohort. The Cox–Oakes tests of exponentiality for the low and high receptor level cohorts are $X_{co}^2 = 2.31$ ($p = .13$) and $X_{co}^2 = 4.44$ ($p = .04$), respectively. These results are a bit surprising since, from Figure 10.7, the low receptor level cohort is the one that appears to exhibit the greatest departure from exponentiality.

The stratum-specific hazard rate estimates are $\hat{\lambda}_1 = 10.74 \times 10^{-3}$ and $\hat{\lambda}_2 = 3.64 \times 10^{-3}$, which suggests that having low receptor level increases mortality from breast cancer. Based on the implicit approach, the 95% confidence intervals for λ_1 and λ_2 are $[7.09, 16.26] \times 10^{-3}$ and $[2.50, 5.29] \times 10^{-3}$, respectively. The confidence intervals do not overlap, suggesting that λ_1 and λ_2 are unequal. The estimate of the hazard ratio is $\widehat{HR} = (22 \times 7422)/(27 \times 2049) = 2.95$, the 95% confidence interval for HR is $[1.68, 5.18]$, and the Wald and likelihood ratio tests of association are $X_w^2 = 9.73$ ($p = .002$) and $X_{lr}^2 = 13.11$ ($p < .001$). These results are similar to the findings in Example 9.2, which was based on the much more complicated odds ratio approach. It is interesting that the expected counts are nearly identical for the Poisson and odds ratio methods.

TABLE 10.6   Observed and Expected Counts and Person-Months: Receptor Level–Breast Cancer

	Receptor level		
	low	high	
observed	22	27	49
expected	10.60	38.40	49
person-months	2049	7422	9471

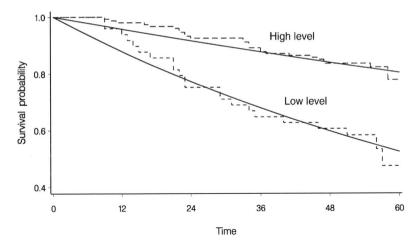

FIGURE 10.7 Exponential and Kaplan–Meier survival curves: Breast cancer cohort stratified by receptor level

Example 10.9 (Receptor Level–Breast Cancer: Stage III) Table 10.7 corresponds to Table 10.6 except that now attention is restricted to subjects with stage III disease. The estimated hazard ratio is $\widehat{HR} = 2.55$ and the 95% confidence interval for HR is $[1.04, 6.24]$. Note that the lower bound of the confidence interval is only slightly larger than 1. The Wald and likelihood ratio tests of association are $X_{w}^{2} = 4.09$ ($p = .04$) and $X_{lr}^{2} = 4.32$ ($p = .04$). So there is moderate evidence for an association between receptor level and survival in the stage III cohort.

10.2.2 Exact Conditional Methods for a Single 1 × 2 Table

Conditional Poisson Distribution
In the unconditional case, D_1 and D_2 are independent Poisson random variables with parameters $\nu_1 = \lambda_1 n_1$ and $\nu_2 = \lambda_2 n_2$. According to the conditional approach, we assume that the total number of deaths m is a known constant. As a result, D_1 and D_2 satisfy the constraint $D_1 + D_2 = m$ and are no longer independent. We choose the

TABLE 10.7 Observed and Expected Counts and Person-Months: Receptor Level–Breast Cancer (Stage III)

	Receptor level		
	low	high	
observed	12	8	20
expected	7.41	12.59	20
person-months	384	653	1037

left cell of Table 10.2 to be the index cell for the conditional analysis and continue to denote the corresponding random variable by D_1. As shown in Appendix C, D_1 has the probability function

$$P(D_1 = d_1|HR) = \binom{m}{d_1} \pi^{d_1}(1 - \pi)^{m-d_1} \tag{10.17}$$

where

$$\pi = \frac{v_1}{v_1 + v_2} = \frac{HRn_1}{HRn_1 + n_2}. \tag{10.18}$$

So D_1 is binomial with parameters (π, m). Observe that by conditioning on m, the nuisance parameter λ_2 has been eliminated, leaving HR as the only unknown parameter in (10.17). Solving (10.18) for HR yields

$$HR = \frac{\pi n_2}{(1 - \pi)n_1}. \tag{10.19}$$

The binomial mean and variance of D_1 are

$$E(D_1|HR) = \pi m = \frac{HRn_1 m}{HRn_1 + n_2} \tag{10.20}$$

and

$$\text{var}(D_1|HR) = \pi(1 - \pi)m = \frac{HRn_1 n_2 m}{(HRn_1 + n_2)^2}. \tag{10.21}$$

Confidence Interval
From (3.3) and (3.4), an exact $(1 - \alpha) \times 100\%$ confidence interval for π is obtained by solving the equations

$$\frac{\alpha}{2} = P(D_1 \geq d_1|\underline{\pi}) = \sum_{x=d_1}^{m} \binom{m}{x} \underline{\pi}^x (1 - \underline{\pi})^{m-x}$$

and

$$\frac{\alpha}{2} = P(D_1 \leq d_1|\overline{\pi}) = \sum_{x=0}^{d_1} \binom{m}{x} \overline{\pi}^x (1 - \overline{\pi})^{m-x}$$

for $\underline{\pi}$ and $\overline{\pi}$. A confidence interval for HR results after transforming $\underline{\pi}$ and $\overline{\pi}$ using (10.19).

Exact Test of Association
From (10.18), $H_0 : HR = 1$ is equivalent to $H_0 : \pi = \pi_0$, where $\pi_0 = n_1/n$. From (3.1) and (3.2), an exact test of association is based on the tail probabilities

$$P(D_1 \leq d_1 | \pi_0) = \sum_{x=0}^{d_1} \binom{m}{x} \pi_0^x (1 - \pi_0)^{m-x}$$

and

$$P(D_1 \geq d_1 | \pi_0) = \sum_{x=d_1}^{m} \binom{m}{x} \pi_0^x (1 - \pi_0)^{m-x}.$$

Example 10.10 (Receptor Level–Breast Cancer: Stage III)   From

$$\frac{\alpha}{2} = \sum_{x=12}^{20} \binom{20}{x} \underline{\pi}^x (1 - \underline{\pi})^{20-x}$$

and

$$\frac{\alpha}{2} = \sum_{x=0}^{12} \binom{20}{x} \overline{\pi}^x (1 - \overline{\pi})^{20-x}$$

the 95% confidence interval for π is [.361, .809]. Applying (10.19), the 95% confidence interval for HR is [.959, 7.19]. Under $H_0 : HR = 1$, we have $\pi_0 = 384/1037 = .370$. Table 10.8 gives a portion of the probability function for the binomial distribution with parameters (.370, 20). Based on the doubling method, the p-value for the exact test of association is $p = 2(.031) = .062$. Observe that the exact results provide less evidence for an association between receptor level and breast cancer survival than the asymptotic results of Example 10.9. In this case it is prudent to rely on the exact findings.

TABLE 10.8   Probability Function (%) for the Binomial Distribution with Parameters (.370, 20)

d	$P(D_1 = d|.370)$	$P(D_1 \leq d|.370)$	$P(D_1 \geq d|.370)$
$\vdots$	$\vdots$	$\vdots$	$\vdots$
2	.63	.75	99.88
3	2.23	2.98	99.25
4	5.57	8.55	97.02
$\vdots$	$\vdots$	$\vdots$	$\vdots$
11	4.69	96.90	7.80
12	2.07	98.97	3.10
13	.75	99.72	1.03
$\vdots$	$\vdots$	$\vdots$	$\vdots$

10.2.3   Asymptotic (Conditional) Methods for a Single 1 × 2 Table

Point Estimate

The conditional maximum likelihood equation is

$$d_1 = E(D_1|\widehat{HR}) = \frac{\widehat{HR}n_1 m}{\widehat{HR}n_1 + n_2}. \tag{10.22}$$

So the asymptotic conditional estimate of HR is

$$\widehat{HR} = \frac{d_1 n_2}{d_2 n_1}$$

which is the same as (10.15), the asymptotic unconditional estimate.

Confidence Interval

From (10.21) an estimate of $\mathrm{var}(D_1|\widehat{HR})$ is

$$\hat{v} = \frac{\widehat{HR}n_1 n_2 m}{(\widehat{HR}n_1 + n_2)^2} = \left(\frac{1}{d_1} + \frac{1}{d_2}\right)^{-1}. \tag{10.23}$$

As shown in Appendix C, an asymptotic conditional estimate of $\mathrm{var}(\log \widehat{HR})$ is

$$\widehat{\mathrm{var}}(\log \widehat{HR}) = \frac{1}{\hat{v}} = \frac{1}{d_1} + \frac{1}{d_2}$$

which is the same as (10.16), the asymptotic unconditional estimate (Tarone et al., 1983).

Mantel–Haenszel Test of Association for Person-Time Data

Under the hypothesis of no association $H_0 : HR = 1$, it follows from (10.22) and (10.23) that

$$\hat{e}_1 = \frac{n_1 m}{n}$$

and

$$\hat{v}_0 = \frac{n_1 n_2 m}{n^2}.$$

Following Rothman and Greenland (1998, p. 274) we refer to

$$X_{\mathrm{pt}}^2 = \frac{(d_1 - \hat{e}_1)^2}{\hat{v}_0} \qquad (\mathrm{df} = 1)$$

as the Mantel–Haenszel test of association for person-time data (Oleinick and Mantel, 1970). This lengthy title will be shortened to the Mantel–Haenszel test when there is no possibility of confusion with the corresponding test for the odds ratio. Setting

$$\hat{e}_2 = \frac{n_2 m}{n}$$

it is readily demonstrated that

$$X_{\text{pt}}^2 = \frac{(d_1 - \hat{e}_1)^2}{\hat{e}_1} + \frac{(d_2 - \hat{e}_2)^2}{\hat{e}_2}.$$

The normal approximation underlying the Mantel–Haenszel test should be satisfactory provided $\hat{e}_1$ and $\hat{e}_2$ are greater than or equal to 5 (Rothman and Greenland, 1998, p. 239).

Example 10.11 (Receptor Level–Breast Cancer: Stage III)   The Mantel–Haenszel test is $X_{\text{pt}}^2 = (12 - 7.41)^2/4.66 = 4.53$ ($p = .03$).

10.2.4 Asymptotic Methods for a Single 1 × *I* Table

The data layout for the case of $I \geq 2$ exposure categories is given in Table 10.9. We model the ith exposure category using the Poisson distribution with parameters (λ_i, n_i) ($i = 1, 2, \ldots, I$). With $i = 1$ as the reference category, the hazard ratio for the ith exposure category is $HR_i = \lambda_i / \lambda_1$.

The maximum likelihood estimate of HR_i is

$$\widehat{HR}_i = \frac{d_i n_1}{d_1 n_i}$$

where we note that $\widehat{HR}_1 = 1$. A confidence interval for HR_i can be estimated using (10.16). We say there is no association between exposure and disease if $\lambda_1 = \lambda_2 = \cdots = \lambda_I$. The expected count for the ith exposure category is

$$\hat{e}_i = \frac{n_i m}{n}.$$

It is readily verified that $\hat{e}_\bullet = d_\bullet = m$. Conditioning on the total number of cases m results in the multinomial distribution (Appendix E). The Mantel–Haenszel test for a 1 × *I* table is

TABLE 10.9 Observed Counts and Person-Time: Censored Survival Data

	Exposure category						
	1	2	$\cdots$	i	$\cdots$	I	
deaths	d_1	d_2	$\cdots$	d_i	$\cdots$	d_I	m
person-time	n_1	n_2	$\cdots$	n_i	$\cdots$	n_I	n

TABLE 10.10   Observed and Expected Counts
and Person-Months: Stage–Breast Cancer

	Stage			
	I	II	III	
observed	6	23	20	49
expected	18.82	24.82	5.37	49
person-months	3637	4797	1037	9471

$$X_{\text{pt}}^2 = \sum_{i=1}^{I} \frac{(d_i - \hat{e}_i)^2}{\hat{e}_i} \qquad (\text{df} = I - 1) \tag{10.24}$$

(Breslow and Day, 1987, p. 96).

Let s_i be the exposure level for the ith category with $s_1 < s_2 < \cdots < s_I$. Consider the scatter plot of $\log(\hat{\lambda}_i)$ against s_i ($i = 1, 2, \ldots, I$) and let $\log(\hat{\lambda}_i) = \hat{\alpha} + \hat{\beta}s_i$ be the best-fitting straight line for these points, where α and β are constants. As shown in Appendix E, the score test of $H_0 : \beta = 0$, which will be referred to as the test for linear trend (in log-hazards), is

$$X_{\text{t}}^2 = \frac{\left[\sum_{i=1}^{I} s_i(d_i - \hat{e}_i)\right]^2}{\sum_{i=1}^{I} s_i^2 \hat{e}_i - \left(\sum_{i=1}^{I} s_i \hat{e}_i\right)^2 \bigg/ \hat{e}_\bullet} \qquad (\text{df} = 1) \tag{10.25}$$

(Armitage, 1966; Clayton, 1982; Breslow and Day, 1987, p. 96). Although X_{t}^2 has been presented in terms of log-hazards, it has an equivalent interpretation as a test for linear trend in hazards or hazard ratios.

Example 10.12 (Stage–Breast Cancer)   Table 10.10 gives the observed and expected counts and person-months for the breast cancer cohort according to stage of disease. Figure 10.8 shows the graphs of the Kaplan–Meier and exponential survival curves. The fit for stage III is less than might be desired, but overall the exponential (Poisson) model performs reasonably well.

Table 10.11 gives the hazard ratio estimates and 95% confidence intervals with stage I taken as the reference category. An increasing trend across stage is evident (where $\widehat{HR}_1 = 1$), but the confidence intervals exhibit substantial overlap.

TABLE 10.11   Hazard Ratio Estimates and 95%
Confidence Intervals: Stage–Breast Cancer

Stage	$\widehat{HR}$	$\underline{HR}$	$\overline{HR}$
II	2.91	1.18	7.14
III	11.69	4.70	29.11

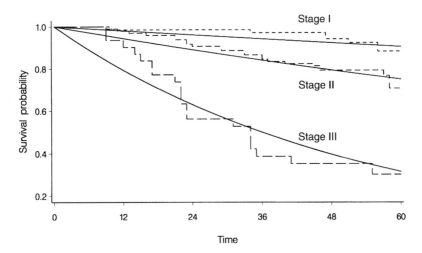

FIGURE 10.8   Exponential and Kaplan–Meier survival curves: Breast cancer cohort stratified by stage

The Mantel–Haenszel test provides considerable evidence for an association between stage and survival:

$$X_{pt}^2 = \frac{(6 - 18.82)^2}{18.82} + \frac{(23 - 24.82)^2}{24.82} + \frac{(20 - 5.37)^2}{5.37} = 48.78 \ (p < .001).$$

Setting $s_1 = 1$, $s_2 = 2$, and $s_3 = 3$, the test for linear trend is

$$X_t^2 = \frac{(27.45)^2}{166.4 - (84.55)^2/49} = 36.78 \ (p < .001).$$

These results are similar to those of Example 9.6, where the analysis was based on odds ratio methods.

10.2.5   Assessment of the Poisson-Exponential Assumption

We now present a method of assessing the Poisson-exponential assumption which is based on the Mantel–Haenszel test of association for $1 \times I$ tables. The key idea is that the period of follow-up is partitioned into time periods that are used to stratify the data. This is reminiscent of the odds ratio analysis of censored survival data in Chapter 9, where death times were used for a similar purpose. Suppose that survival during the ith time period is exponential with hazard rate λ_i ($i = 1, 2, \ldots, I$). We say there is no association between "time period" and survival if $\lambda_1 = \lambda_2 = \ldots = \lambda_I$. If the hypothesis of no association is not rejected, we conclude that there is a common hazard rate across time periods; that is, there is overall exponentiality. It is convenient to give the remaining details of the method using a specific example rather than provide a description in complete generality.

Example 10.13 (Breast Cancer)   Consider Figure 10.2(a), which shows the Kaplan–Meier survival curve for the entire breast cancer cohort. To create a partition of the period of follow-up we inspect the Kaplan–Meier survival curve and determine, on an empirical basis, a series of time periods where survival seems to be exponential and where the individual hazard rates may be unequal. Evidently this introduces an element of subjectivity into the procedure. There is a plateau in the survival curve until just prior to 12 months, after which there is a gradual decline. Based on this observation we partition the 60-month period of follow-up into the two time periods, $[0, 12)$ and $[12, 60]$.

For the first time period, denote the number of deaths, number of person-months, and hazard rate by d_1, n_1, and λ_1; for the second time period, the corresponding notation is d_2, n_2, and λ_2. The number of person-months in each time period is calculated as follows. Let t_i be the survival time for the ith subject. If $t_i < 12$, this individual contributes t_i person-months to n_1 and 0 person-months to n_2. If $t_i \geq 12$, the contribution is 12 person-months to n_1 and $t_i - 12$ person-months to n_2. With these definitions, $d_1 + d_2 = d$ and $n_1 + n_2 = n$. From Table 10.12,

$$X_{\mathrm{pt}}^2 = \frac{(5 - 12.23)^2}{12.23} + \frac{(44 - 36.77)^2}{36.77} = 5.69 \ (p = .02)$$

which provides moderate evidence that λ_1 and λ_2 are unequal. We note from Example 10.1 that the Cox–Oakes test of exponentiality gives an almost identical result. As discussed in Example 10.1, when there is reason to reject the assumption of overall exponentiality, a decision must be made as to whether there is a practical advantage to considering a more complicated parametric model.

10.3   POISSON METHODS FOR STRATIFIED SURVIVAL DATA

In this section we present methods for comparing cohorts across two or more categories of exposure when the data are stratified. The techniques to be described correspond closely to the odds ratio methods of Chapter 5, which were subsequently adapted to the analysis of censored survival data in Chapter 9. In order to avoid confusion with Chapter 9, where j was used to index death times, we let k index the stratifying variable. In the odds ratio setting we distinguished between large-strata and sparse-strata conditions. A corresponding contrast is made here, except that now the distinction rests on the number of deaths in each stratum (Greenland and Robins,

TABLE 10.12   Stratification by Time Period: Breast Cancer

Time period	d_i	n_i	$\hat{\lambda}_i \times 10^3$	$\hat{e}_i$
$[0, 12)$	5	2363	2.12	12.23
$[12, 60]$	44	7108	6.19	36.77

1985b). References for this section are Clayton (1982), Breslow (1984a), Greenland and Robins (1985b), and Breslow and Day (1987).

10.3.1 Asymptotic (Unconditional) Methods for K (1 × 2) Tables

We now consider the case of a dichotomous exposure variable with the data stratified into K strata. Suppose that, in the kth stratum, the development of disease in the exposed and unexposed cohorts is governed by Poisson random variables D_{1k} and D_{2k} with parameters (λ_{1k}, n_{1k}) and (λ_{2k}, n_{2k}), respectively $(k = 1, 2, \ldots, K)$. For the kth stratum, the data layout is given in Table 10.13, where the hazard ratio is $HR_k = \lambda_{1k}/\lambda_{2k}$. When the hazard ratios are homogeneous we denote the common stratum-specific value by HR.

Point Estimates and Fitted Counts
The unconditional maximum likelihood equations are

$$\sum_{k=1}^{K} d_{1k} = \sum_{k=1}^{K} \widehat{HR}\hat{\lambda}_{2k}n_{1k} \qquad (10.26)$$

and

$$m_k = \widehat{HR}\hat{\lambda}_{2k}n_{1k} + \hat{\lambda}_{2k}n_{2k} \qquad (k = 1, 2, \ldots, K). \qquad (10.27)$$

Solving (10.27) for $\hat{\lambda}_{2k}$ gives

$$\hat{\lambda}_{2k} = \frac{m_k}{\widehat{HR}n_{1k} + n_{2k}} \qquad (10.28)$$

which can be substituted in (10.26) to yield

$$\sum_{k=1}^{K} d_{1k} = \sum_{k=1}^{K} \frac{\widehat{HR}m_k n_{1k}}{\widehat{HR}n_{1k} + n_{2k}}. \qquad (10.29)$$

This is an equation in the single unknown $\widehat{HR}$, which can be solved by trial and error. Alternatively we can use an iterative procedure due to Clayton (1982). Rewriting (10.29) as

TABLE 10.13 Observed
Counts and Person-Time:
Censored Survival Data

	Exposure		
	yes	no	
deaths	d_{1k}	d_{2k}	m_k
person-time	n_{1k}	n_{2k}	n_k

$$0 = \sum_{k=1}^{K} \left(d_{1k} - \frac{\widehat{HR}m_k n_{1k}}{\widehat{HR}n_{1k} + n_{2k}} \right)$$

$$= \sum_{k=1}^{K} \frac{d_{1k}n_{2k} - \widehat{HR}d_{2k}n_{1k}}{\widehat{HR}n_{1k} + n_{2k}}$$

$$= \sum_{k=1}^{K} \frac{d_{1k}n_{2k}}{\widehat{HR}n_{1k} + n_{2k}} - \widehat{HR} \sum_{k=1}^{K} \frac{d_{2k}n_{1k}}{\widehat{HR}n_{1k} + n_{2k}}$$

and solving for the $\widehat{HR}$ preceding the second summation, we have

$$\widehat{HR} = \sum_{k=1}^{K} \frac{d_{1k}n_{2k}}{\widehat{HR}n_{1k} + n_{2k}} \bigg/ \sum_{k=1}^{K} \frac{d_{2k}n_{1k}}{\widehat{HR}n_{1k} + n_{2k}}. \tag{10.30}$$

The iterative process begins by substituting $\widehat{HR}^{(1)} = 1$ in the right-hand side of (10.30) and performing the calculations to get an updated value $\widehat{HR}^{(2)}$. Then $\widehat{HR}^{(2)}$ is substituted in the right-hand side of (10.30) to get the next updated value $\widehat{HR}^{(3)}$, and so on. This process is repeated until the desired accuracy is obtained. The algorithm is very efficient, and typically only three or four iterations are required to obtain an accurate estimate of HR. From $\hat{\lambda}_{1k} = \widehat{HR}\hat{\lambda}_{2k}$ and (10.28) we obtain the fitted counts

$$\hat{d}_{1k} = \hat{\lambda}_{1k}n_{1k}$$

and

$$\hat{d}_{2k} = \hat{\lambda}_{2k}n_{2k}.$$

Confidence Interval
Let

$$\hat{v}_k = \left(\frac{1}{\hat{d}_{1k}} + \frac{1}{\hat{d}_{2k}} \right)^{-1} \tag{10.31}$$

and let $\hat{V} = \sum_{k=1}^{K} \hat{v}_k$. As shown in Appendix C, an estimate of $\text{var}(\log \widehat{HR})$ is

$$\widehat{\text{var}}(\log \widehat{HR}) = \frac{1}{\hat{V}}$$

and a $(1 - \alpha) \times 100\%$ confidence interval for HR is obtained by exponentiating

$$\left[\log \underline{HR}, \log \overline{HR} \right] = \log(\widehat{HR}) \pm \frac{z_{\alpha/2}}{\sqrt{\hat{V}}}$$

(Tarone et al., 1983).

Tests of Association, Homogeneity, and Trend

The hypothesis of no association is $H_0 : \log(HR) = 0$ and, for the kth stratum, the expected counts are

$$\hat{e}_{1k} = \frac{n_{1k} m_k}{n_k} \quad \text{and} \quad \hat{e}_{2k} = \frac{n_{2k} m_k}{n_k}.$$

Let

$$\hat{v}_{0k} = \left(\frac{1}{\hat{e}_{1k}} + \frac{1}{\hat{e}_{2k}} \right)^{-1} = \frac{n_{1k} n_{2k} m_k}{n_k^2}$$

and let $\hat{V}_0 = \sum_{k=1}^{K} \hat{v}_{0k}$. Under H_0 an estimate of $\mathrm{var}(\log \widehat{HR})$ is

$$\widehat{\mathrm{var}}_0(\log \widehat{HR}) = \frac{1}{\hat{V}_0}.$$

The Wald and likelihood ratio tests of association are

$$X_{\mathrm{w}}^2 = (\log \widehat{HR})^2 \hat{V}_0 \qquad (\mathrm{df} = 1)$$

and

$$X_{\mathrm{lr}}^2 = 2 \sum_{k=1}^{K} \left[d_{1k} \log \left(\frac{d_{1k}}{\hat{e}_{1k}} \right) + d_{2k} \log \left(\frac{d_{2k}}{\hat{e}_{2k}} \right) \right] \qquad (\mathrm{df} = 1)$$

respectively. The likelihood ratio test of homogeneity is

$$X_{\mathrm{h}}^2 = 2 \sum_{k=1}^{K} \left[d_{1k} \log \left(\frac{d_{1k}}{\hat{d}_{1k}} \right) + d_{2k} \log \left(\frac{d_{2k}}{\hat{d}_{2k}} \right) \right] \qquad (\mathrm{df} = K - 1).$$

Let s_k be the exposure level for the kth stratum with $s_1 < s_2 < \cdots < s_K$. Consider the scatter plot of $\log(\widehat{HR}_k)$ against s_k $(k = 1, 2, \ldots, K)$ and let $\log(\widehat{HR}_k) = \hat{\alpha} + \hat{\beta} s_k$ be the "best-fitting straight line" for these points, where α and β are constants. The score test of $H_0 : \beta = 0$, which we refer to as the test for linear trend (in log-hazard ratios), is

$$X_{\mathrm{t}}^2 = \frac{\left[\sum_{k=1}^{K} s_k (d_{1k} - \hat{d}_{1k}) \right]^2}{\sum_{k=1}^{K} s_k^2 \hat{v}_k - \left(\sum_{k=1}^{K} s_k \hat{v}_k \right)^2 \Big/ \hat{v}_\bullet} \qquad (\mathrm{df} = 1) \qquad (10.32)$$

where $\hat{v}_k$ is given by (10.31) (Breslow, 1984a). Although X_{t}^2 has been presented in terms of log-hazard ratios, it has an equivalent interpretation as a test for linear trend in hazards ratios.

Example 10.14 (Receptor Level–Breast Cancer)   In this example we extend the analysis of Example 10.8 by stratifying by stage of disease. Table 10.14 gives the ob-

TABLE 10.14 Observed, Expected, and Fitted Counts, and Person-Months: Receptor Level–Breast Cancer

	Stage I			Stage II			Stage III		
	Receptor level			Receptor level			Receptor level		
	low	high		low	high		low	high	
observed	2	4	6	8	15	23	12	8	20
expected	1.07	4.93	6	4.87	18.13	23	7.41	12.59	20
fitted	1.97	4.03	6	8.65	14.35	23	11.38	8.62	20
person-months	650	2987	3637	1015	3782	4797	384	653	1037

served, expected, and fitted counts as well as the person-months, stratified by stage. We note that the observed and fitted counts are quite close in value, and so the model based on homogeneity appears to provide a reasonably good fit to the data. Table 10.15 gives the stage-specific analysis based on the methods described above for 1×2 tables. There is considerable overlap among the confidence intervals and no apparent trend across strata. With $\widehat{HR}^{(1)} = 1$, the first few iterations based on (10.30) are $\widehat{HR}^{(2)} = 2.263$, $\widehat{HR}^{(3)} = 2.245$, and $\widehat{HR}^{(4)} = 2.246$, and so we take $\widehat{HR} = 2.25$. This estimate is quite close to the stage-adjusted estimates in Table 9.10 based on the odds ratio approach. From $\hat{V} = 1.32 + 5.40 + 4.90 = 11.62$, the 95% confidence interval for HR is $[1.26, 3.99]$. The Wald and likelihood ratio tests of association are $X_w^2 = (\log 2.25)^2 (9.38) = 6.14$ ($p = .01$) and $X_{lr}^2 = 7.41$ ($p = .01$). The likelihood ratio test of homogeneity is $X_h^2 = .158$ ($p = .92$). Setting $s_1 = 1$, $s_2 = 2$, and $s_3 = 3$, the test for linear trend is

$$X_t^2 = \frac{(.584)^2}{67.05 - (26.83)^2/11.62} = .067 \ (p = .80).$$

10.3.2 Asymptotic (Conditional) Methods for K (1×2) Tables

We now consider asymptotic conditional methods for analyzing K (1×2) tables. From (10.20) and (10.21) the binomial mean and variance of D_{1k} are

$$E(D_{1k}|HR) = \frac{HRn_{1k}m_k}{HRn_{1k} + n_{2k}} \qquad (10.33)$$

and

TABLE 10.15 Hazard Ratio Estimates and 95% Confidence Intervals: Receptor Level–Breast Cancer

Stage	$\widehat{HR}$	$\underline{HR}$	$\overline{HR}$
I	2.30	.42	12.54
II	1.99	.84	4.69
III	2.55	1.04	6.24

$$\text{var}(D_{1k}|HR) = \frac{HRn_{1k}n_{2k}m_k}{(HRn_{1k} + n_{2k})^2}.$$ (10.34)

The conditional maximum likelihood equation is

$$\sum_{k=1}^{K} d_{1k} = \sum_{k=1}^{K} E(D_{1k}|\widehat{HR}) = \sum_{k=1}^{K} \frac{\widehat{HR}n_{1k}m_k}{\widehat{HR}n_{1k} + n_{2k}}$$

which is the same as the unconditional maximum likelihood equation (10.29). It follows that the asymptotic conditional estimate of HR is identical to the asymptotic unconditional estimate. When $HR = 1$, (10.33) and (10.34) simplify to

$$\hat{e}_{1k} = \frac{n_{1k}m_k}{n_k}$$

and

$$\hat{v}_{0k} = \frac{n_{1k}n_{2k}m_k}{n_k^2}.$$ (10.35)

The Mantel–Haenszel test of association for person-time data is

$$X_{\text{pt}}^2 = \frac{(d_{1\bullet} - \hat{e}_{1\bullet})^2}{\hat{v}_{0\bullet}} \qquad (\text{df} = 1)$$

(Shore et al., 1976; Breslow, 1984a; Breslow and Day, 1987, p. 108). The normal approximation underlying the Mantel–Haenszel test should be satisfactory provided $\hat{e}_{1\bullet}$, $\hat{e}_{2\bullet}$, $\hat{d}_{1\bullet}$, and $\hat{d}_{2\bullet}$ are all greater than or equal to 5 (Rothman and Greenland, 1998, p. 274). A test of homogeneity is

$$X_{\text{h}}^2 = \sum_{k=1}^{K} \frac{(d_{1k} - \hat{d}_{1k})^2}{\hat{v}_k} \qquad (\text{df} = K - 1)$$ (10.36)

(Breslow, 1984a; Breslow and Day, 1987, p. 112).

Example 10.15 (Receptor Level–Breast Cancer)   The Mantel–Haenszel test is $X_{\text{pt}}^2 = (22 - 13.34)^2/9.38 = 7.99$ ($p = .01$), and the test of homogeneity is

$$X_{\text{h}}^2 = \frac{(2 - 1.97)^2}{1.32} + \frac{(8 - 8.65)^2}{5.40} + \frac{(12 - 11.38)^2}{4.90} = .158 \ (p = .92).$$

10.3.3   Mantel–Haenszel Estimate of the Hazard Ratio

The Mantel–Haenszel estimate of the hazard ratio is

$$\widehat{HR}_{\text{mh}} = \frac{R_{\bullet}}{S_{\bullet}}$$ (10.37)

where

$$R_k = \frac{d_{1k} n_{2k}}{n_k}$$

and

$$S_k = \frac{d_{2k} n_{1k}}{n_k}$$

(Rothman and Boice, 1979). Interestingly, with $HR^{(1)} = 1$, the first iteration of (10.30) produces the Mantel–Haenszel estimate, that is, $HR^{(2)} = \widehat{HR}_{mh}$ (Tarone, 1981; Clayton, 1982). Greenland and Robins (1985b) give an estimate of $var(\log \widehat{HR}_{mh})$ which is valid under both large-strata and sparse-strata conditions:

$$\widehat{var}(\log \widehat{HR}_{mh}) = \frac{\hat{v}_{0\bullet}}{(R_\bullet)(S_\bullet)}$$

where $\hat{v}_{0k}$ is given by (10.35). A $(1 - \alpha) \times 100\%$ confidence interval for HR is obtained by exponentiating

$$\left[\log \underline{HR}_{mh}, \log \overline{HR}_{mh}\right] = \log(\widehat{HR}_{mh}) \pm z_{\alpha/2}\sqrt{\widehat{var}(\log \widehat{HR}_{mh})}.$$

Example 10.16 (Receptor Level–Breast Cancer) The Mantel–Haenszel estimate is $\widehat{HR}_{mh} = 15.51/6.85 = 2.26$. From $\widehat{var}(\log \widehat{HR}_{mh}) = 9.38/(15.51 \times 6.85) = (.297)^2$, the 95% confidence interval for HR is $[1.26, 4.05]$.

10.3.4 Weighted Least Squares Methods for K (1 × 2) Tables

For the weighted least squares methods, the weight for the kth stratum is defined to be

$$\hat{w}_k = \frac{1}{\widehat{var}(\log \widehat{HR}_k)} = \left(\frac{1}{d_{1k}} + \frac{1}{d_{2k}}\right)^{-1}.$$

The hazard ratio formulas are the same as (5.33)–(5.37) except that $\hat{w}_k$ is defined as above and HR replaces OR.

Example 10.17 (Receptor Level–Breast Cancer)   From

$$\log(\widehat{HR}_{ls}) = \frac{(1.33 \times .832) + (5.22 \times .687) + (4.80 \times .936)}{11.35} = .809$$

the WLS estimate of the hazard ratio is $\widehat{HR}_{ls} = \exp(.809) = 2.25$. From $\widehat{var}(\log \widehat{HR}_{ls}) = 1/11.35 = (.297)^2$, the 95% confidence interval for HR is $[1.26, 4.02]$. The test of association is $X_{ls}^2 = (\log 2.25)^2 (9.38) = 6.15$ ($p = .01$), and the test of homogeneity is

$$X_h^2 = 1.33(.832-.809)^2 + 5.22(.687-.809)^2 + 4.80(.936-.809)^2 = .157 \, (p = .92).$$

10.3.5 Standardized Hazard Ratio

Following Section 2.5.4, the observed and standardized expected counts are defined to be

$$O = d_{1\bullet} = \sum_{k=1}^{K} \lambda_{1k} n_{1k}$$

and

$$sE = \sum_{k=1}^{K} \lambda_{2k} n_{1k}$$

and the standardized hazard ratio is defined to be

$$sHR = \frac{O}{sE} = \frac{\sum_{k=1}^{K} \lambda_{1k} n_{1k}}{\sum_{k=1}^{K} \lambda_{2k} n_{1k}}. \tag{10.38}$$

Note the similarity between (10.38) and the first equality in (2.22).

10.3.6 Summary of Examples and Recommendations

Table 10.16 summarizes the results of the receptor level–breast cancer analyses based on the asymptotic unconditional (AU), asymptotic conditional (AC), Mantel–Haenszel (MH), and weighted least squares (WLS) methods. Recall that the AU and AC methods are identical, the nominal distinction serving only to represent the organization of material in this chapter. As can be seen, the various methods produce remarkably similar results.

TABLE 10.16 Summary of Receptor Level–Breast Cancer Results

Result	AU	AC	MH	WLS
$\widehat{HR}$	2.25	—	2.26	2.25
$[\underline{HR}, \overline{HR}]$	[1.26, 3.99]	—	[1.26, 4.05]	[1.26, 4.02]
Association p-value	$.01^a$	.01	—	.01
Homogeneity p-value	$.92^b$	.92	—	.92
Trend p-value	.80	—	—	—

$^a X_{lr}^2$
^{b} Likelihood ratio

Recommendations for the analysis of censored survival data based on the Poisson distribution are similar to those made in Section 5.6 for the analysis of closed cohort data using odds ratio methods. A difference is that in the Poisson setting we do not need to distinguish between asymptotic unconditional and asymptotic conditional estimates. $\widehat{HR}_{\mathrm{mh}}$, $\widehat{\mathrm{var}}(\log \widehat{HR}_{\mathrm{mh}})$, and X_{pt}^2 are easily calculated and have good asymptotic properties (Tarone et al., 1983; Walker, 1985; Greenland and Robins, 1985b). These methods are recommended for the analysis of censored survival data, provided the Poisson-exponential assumption is satisfied and asymptotic conditions are met.

10.3.7   Methods for K (1 × I) Tables

We now consider methods for analyzing stratified data when the exposure variable is polychotomous. The data layout for the kth stratum is given in Table 10.17. We say there is no association between exposure and disease if $\lambda_{1k} = \lambda_{2k} = \cdots = \lambda_{Ik}$ for all k. The expected count for the ith exposure category in the kth stratum is

$$\hat{e}_{ik} = \frac{n_{ik} m_k}{n_k}.$$

With $i = 1$ as the reference category, let $\widehat{HR}_{\mathrm{mh}i}$ denote the Mantel–Haenszel hazard ratio estimate comparing the ith exposure category to the first category.

The Mantel–Haenszel test X_{pt}^2 has a generalization to the K (1 × I) setting, but the formula involves matrix algebra (Appendix E; Breslow and Day, 1987, p. 113). As shown in Appendix E, a conservative approximation to X_{pt}^2 is

$$X_{\mathrm{oe}}^2 = \sum_{i=1}^{I} \frac{(d_{i\bullet} - \hat{e}_{i\bullet})^2}{\hat{e}_{i\bullet}} \qquad (\mathrm{df} = I - 1). \qquad (10.39)$$

that is, $X_{\mathrm{oe}}^2 \le X_{\mathrm{pt}}^2$ (Clayton, 1982). Let s_i be the exposure level for the ith category with $s_1 < s_2 < \cdots < s_I$. For each k define

$$U_k = \sum_{i=1}^{I} s_i (d_{ik} - \hat{e}_{ik})$$

TABLE 10.17   Observed Counts and Person-Time: Censored Survival Data

	Exposure category						
	1	2	$\cdots$	i	$\cdots$	I	
deaths	d_{1k}	d_{2k}	$\cdots$	d_{ik}	$\cdots$	d_{Ik}	m_k
person-time	n_{1k}	n_{2k}	$\cdots$	n_{ik}	$\cdots$	n_{Ik}	n_k

and

$$V_k = \sum_{i=1}^{I} s_i^2 \hat{e}_{ik} - \left(\sum_{i=1}^{I} s_i \hat{e}_{ik} \right)^2 \bigg/ \hat{e}_{\bullet k}.$$

An overall test for linear trend is

$$X_t^2 = \frac{(U_\bullet)^2}{V_\bullet} \quad (df = 1). \tag{10.40}$$

A conservative approximation to (10.40) is

$$X_t^2 = \frac{\left[\sum_{i=1}^{I} s_i (d_{i\bullet} - \hat{e}_{i\bullet}) \right]^2}{\sum_{i=1}^{I} s_i^2 \hat{e}_{i\bullet} - \left(\sum_{i=1}^{I} s_i \hat{e}_{i\bullet} \right)^2 \bigg/ \hat{e}_{\bullet\bullet}} \quad (df = 1). \tag{10.41}$$

As illustrated in the following example, for censored survival data, X_{oe}^2 and (10.41) are usually sufficiently accurate approximations to X_{pt}^2 and (10.40) for practical purposes. There is an obvious similarity between (10.39) and (9.10) and between (10.41) and (9.11). We note, however, that for (9.10) and (9.11) the stratifying variable is "time."

Example 10.18 (Stage–Breast Cancer)   In this example we extend the analysis of Example 10.12 by stratifying by receptor level. The observed and expected counts and person-months are given in Table 10.18. Table 10.19 gives the Mantel–Haenszel

TABLE 10.18   Observed and Expected Counts and Person-Months: Stage–Breast Cancer

	Low receptor level				High receptor level			
	Stage				Stage			
	I	II	III		I	II	III	
observed	2	8	12	22	4	15	8	27
expected	6.98	10.90	4.12	22	10.87	13.76	2.38	27
person-months	650	1015	384	2049	2987	3782	653	7422

TABLE 10.19   Mantel–Haenszel Hazard Ratio Estimates and Greenland–Robins 95% Confidence Intervals: Stage–Breast Cancer

Stage	$\widehat{HR}_{mhi}$	$\underline{HR}_{mhi}$	$\overline{HR}_{mhi}$
II	2.82	1.15	6.92
III	9.66	3.67	25.46

hazard ratio estimates and Greenland–Robins 95% confidence intervals, with stage I as the reference category and with adjustment for receptor level. The adjusted estimates in Table 10.19 are close to the crude estimates in Table 10.11, suggesting that receptor level may not be an important confounder. The tests of association are $X^2_{pt} = 36.89$ ($p < .001$) and

$$X^2_{oe} = \frac{(6 - 17.85)^2}{17.85} + \frac{(23 - 24.66)^2}{24.66} + \frac{(20 - 6.50)^2}{6.50} = 36.03 \ (p < .001).$$

Setting $s_1 = 1$, $s_2 = 2$, and $s_3 = 1$, the tests for linear trend are (10.40) $= (25.35)^2/21.30 = 30.16$ ($p < .001$) and

$$(10.41) = \frac{(25.35)^2}{174.96 - (86.65)^2/49} = 29.58 \ (p < .001).$$

CHAPTER 11

Odds Ratio Methods for
Case-Control Data

11.1 JUSTIFICATION OF THE ODDS RATIO APPROACH

Cohort studies have a design that is intuitively appealing in that subjects are followed forward in time from exposure to the onset of disease, a temporal relationship that parallels causal mechanisms. In case-control studies, subjects with the disease (cases) and subjects who do not have the disease (controls) are sampled, and a history of exposure is determined retrospectively. It is sometimes said that in a case-control study, subjects are followed "backwards" in time from disease onset to exposure. The case-control design was developed in order to provide a method of studying diseases that are so rare that a cohort study would not be feasible. Due in large part to its retrospective nature, the case-control design is generally regarded as being methodologically complex (Austin et al., 1994). A few examples of the challenges inherent in the case-control design are described below. References for further reading on case-control studies are Schlesselman (1982) and Rothman and Greenland (1998).

11.1.1 Methodologic Issues in Case-Control Studies

In order for the results of a case-control study to be generalizable to the population as a whole, it is necessary for the sample of cases to be representative of individuals in the population who develop the disease, and likewise it is necessary for the sample of controls to be representative of those who do not. The point in the disease process when cases are sampled has implications for the validity of study findings. Accordingly, we distinguish between cases who are newly diagnosed (incident) and those who currently have the disease regardless of when onset occurred (prevalent). Consider a case-control study of an exposure that, unknown to the investigator, causes a particularly lethal form of the disease. If cases are recruited into the study at any time after the disease has developed, it is possible that individuals who would have been enrolled in the study if they had been contacted early in the course of their illness will be unable to participate due to debility or death. This means that the sample of cases will have fewer subjects with a history of the exposure of interest than there would have been if recruitment had been initiated immediately after diagnosis. As a result, the relationship between exposure and disease will appear weaker (biased toward the null) in the study data compared to the population. For this reason it is desirable to base case-control studies on incident rather than prevalent cases.

Collecting exposure data retrospectively is another of the methodologic challenges associated with the case-control design. It is easy to imagine that when exposure has occurred in the remote past, it may be difficult to ensure that details regarding exposure history (onset, duration, intensity, etc.) will be recalled accurately. Another problem is that individuals who develop the disease may be inclined to reflect on why this has occurred and, in particular, to search for past exposures that may have led to illness. Due to this aspect of human nature, cases are likely to provide a more complete exposure history than controls. In this situation the relationship between exposure and disease will appear stronger (biased away from the null) in the study data compared to the population.

In previous chapters we showed that it is possible to estimate epidemiologically meaningful parameters from cohort data. The case-control and cohort designs are so different it is reasonable to ask whether a useful measure of effect can be estimated from case-control data. There are a number of case-control study designs, two of which are described below. For the moment we assume that data have been collected on m_1 cases and m_2 controls using simple random sampling. Analogous to the closed cohort setting, we model outcomes using binomial distributions with parameters (ϕ_1, m_1) and (ϕ_2, m_2), where ϕ_1 is the probability that a case has a history of exposure, and ϕ_2 is the corresponding probability for controls. The expected values for the case-control study are given in Table 11.1. The odds ratio for the case-control study is

$$OR^* = \frac{\phi_1/(1 - \phi_1)}{\phi_2/(1 - \phi_2)} = \frac{\phi_1(1 - \phi_2)}{\phi_2(1 - \phi_1)} \tag{11.1}$$

where the asterisk is a reminder that the study has a case-control design. The intermediate equality in (11.1) is shown to emphasize that the odds are defined "across the rows" of Table 11.1. The interpretation of OR^* is quite different from the corresponding odds ratio from a closed cohort study: OR^* is the factor by which the odds of *exposure* increases or decreases when there is a history of disease. In fact we are primarily interested in estimating the factor by which the odds of *disease* increases

TABLE 11.1   Expected Values: Case-Control Study

Disease	Exposure		
	yes	no	
case	$\phi_1 m_1$	$(1 - \phi_1)m_1$	m_1
control	$\phi_2 m_2$	$(1 - \phi_2)m_2$	m_2

TABLE 11.2   Observed Counts: Oral Contraceptives–Myocardial Infarction

Myocardial infarction	Oral contraceptive yes	Oral contraceptive no	
case	29	205	234
control	135	1607	1742
	164	1812	1976

or decreases when there is a history of exposure. At this point it seems that OR^* is of little or no epidemiologic interest.

Example 11.1 (Oral Contraceptives–Myocardial Infarction)   Table 11.2 gives data from a case-control study investigating oral contraceptives as a risk factor for myocardial infarction (Shapiro et al., 1979). These data have been analyzed by Schlesselman (1982, p. 186). At the time this study was conducted, oral contraceptives contained relatively large amounts of estrogen, a female hormone that tends to elevate serum lipids and raise blood pressure, thereby increasing the risk of myocardial infarction (heart attack). For these data, $29/234 = 12.3\%$ of cases have a history of exposure compared to $135/1742 = 7.75\%$ of controls. The fact that oral contraceptive use is more common in cases than controls suggests that this medication may be associated with myocardial infarction. The estimated odds ratio is $\widehat{OR}^* = 1.68$, which has the interpretation that a history of oral contraceptive use is more likely in women who have had a myocardial infarction than in those who have remained well. This finding is of some interest, but it is not yet clear whether these data can be used to estimate the increase in the risk of myocardial infarction associated with using oral contraceptives.

11.1.2   Case-Control Study Nested in a Closed Cohort Study

Consider a closed cohort study in which the exposure is dichotomous, as depicted in Tables 2.1(a) and 2.1(b). We now describe a case-control design that is said to be nested in the closed cohort study. The cases for the case-control study are a simple random sample of subjects in the cohort study who develop the disease, and the controls are a simple random sample of subjects who remain well. Denote the sampling fractions for the cases and controls by γ_1 and γ_2, respectively, where $0 < \gamma_1 \leq 1$ and $0 < \gamma_2 \leq 1$. So, for example, the number of cases is $\gamma_1 m_1$. Since the cohort study is closed, all subjects who do not develop the disease remain under observation until the end of the period of follow-up. It is convenient to sample the controls at that time point in order to avoid the problem of selecting someone to be a control early in the study, only to have that person become a case later on. The expected values for the nested case-control study are given in Table 11.3.

TABLE 11.3   Expected Values: Case-Control
Study Nested in a Closed Cohort Study

Disease	Exposure		
	yes	no	
case	$\gamma_1 \pi_1 r_1$	$\gamma_1 \pi_2 r_2$	$\gamma_1 m_1$
control	$\gamma_2(1 - \pi_1)r_1$	$\gamma_2(1 - \pi_2)r_2$	$\gamma_2 m_2$

The odds ratio for the nested case-control study is

$$OR^* = \frac{(\gamma_1 \pi_1 r_1)/(\gamma_1 \pi_2 r_2)}{[\gamma_2(1 - \pi_1)r_1]/[\gamma_2(1 - \pi_2)r_2]} = \frac{\pi_1(1 - \pi_2)}{\pi_2(1 - \pi_1)}$$

which is precisely the odds ratio (2.1) for the closed cohort study. If we ignore the case-control design and treat Table 11.3 as if the "data" had been collected using a closed cohort design, the odds ratio is unchanged:

$$OR = \frac{(\gamma_1 \pi_1 r_1)/[\gamma_2(1 - \pi_1)r_1]}{(\gamma_1 \pi_2 r_2)/[\gamma_2(1 - \pi_2)r_2]} = \frac{\pi_1(1 - \pi_2)}{\pi_2(1 - \pi_1)}.$$

This means that we can use the odds ratio methods developed in Chapters 4 and 5 to analyze data from a case-control study that is nested in a closed cohort study. In Section 2.2.2 we observed that, if the disease is rare, the odds ratio and risk ratio from a closed cohort are approximately equal. In this situation, data from a nested case-control study can be used to estimate the risk ratio for the closed cohort study.

The above argument is often put forward as a justification for using odds ratio methods to analyze case-control data. The problem with this rationale is that the above study design is seldom used in practice. In particular, if a closed cohort study has been completed, it would be wasteful to analyze only a portion of the data using the nested approach. However, nesting a case-control study in a closed cohort study can be efficient when the disease is especially rare. In most cohort studies, detailed information on exposure and other variables is collected from all subjects at the time of enrollment into the cohort. When the disease is rare, only a few of these individuals will eventually develop the disease. Beyond a certain point, data on subjects who do not develop the disease contributes little to the substance of the study. For a rare disease, an alternative is to collect a minimum of information from each subject at the time of enrollment, conduct a nested case-control study with a small sampling fraction for controls, and then administer extensive questionnaires only to subjects in the case and control samples (Mantel, 1973; Langholz and Goldstein, 1996).

11.1.3   Case-Control Study Nested in an Open Cohort Study

Most case-control studies are conducted over a specific (calendar) time period, with incident cases and controls sampled from a well-defined population. For example, cases might be identified through a population-based registry such as a cancer registry, from the employment records of a large workforce, or through a network of medical clinics serving a defined catchment area. Controls are usually sampled from the population on an ongoing basis during the course of the study. We refer to such a case-control study as having an incidence design. We can think of the underlying population as the cohort in an open cohort study which is conducted over the time period of the case-control study (Rothman and Greenland, 1998, Chapter 7). Thus the cohort for the study consists of all individuals living in the population at the beginning of the time period as well as those entering the population through birth and in-migration. Censoring is permitted as a result of out-migration, death, and survival to the end of the time period without developing the disease of interest. In this way an incidence case-control study can be viewed as "nested" in an open cohort study conducted on the underlying population over a defined time period. The open cohort study considered here differs from the one described in Section 8.1 in that calendar time is retained as a time dimension. So the open cohort study we are discussing corresponds to Figure 8.1(a) rather than Figure 8.1(b). Since there is constant movement in and out of the population, the term "dynamic" cohort is sometimes used as an alternative to open cohort.

The hazard function occupied a central place in our earlier discussion of survival analysis. In that context the hazard function was expressed in terms of a single "time" variable that measured duration from the beginning of follow-up. A hazard function can be defined for an entire population, but now we must consider two time dimensions—calendar time and age. We define the hazard function for the population as follows: $r(x, t)$ is the instantaneous probability per unit time that a member of the population who is free of disease at age x and at time t will develop the disease in the next instant. For a given time t we can take the average of $r(x, t)$ across all ages x to get an overall hazard function for the population at time t, a quantity we denote by $r(t)$. We interpret $r(t)$ as the instantaneous probability per unit time that a randomly selected member of the population who is free of disease at time t will develop the disease in the next instant. In a similar fashion we can define hazard functions $r_1(t)$ and $r_2(t)$, which are specific to those with and without a history of exposure, respectively. Let $N_1(t)$ be the number of individuals in the population at time t with a history of exposure who are free of disease. Similarly, let $N_2(t)$ be the number of individuals in the population at time t without a history of exposure who are free of disease. So, at time t, there are $N_1(t) + N_2(t)$ individuals in the population "at risk" of disease.

We now invoke the stationary population assumption (Keyfitz, 1977). The nature of the stationary population assumption varies somewhat depending on the context, but in general it requires that specified features of the population be independent of (calendar) time. In the present setting, the stationary population assumption is taken to mean that $r_1(t)$, $r_2(t)$, $N_1(t)$, and $N_2(t)$ are each independent of t. We denote the constant values of these functions by R_1, R_2, N_1, and N_2, respectively. Suppose that the case-control study begins at time t_0 and continues until a later time $t_0 + \Delta$. Ordinarily Δ is no more than, say, 2 or 3 years, which is usually not enough time for the population to undergo a significant shift in demographic composition or a major change in the relationship between exposure and disease. Therefore it may be reasonable to regard a population as being approximately stationary for the duration of a case-control study. However, when the history of exposure goes back many years, the stationary population assumption is harder to justify.

As discussed in Appendix G, when there is no confounding, the population parameter of epidemiologic interest is R_1/R_2, which we refer to as the ratio of hazard rates. Suppose that the hazard functions in the exposed and unexposed populations which give rise to R_1 and R_2 satisfy the proportional hazards assumption, and denote the hazard ratio by HR. In Appendix G we show that, although R_1/R_2 does not generally equal HR, in practice, R_1/R_2 and HR will be very close in value. In light of results described below, this provides a link between the methods used for incidence case-control studies and those described for open cohort studies in Chapters 8–10.

During the time period from t_0 to $t_0 + \Delta$, the number of person-years experienced by cohort members with a history of exposure who are at risk of disease is $N_1\Delta$. It follows that the (expected) number of incident cases among these individuals is $R_1 N_1 \Delta$. Likewise, the number of incident cases among cohort members without a history of exposure who are at risk of disease is $R_2 N_2 \Delta$. Let γ_1 and γ_2 be the case and control sampling fractions, respectively, where $0 < \gamma_1 \leq 1$ and $0 < \gamma_2 \leq 1$. Then the (expected) numbers of exposed and unexposed cases are $\gamma_1 R_1 N_1 \Delta$ and $\gamma_1 R_2 N_2 \Delta$. This gives the top row of Table 11.4. Since the population is stationary, at any time t there are $N_1 + N_2$ subjects in the population who do not have the disease and are therefore eligible to be controls. For simplicity, we assume that all controls are sampled at the end of the cohort study in order to avoid the previously noted complication of a control selected early in the study becoming a case later on. This gives the second row of Table 11.4.

Therefore the odds ratio for the incidence case-control study is

$$OR^* = \frac{(\gamma_1 R_1 N_1 \Delta)/(\gamma_1 R_2 N_2 \Delta)}{(\gamma_2 N_1)/(\gamma_2 N_2)} = \frac{(R_1/R_2)(N_1/N_2)}{N_1/N_2} = \frac{R_1}{R_2} \tag{11.2}$$

(Miettinen, 1976). Note that nowhere in this derivation have we assumed that the disease is rare (Greenland and Thomas, 1982). From (11.2) we see that it is the odds for the cases $(R_1 N_1)/(R_2 N_2)$ that contains the crucial information about the ratio of hazard rates. The purpose of dividing by the odds for controls is to eliminate the factor N_1/N_2. If we ignore the case-control design and treat Table 11.4 as if the "data" had been collected using a closed cohort design, the odds ratio is unchanged:

TABLE 11.4   Expected Values: Case-Control Study Nested in an Open Cohort Study

Disease	Exposure		
	yes	no	
case	$\gamma_1 R_1 N_1 \Delta$	$\gamma_1 R_2 N_2 \Delta$	$\gamma_1 (R_1 N_1 + R_2 N_2)\Delta$
control	$\gamma_2 N_1$	$\gamma_2 N_2$	$\gamma_2 (N_1 + N_2)$

$$OR = \frac{(\gamma_1 R_1 N_1 \Delta)/(\gamma_2 N_1)}{(\gamma_1 R_2 N_2 \Delta)/(\gamma_2 N_2)} = \frac{R_1}{R_2}. \tag{11.3}$$

The fact that $OR = R_1/R_2$ leads to the following strategy for analyzing data from an incidence case-control study: Treat the data as if collected using a closed cohort design, use odds ratio methods for closed cohort studies to analyze the data, and interpret the results in terms of the ratio of hazard rates in the population. In practice it is usual to retain odds ratio terminology and let the interpretation in terms of the ratio of hazard rates remain implicit. It should be emphasized that the preceding strategy does not extend to analyses based on either the risk ratio or the risk difference. This accounts for the popularity of odds ratio methods in the analysis of case-control data. As mentioned earlier, when the proportional hazards assumption is satisfied, R_1/R_1 is generally very close in value to HR, and so from (11.3) we have the approximate equality $OR = HR$. In practice it is usual to identify the odds ratio from an incidence case-control study with the hazard ratio in the population. In Appendix G it is pointed out that under the stationary population assumption the counterfactual definition of confounding in an open cohort study can be adapted to the case-control setting.

The strategy presented above for analyzing data from an incidence case-control study does not extend to the standardized measures of effect that were discussed in Section 2.5.4. The reason is that the formulas presented there use marginal totals rather than only interior cell counts. However, it is possible to define a standardized measure of effect for the incidence case-control design. We return for the moment to the deterministic model of Chapter 2. In the notation of Table 5.1 the observed count is $O = a_{1\bullet}$ and the standardized expected count is defined to be

$$sE = \sum_{j=1}^{J} \frac{a_{2j} b_{1j}}{b_{2j}}$$

(Miettinen, 1972b; Greenland, 1982). From a stratified version of Table 11.4 and assuming that the same sampling fractions are used in each stratum, we have

$$O = \sum_{j=1}^{J} \gamma_1 R_{1j} N_{1j} \Delta = \gamma_1 \Delta \sum_{j=1}^{J} R_{1j} N_{1j}$$

and

$$sE = \sum_{j=1}^{J} \frac{(\gamma_1 R_{2j} N_{2j} \Delta)(\gamma_2 N_{1j})}{\gamma_2 N_{2j}} = \gamma_1 \Delta \sum_{j=1}^{J} R_{2j} N_{1j}.$$

The standardized hazard ratio for an incidence case-control study is defined to be

$$sHR = \frac{O}{sE} = \frac{\sum_{j=1}^{J} R_{1j} N_{1j}}{\sum_{j=1}^{J} R_{2j} N_{1j}}$$

which has an obvious similarity to (2.22) and (10.38).

TABLE 11.5(a)   Observed Counts: Oral Contraceptives–Myocardial Infarction

Disease	25–34 OC			35–44 OC			45–49 OC		
	yes	no		yes	no		yes	no	
case	13	14	27	10	98	108	6	93	99
control	95	614	709	35	692	727	5	301	306
	108	628	736	45	790	835	11	394	405

TABLE 11.5(b)   Asymptotic Unconditional Odds Ratio Estimates and 95% Confidence Intervals: Oral Contraceptives–Myocardial Infarction

Age group	$\widehat{OR}_{uj}$	$\underline{OR}_{uj}$	$\overline{OR}_{uj}$
25–34	6.00	2.74	13.16
35–44	2.02	.97	4.20
45–49	3.88	1.16	13.02

Example 11.2 (Oral Contraceptives–Myocardial Infarction)   The case-control study considered in Example 11.1 has an incidence design, and so we are free to apply the methods of Chapters 4 and 5. Our earlier misgivings about the interpretation of the odds ratio appear to have been unfounded. Provided there is no confounding, we can interpret $\widehat{OR}_u = 1.68$ as a crude estimate of the hazard ratio. However, confounding by age is a distinct possibility because this variable is a major risk factor for myocardial infarction and is also associated with oral contraceptive use. Table 11.5(a) gives the case-control data stratified by age group, and Table 11.5(b) gives the age-specific asymptotic unconditional analysis. The Mantel–Haenszel estimate of the odds ratio is $\widehat{OR}_{mh} = 3.34$, which is quite a bit larger than the crude estimate. Based on the RBG estimate of $\widehat{var}(\widehat{OR}_{mh})$, the 95% confidence interval for OR is [2.07, 5.38]. The Mantel–Haenszel test is $X^2_{mh} = 27.21$ ($p < .001$) and the Breslow–Day test of homogeneity is $X^2_{bd} = 4.09$ ($p = .13$), which includes the correction term of .00015. If we accept that age is a confounder, $\widehat{OR}_{mh} = 3.34$ can be interpreted as a summary estimate of the hazard ratio.

11.2   ODDS RATIO METHODS FOR MATCHED-PAIRS CASE-CONTROL DATA

When few cases are available for a case-control study, selecting controls using simple random sampling may be inefficient, especially when there are multiple confounders. For example, consider an incidence case-control study in which 50 cases are available and where the confounders are age (4), sex (2), socioeconomic status (3), and past medical history (2). The numbers in parentheses are the number of categories corresponding to each variable. After cross-classifying the confounders, there are $4 \times 2 \times 3 \times 2 = 48$ categories, almost as many as the number of cases. Suppose that the cases are thinly distributed across categories. A simple random sample of controls might have no subjects in several of the strata, even if the control sample is relatively large. When this occurs, strata in which there are cases but no controls are effectively dropped from the odds ratio analysis.

A way to avoid the problem of wasted cases is to match controls to cases based on the confounder profile of cases. In the preceding example, consider an incident case with a particular age, sex, socioeconomic status, and past medical history. With a matched design, one or more controls with the same confounder profile would be sampled from the population and linked (matched) to the case to create a matched set. We can think of the population from which the controls are selected as having been stratified according to confounder categories, thereby making the controls a stratified random sample. The distinguishing feature of the matched case-control design is that stratification is incorporated into the study at the sampling stage rather than at the time of data analysis. As a result of matching, cases and controls necessarily have the same distribution with respect to the matching variables, and so the matching variables are eliminated as sources of confounding. Unfortunately, this also means that the matching variables cannot be examined as risk factors in the data analysis (although they can still be assessed for effect modification). When matching is included as part of a case-control study, an already complicated design is made that much more complex. As an illustration of the problems that can result, consider that a matching variable that is not a confounder in the population can be turned into a confounder "in the data" as a result of matching (Rothman and Greenland, 1998, Chapter 10).

Matching brings a potential improvement in efficiency in the sense that the variance of the odds ratio estimate may be reduced compared to simple random sampling of cases and controls. However, whether the anticipated gain in efficiency is realized depends on a number of considerations: the exposure–disease–confounder associations (Kupper et al., 1981; Thomas and Greenland, 1983, 1985), the way matched sets are formed (Brookmeyer et al., 1986), and the relative costs associated with gathering information on cases and controls (Miettinen, 1969; Walter, 1980a; Thompson et al., 1982). One of the determinants of the success of a matched case-control study is the feasibility of finding controls with the desired confounder profiles. Methods are available for estimating the expected number of matches (McKinlay, 1974; Walter, 1980b). For further reading on matching see, for example, Anderson et al. (1980), Breslow and Day (1987), and Rothman and Greenland (1998).

When each case is matched to a single control, the case-control study is said to have a matched-pairs design. Pair-matching can be thought of as an extreme form of stratification in which each stratum consists of a single case and a single control. In keeping with the notation of Chapter 5, we denote the number of matched-pairs by J. If J is sufficiently large the sparse-strata conditions discussed in Section 5.2 are satisfied and so asymptotic conditional and MH–RBG methods can be used to

TABLE 11.6   Configurations: Matched-Pairs Case-Control Study

$f_{(1,1)}$				$f_{(1,0)}$			
Disease	Exposure			Disease	Exposure		
	yes	no			yes	no	
case	1	0	1	case	1	0	1
control	1	0	1	control	0	1	1
	2	0	2		1	1	2

$f_{(0,1)}$				$f_{(0,0)}$			
Disease	Exposure			Disease	Exposure		
	yes	no			yes	no	
case	0	1	1	case	0	1	1
control	1	0	1	control	0	1	1
	1	1	2		0	2	2

analyze the data. The formulas given in Chapter 5 can be applied directly, but the matched-pairs design results in certain simplifications, as demonstrated below.

Corresponding to each matched pair, there is a 2×2 table of the form of Table 5.1 with $m_{1j} = m_{2j} = 1$. Since each case and each control is either exposed or unexposed, there are four possible configurations as shown in Table 11.6. For example, the upper right configuration corresponds to a matched pair in which the case has a history of exposure but the control does not. We refer to this configuration as being of type $(1, 0)$ and denote the number of matched-pairs having this configuration by $f_{(1,0)}$. Similar definitions apply to the remaining configurations. The configurations of type $(1, 0)$ and $(0, 1)$ are said to be discordant because the members of each matched pair have different exposure histories. The configurations of type $(1, 1)$ and $(0, 0)$ are referred to as concordant. The configurations are depicted more compactly in Table 11.7, and the numbers of configurations are given in Table 11.8. Since there are J strata (matched pairs), we have $J = f_{(1,1)} + f_{(1,0)} + f_{(0,1)} + f_{(0,0)}$.

TABLE 11.7   Configurations: Matched-Pairs Case-Control Study $(m = 0, 1)$

$f_{(1,m)}$				$f_{(0,m)}$			
Disease	Exposure			Disease	Exposure		
	yes	no			yes	no	
case	1	0	1	case	0	1	1
control	m	$1 - m$	1	control	m	$1 - m$	1
	$1 + m$	$1 - m$	2		m	$2 - m$	2

TABLE 11.8 Observed Numbers
of Configurations: Matched-Pairs
Case-Control Study

Case	Control	
	exposed	unexposed
exposed	$f_{(1,1)}$	$f_{(1,0)}$
unexposed	$f_{(0,1)}$	$f_{(0,0)}$

11.2.1 Asymptotic Conditional Analysis

In this section we apply the asymptotic conditional methods of Section 5.2 to the matched-pairs design (Miettinen, 1970). We assume in what follows that the odds ratio is homogeneous across strata. For a configuration of type $(1,1)$, denote the mean (5.21) and variance (5.22) of the corresponding hypergeometric distribution by $E_{(1,1)}$ and $V_{(1,1)}$, and likewise for the other configurations. We then have

$$
\begin{aligned}
E_{(1,1)} &= 1 & V_{(1,1)} &= 0 \\
E_{(1,0)} &= \frac{OR}{OR+1} & V_{(1,0)} &= \frac{OR}{(OR+1)^2} \\
E_{(0,1)} &= \frac{OR}{OR+1} & V_{(0,1)} &= \frac{OR}{(OR+1)^2} \\
E_{(0,0)} &= 0 & V_{(0,0)} &= 0.
\end{aligned}
\tag{11.4}
$$

It is not surprising that $E_{(1,0)} = E_{(0,1)}$ and $V_{(1,0)} = V_{(0,1)}$ because hypergeometric means and variances are determined by marginal totals, and the discordant pairs have the same marginal totals. From (11.4), the left-hand side of the conditional maximum likelihood equation (5.23) is

$$
\begin{aligned}
a_{1\bullet} &= \left[f_{(1,1)} \times 1 \right] + \left[f_{(1,0)} \times 1 \right] + \left[f_{(0,1)} \times 0 \right] + \left[f_{(0,0)} \times 0 \right] \\
&= f_{(1,1)} + f_{(1,0)}
\end{aligned}
$$

and the right-hand side is

$$
\begin{aligned}
&\left[f_{(1,1)} \hat{E}_{(1,1)} \right] + \left[f_{(1,0)} \hat{E}_{(1,0)} \right] + \left[f_{(0,1)} \hat{E}_{(0,1)} \right] + \left[f_{(0,0)} \hat{E}_{(0,0)} \right] \\
&= f_{(1,1)} + \frac{\left[f_{(1,0)} + f_{(0,1)} \right] \widehat{OR}_c}{\widehat{OR}_c + 1}.
\end{aligned}
$$

So the conditional maximum likelihood equation is

$$f_{(1,0)} = \frac{\left[f_{(1,0)} + f_{(0,1)} \right] \widehat{OR}_c}{\widehat{OR}_c + 1} \qquad (11.5)$$

which can be solved for $\widehat{OR}_c$ to give

$$\widehat{OR}_c = \frac{f_{(1,0)}}{f_{(0,1)}}$$

(Kraus, 1960). In Appendix H we show that, if the unconditional maximum likelihood approach is used, the estimate is $\widehat{OR}_u = \left[f_{(1,0)}/f_{(0,1)} \right]^2$ (Andersen, 1973, p. 69; Breslow, 1981). This demonstrates that unconditional methods can lead to bias when applied in the sparse-strata setting. From (5.25) and (11.4),

$$\hat{V}_c = \left[f_{(1,1)} \hat{V}_{(1,1)} \right] + \left[f_{(1,0)} \hat{V}_{(1,0)} \right] + \left[f_{(0,1)} \hat{V}_{(0,1)} \right] + \left[f_{(0,0)} \hat{V}_{(0,0)} \right]$$

$$= \frac{\left[f_{(1,0)} + f_{(0,1)} \right] \widehat{OR}_c}{(\widehat{OR}_c + 1)^2} = \frac{f_{(1,0)} f_{(0,1)}}{f_{(1,0)} + f_{(0,1)}}.$$

So, from (5.26), an estimate of $\mathrm{var}(\log \widehat{OR}_c)$ is

$$\widehat{\mathrm{var}}(\log \widehat{OR}_c) = \frac{1}{f_{(1,0)}} + \frac{1}{f_{(0,1)}}$$

and a $(1 - \alpha) \times 100\%$ confidence for OR is obtained by exponentiating

$$\log \left[\frac{f_{(1,0)}}{f_{(0,1)}} \right] \pm z_{\alpha/2} \sqrt{\frac{1}{f_{(1,0)}} + \frac{1}{f_{(0,1)}}}.$$

From (11.4), under the hypothesis of no association $H_0 : OR = 1$, the expected counts and variance estimates are

$$\hat{e}_{(1,1)} = 1 \qquad \hat{v}_{0(1,1)} = 0$$

$$\hat{e}_{(1,0)} = \frac{1}{2} \qquad \hat{v}_{0(1,0)} = \frac{1}{4}$$

$$\hat{e}_{(0,1)} = \frac{1}{2} \qquad \hat{v}_{0(0,1)} = \frac{1}{4}$$

$$\hat{e}_{(0,0)} = 0 \qquad \hat{v}_{0(0,0)} = 0.$$

It follows that

$$\hat{e}_{1\bullet} = \left[f_{(1,1)} \times 1 \right] + \left[f_{(1,0)} \times \frac{1}{2} \right] + \left[f_{(0,1)} \times \frac{1}{2} \right] + \left[f_{(0,0)} \times 0 \right]$$

$$= f_{(1,1)} + \frac{f_{(1,0)} + f_{(0,1)}}{2}$$

and

$$\hat{v}_{0\bullet} = \left[f_{(1,1)} \times 0 \right] + \left[f_{(1,0)} \times \frac{1}{4} \right] + \left[f_{(0,1)} \times \frac{1}{4} \right] + \left[f_{(0,0)} \times 0 \right]$$

$$= \frac{f_{(1,0)} + f_{(0,1)}}{4}.$$

So the Mantel–Haenszel test of association (5.29) is

$$X^2_{mh} = \frac{\left[f_{(1,0)} - f_{(0,1)} \right]^2}{f_{(1,0)} + f_{(0,1)}} \qquad (df = 1).$$

An important observation is that the formulas for $\widehat{OR}_c$, $\widehat{\text{var}}(\log \widehat{OR}_c)$, and X^2_{mh} use data from discordant pairs only. This means that information collected from concordant pairs, which may represent much of the effort going into the study, is ignored. Estimates of the odds ratio have been developed which make use of data on concordant as well as discordant pairs (Liang and Zeger, 1988; Kalish, 1990). Since all the data are used, the variance of the resulting odds ratio estimate is reduced compared to $\widehat{OR}_c$. However, this gain in efficiency comes at the cost of introducing a degree of bias into the odds ratio estimate.

Suppose that the pair-matching is broken and that we collapse the data into a single 2×2 table. From Table 11.8 the number of cases with a history of exposure is $f_{(1,1)} + f_{(1,0)}$, the number of controls with a history of exposure is $f_{(1,1)} + f_{(0,1)}$, and so on. The resulting crude table is given in Table 11.9. Note that the sum over all interior cells in Table 11.9 is $2J$, the number of subjects in the study. The crude asymptotic unconditional estimate of the odds ratio is

$$\widehat{OR} = \frac{\left[f_{(1,1)} + f_{(1,0)} \right]\left[f_{(1,0)} + f_{(0,0)} \right]}{\left[f_{(0,1)} + f_{(0,0)} \right]\left[f_{(1,1)} + f_{(0,1)} \right]}.$$

It can be shown that $\widehat{OR}$ is biased toward the null compared to the stratified estimate $\widehat{OR}_c$ (Siegel and Greenhouse, 1973; Armitage, 1975; Breslow and Day, 1980, §7.6). An illustration is provided in Example 13.3. This is another manifestation of

TABLE 11.9 Observed Counts after Breaking Matches: Matched-Pairs Case-Control Study

Disease	Exposure		
	yes	no	
case	$f_{(1,1)} + f_{(1,0)}$	$f_{(0,1)} + f_{(0,0)}$	J
control	$f_{(1,1)} + f_{(0,1)}$	$f_{(1,0)} + f_{(0,0)}$	J

the inequalities relating OR and θ that were discussed in Section 2.4.5, where OR and θ now represent the crude and pair-matched odds ratios, respectively. To translate the results of Section 2.4.5 into the present context, the roles of E and D must be reversed, so that π_{1j} and π_{2j} become probabilities of exposure. Note that for the matched-pairs design, $p_{1j} = p_{2j} = 1/J$ and so (2.15) simplifies accordingly. Liang (1987) describes a test of homogeneity which is applicable to the matched-pairs design.

11.2.2 Mantel–Haenszel and Robins–Breslow–Greenland Estimates

To derive the Mantel–Haenszel estimate of the odds ratio and the RBG variance estimate for matched-pairs case-control data, we argue as in the preceding section and obtain

$$R_{\bullet} = \frac{f_{(1,0)}}{2} \qquad S_{\bullet} = \frac{f_{(0,1)}}{2} \qquad T_{\bullet} = \frac{f_{(1,0)}}{2}$$

$$U_{\bullet} = 0 \qquad V_{\bullet} = 0 \qquad W_{\bullet} = \frac{f_{(0,1)}}{2}.$$

It follows that

$$\widehat{OR}_{\mathrm{mh}} = \frac{f_{(1,0)}}{f_{(0,1)}}$$

and

$$\widehat{\mathrm{var}}(\log\widehat{OR}_{\mathrm{mh}}) = \frac{1}{f_{(1,0)}} + \frac{1}{f_{(0,1)}}. \tag{11.6}$$

These are precisely the estimates based on the asymptotic conditional approach.

11.2.3 Conditional Methods for Discordant Pairs

The asymptotic conditional, Mantel–Haenszel, and RBG estimates considered above are based exclusively on discordant pairs. Another method of analyzing matched-pairs case-control data begins by conditioning on the observed number of discordant pairs $f_{(1,0)} + f_{(0,1)}$ (Miettinen, 1970). For a given discordant pair, either the case or the control has a history of exposure. Let Π denote the probability that in a discordant pair it is the case who has been exposed. From (5.20) and in the notation of (11.4), the hypergeometric probabilities are $P_{(1,0)} = OR/(OR+1)$ and $P_{(0,1)} = 1/(OR+1)$. Therefore

$$\Pi = \frac{P_{(1,0)}}{P_{(1,0)} + P_{(0,1)}} = \frac{OR}{OR + 1} \tag{11.7}$$

and so

$$OR = \frac{\Pi}{1 - \Pi}.$$ (11.8)

Therefore the odds ratio we wish to estimate is equal to the odds from a binomial distribution with parameters (Π, r), where $r = f_{(1,0)} + f_{(0,1)}$. With $a = f_{(1,0)}$ we have the estimate

$$\hat{\Pi} = \frac{a}{r} = \frac{f_{(1,0)}}{f_{(1,0)} + f_{(0,1)}}$$

and so

$$\widehat{OR} = \frac{\hat{\Pi}}{1 - \hat{\Pi}} = \frac{f_{(1,0)}}{f_{(0,1)}}.$$

It follows from (3.12) that an estimate of var(log $\widehat{OR}$) is

$$\widehat{\text{var}}(\log\widehat{OR}) = \frac{1}{\hat{\Pi}(1 - \hat{\Pi})r} = \frac{1}{f_{(1,0)}} + \frac{1}{f_{(0,1)}}.$$

From (11.7), $OR = 1$ is equivalent to $\Pi_0 = 1/2$. Based on (3.9), a test of association is

$$X_m^2 = \frac{(\hat{\Pi} - 1/2)^2}{1/(4r)} = \frac{\left[f_{(1,0)} - f_{(0,1)}\right]^2}{f_{(1,0)} + f_{(0,1)}} \qquad (\text{df} = 1)$$

which is referred to as McNemar's test (McNemar, 1947). It is of note that the above formulas are identical to those based on the asymptotic conditional, Mantel–Haenszel, and RBG methods. A feature of the present approach is that it is amenable to exact binomial calculations, an option that is useful when the number of discordant pairs is small.

Example 11.3 (Estrogen–Endometrial Cancer)   Table 11.10 gives data from a matched-pairs case-control study investigating estrogen use as a risk factor for en-

TABLE 11.10   Observed Counts of Matched-Pairs: Estrogen–Endometrial Cancer

Case	Control	
	exposed	unexposed
exposed	12	43
unexposed	7	121

TABLE 11.11 Observed Counts
after Breaking Matches:
Estrogen–Endometrial Cancer

Cancer	Estrogen yes	no	
case	55	128	183
control	19	164	183

dometrial cancer (Antunes et al., 1979). These data have been analyzed by Schlessel-man (1982, p. 209). The point estimate is $\widehat{OR} = 43/7 = 6.14$, the variance estimate is $\widehat{\text{var}}(\log \widehat{OR}) = (1/43) + (1/7) = .166$, and the 95% confidence interval for OR is [2.76, 13.66]. The test of association is $X^2 = (43-7)^2/(43+7) = 25.92$ ($p < .001$). So there is considerable evidence that estrogen use is associated with an increased risk of endometrial cancer. Based on (3.3) and (3.4), with $a = 43$ and $r = 50$, the exact 95% confidence interval for Π is [.733, .942]. Transforming using (11.8), the exact 95% confidence interval for OR is [2.74, 16.18], which is somewhat wider than the asymptotic interval. If the pair-matching is broken, we obtain Table 11.11, from which $\widehat{OR} = 3.71$. The crude estimate of the odds ratio is much smaller than the matched estimate, suggesting that the matching variables are important confounders.

11.3 ODDS RATIO METHODS FOR (1 : M) MATCHED CASE-CONTROL DATA

The matched-pairs design for case-control studies can be generalized to (1 : M) matching in which each case is matched to exactly M controls, where $M \geq 1$. Corresponding to Tables 11.7 and 11.8, with (1 : M) matching we have Tables 11.12 and 11.13. In this notation there are $f_{(0,0)} + f_{(1,M)}$ concordant matched sets.

TABLE 11.12 Configurations: (1 : M) Matched Case-Control Study ($m = 0, 1, 2, \ldots, M$)

Disease	$f_{(1,m)}$ Exposure yes	no		Disease	$f_{(0,m)}$ Exposure yes	no	
case	1	0	1	case	0	1	1
control	m	$M - m$	M	control	m	$M - m$	M
	$1 + m$	$M - m$	$M + 1$		m	$M + 1 - m$	$M + 1$

TABLE 11.13   Observed Numbers of Configurations: (1 : M) Matched Case-Control Study

Case	Number of exposed controls				
	0	1	$\cdots$ m $\cdots$		M
exposed	$f_{(1,0)}$	$f_{(1,1)}$	$\cdots$	$f_{(1,m)}$ $\cdots$	$f_{(1,M)}$
unexposed	$f_{(0,0)}$	$f_{(0,1)}$	$\cdots$	$f_{(0,m)}$ $\cdots$	$f_{(0,M)}$

11.3.1   Asymptotic Conditional Analysis

The asymptotic conditional formulas for (1 : M) matching given below are due to Miettinen (1969, 1970); see Appendix H for derivations. The conditional maximum likelihood equation is

$$\sum_{m=1}^{M} f_{(1,m-1)} = \widehat{OR}_c \sum_{m=1}^{M} \frac{\left[f_{(1,m-1)} + f_{(0,m)} \right] m}{m\widehat{OR}_c + M + 1 - m}.$$

An estimate of var(log $\widehat{OR}_c$) is

$$\widehat{var}(\log\widehat{OR}_c) = \frac{1}{\hat{V}_c}$$

where

$$\hat{V}_c = \widehat{OR}_c \sum_{m=1}^{M} \frac{\left[f_{(1,m-1)} + f_{(0,m)} \right] m(M + 1 - m)}{(m\widehat{OR}_c + M + 1 - m)^2}.$$

The Mantel–Haenszel test of association is

$$X_{mh}^2 = \left(\sum_{m=1}^{M} f_{(1,m-1)} - \sum_{m=1}^{M} \frac{\left[f_{(1,m-1)} + f_{(0,m)} \right] m}{M + 1} \right)^2 \Bigg/ \\ \sum_{m=1}^{M} \frac{\left[f_{(1,m-1)} + f_{(0,m)} \right] m(M + 1 - m)}{(M + 1)^2}.$$

11.3.2   Mantel–Haenszel and Robins–Breslow–Greenland Estimates

For (1 : M) matched case-control data, the Mantel–Haenszel odds ratio estimate is

$$\widehat{OR}_{mh} = \frac{R_\bullet}{S_\bullet} = \frac{\sum_{m=1}^{M} f_{(1,m-1)}(M + 1 - m)}{\sum_{m=1}^{M} f_{(0,m)}m}$$

and the RBG variance estimate is

$$\widehat{\text{var}}(\log \widehat{OR}_{\text{mh}}) = \frac{T_\bullet}{2(R_\bullet)^2} + \frac{U_\bullet + V_\bullet}{2(R_\bullet)(S_\bullet)} + \frac{W_\bullet}{2(S_\bullet)^2}$$

where

$$R_\bullet = \frac{1}{M+1} \sum_{m=1}^{M} f_{(1,m-1)}(M+1-m)$$

$$S_\bullet = \frac{1}{M+1} \sum_{m=1}^{M} f_{(0,m)}m$$

$$T_\bullet = \frac{1}{(M+1)^2} \sum_{m=1}^{M} f_{(1,m-1)}(M+1-m)(M+2-m)$$

$$U_\bullet = \frac{1}{(M+1)^2} \sum_{m=1}^{M} f_{(0,m)}m(M-m)$$

$$V_\bullet = \frac{1}{(M+1)^2} \sum_{m=1}^{M} f_{(1,m-1)}(m-1)(M+1-m)$$

$$W_\bullet = \frac{1}{(M+1)^2} \sum_{m=1}^{M} f_{(0,m)}m(m+1).$$

Note that the $f_{(0,0)} + f_{(1,M)}$ concordant matched sets do not contribute terms to the above formulas. The methods of Sections 11.3.1 and 11.3.2 are generalizations of the matched-pairs techniques presented earlier, in that when $M = 1$ the $(1 : M)$ formulas simplify to the corresponding matched-pairs formulas.

When the number of cases is small, matching several controls to each case provides a way of increasing the sample size of the study and thereby reducing random error. The relative efficiency of $(1 : M)$ matching compared to pair-matching equals $2M/(M+1)$ (Ury, 1975; Breslow and Day, 1980, p. 169; Walter, 1980a). As M gets larger, this quantity increases toward a limiting value of 2, but the rate of increase diminishes rapidly once M exceeds 5. So there is a ceiling beyond which little is to be gained from recruiting additional controls into a matched case-control study. In practice it would be unusual to match more than four or five controls to each case.

Even when the study design calls for $(1 : M)$ matching, it may not be possible to match precisely M controls to each case. This can occur, for example, as a consequence of especially stringent matching criteria or because the population from which to sample controls is small. The $(1 : M)$ methods can be generalized further to accommodate variable numbers of controls, but the formulas are even more unwieldy. In this situation it is more convenient to apply the sparse-strata methods of Chapter 5 directly, just as they would be employed in any other stratified analysis.

TABLE 11.14 Observed Numbers of
Configurations: Estrogen–Endometrial Cancer

Case	Number of exposed controls					
	0	1	2	3	4	
exposed	1	10	10	10	2	33
unexposed	0	1	1	1	0	3

Example 11.4 (Estrogen–Endometrial Cancer)   The (1:4) matched case-control data given in Table 11.14 are taken from a study of estrogen use as a risk factor for endometrial cancer (Mack et al., 1976). These data have been analyzed by Breslow and Day (1987, p. 175). The conditional maximum likelihood equation is

$$31 = \widehat{OR}_c \left[\frac{2}{\widehat{OR}_c + 4} + \frac{22}{2\widehat{OR}_c + 3} + \frac{33}{3\widehat{OR}_c + 2} + \frac{40}{4\widehat{OR}_c + 1} \right]$$

which has the solution $\widehat{OR}_c = 9.76$. From

$$\hat{V}_c = 9.76 \left(\frac{8}{[9.76 + 4]^2} + \frac{66}{[2(9.76) + 3]^2} + \frac{66}{[3(9.76) + 2]^2} + \frac{40}{[4(9.76) + 1]^2} \right)$$
$$= 2.58$$

the asymptotic conditional variance estimate is $\widehat{var}(\log \widehat{OR}_c) = 1/2.58 = .387$ and the 95% confidence interval for OR is [2.88, 33.03]. By contrast, the Mantel–Haenszel odds ratio estimate is $\widehat{OR}_{mh} = 12.80/1.20 = 10.67$, the RBG variance estimate is

$$\widehat{var}(\log \widehat{OR}_{mh}) = \frac{8.80}{2(12.80)^2} + \frac{.40 + 4.00}{2(12.80)(1.20)} + \frac{.80}{2(1.20)^2} = .448$$

TABLE 11.15 Observed Counts
after Breaking Matches:
Estrogen–Endometrial Cancer

Cancer	Estrogen		
	yes	no	
case	33	3	36
control	35	1	36

and the 95% confidence interval for OR is [2.87, 39.60]. The Mantel–Haenszel test of association is $X^2_{mh} = (31.00 - 19.40)^2/7.20 = 18.69$ ($p < .001$). If the matching is broken, we obtain Table 11.15, from which $\widehat{OR} = .314$. This estimate is not even on the same side of 1 as the matched estimates, suggesting that the matching variables are important confounders.

CHAPTER 12

Standardized Rates and Age–Period–Cohort Analysis

The analysis of death rates over time and across geographic regions has an important place in the surveillance of disease trends. When crude death rates are compared, it is necessary to account for whatever differences there may be in the age distributions of the populations; otherwise, spurious conclusions may be reached. In this chapter we describe age standardization, a classical approach to adjusting for differences in age and other demographic variables. When age-specific death rates are examined over time, there is usually a mixing of effects due to (time) period and (birth) cohort. Age–period–cohort analysis attempts to disentangle the influences of these three time-related variables on the pattern of death rates. Graphical and multivariate methods of age–period–cohort analysis are briefly described and their properties are discussed.

12.1 POPULATION RATES

In Section 10.1.2 we defined a rate to be a parameter that can be interpreted as the number of events in a cohort divided by the corresponding amount of person-time. The term rate is used throughout epidemiology to denote a variety of different indices, not all of which conform to this usage. Following are a few examples of the use of this term:

$$\text{Case fatality rate} = \frac{\text{Number of cases dying of the disease}}{\text{Number of cases}}$$

$$\text{Point prevalence rate} = \frac{\text{Number of existing cases at a given time point}}{\text{Population at time point}}$$

$$\text{Annual death rate} = \frac{\text{Number of deaths during a calendar year}}{\text{Midyear population}}$$

$$\text{Annual incidence rate} = \frac{\text{Number of incident cases during a calendar year}}{\text{Midyear population}}.$$

For both the case fatality rate and the point prevalence rate, the numerator counts events occurring to subjects in the denominator; that is, the numerator is "contained in" the denominator. Accordingly, these quotients should be thought of as probabilities or proportions, not as rates. At first glance it seems that the annual death rate ought to have a similar interpretation. However, those individuals who die during the first half of the year are not counted in the midyear population, and so the annual death rate is not a probability. As discussed below, under certain conditions the midyear population can be viewed as an estimate of the number of person-years experienced by the population, and so the annual death rate can be interpreted as a rate in the above sense of the term. Similar remarks apply to the annual incidence rate.

Usually population rates are based on data for a given calendar year and therefore are referred to as annual rates. Unless stated otherwise, all rates will be annual rates. As in Section 11.1.3 we consider the population to be an open cohort. Let D be the number of events occurring in the population during the year, such as the number of deaths from a specific cause or the number of incident cases of a particular disease. If we assume that the event is rare, we can treat D as a Poisson random variable. Following Section 11.1.3, let $r(t)$ be the population hazard function and let $N(t)$ be the number of individuals in the population at time t who are at risk of the event. Given that we are considering the population over a single year, it is reasonable to invoke the stationary population assumption. Accordingly, we assume that $N(t)$ and $r(t)$ are independent of t and denote their constant values by N and R, respectively. It follows that the number of person-years experienced by the population during the year is N. So D is Poisson with parameters (R, N). We can now apply the methods of Section 10.1 for the Poisson analysis of a single sample. In particular we have the estimates $\hat{R} = D/N$ and $\widehat{\text{var}}(\hat{R}) = D/N^2$. It is shown below that this same approach can be applied to comparative studies.

The data needed to estimate R depends on the event under consideration. If R is the death rate for all causes, then D is the annual number of deaths. In this case, the entire population is at risk and so N is the (number of individuals in the) midyear population. When the event of interest is more narrowly defined, some modification is necessary. For example, if R is the death rate for a particular age group, the preceding approach can be used, except that now both D and N are restricted to the age group in question. This results in what is referred to as an age-specific death rate. Now suppose that R is the incidence rate for a certain disease. Let I denote the annual number of incident cases in the population as determined, for example, by a disease registry. Those members of the population who have already developed the disease—that is, the prevalent cases—are not included in the population at risk. Let P denote the number of prevalent cases at midyear and, as before, let N be the midyear population. Then $\hat{R} = I/(N - P)$ is an estimate of the annual incidence rate. For most diseases, P will be small compared to N, and so for practical purposes it is sufficient to use the approximation $\hat{R} = I/N$.

For the remainder of this chapter we frame the discussion in terms of annual death rates for all causes, but the concepts carry over immediately to cause-specific death rates and incidence rates. Most of the discussion will be expressed in terms of estimates rather than parameters. For convenience we drop the caret $\widehat{}$ from notation except for variance estimates.

12.2 DIRECTLY STANDARDIZED DEATH RATE

The following discussion is presented in terms of a given population that we refer to as population $\mathcal{A}$. Following the notation of Chapter 9, partition the life span into $K + 1$ age groups: $[x_0, x_1), [x_1, x_2), \ldots, [x_k, x_{k+1}), \ldots, [x_{K-1}, x_K), [x_K, x_{K+1}]$, where $x_0 = 0$ and x_{K+1} is the upper limit of the life span. We refer to $[x_k, x_{k+1})$ as the kth age group. For this age group, let D_{ak} be the annual number of deaths in population $\mathcal{A}$ and let N_{ak} be the midyear population. Evidently, the total number of deaths in population $\mathcal{A}$ over the course of the year is $D_a = \sum_{k=0}^{K} D_{ak}$ and the total midyear population is $N_a = \sum_{k=0}^{K} N_{ak}$. We refer to $R_a = D_a / N_a$ as the crude death rate and to $R_{ak} = D_{ak} / N_{ak}$ as the age-specific death rate for the kth age group. It is readily demonstrated that

$$R_a = \sum_{k=0}^{K} \left(\frac{N_{ak}}{N_a} \right) R_{ak} \qquad (12.1)$$

and so the crude death rate is a weighted average of the age-specific death rates, where the weights N_{ak}/N_a are determined by the age distribution of the population.

We now consider methods for comparing death rates in population $\mathcal{A}$ to those in another population, which we refer to as population $\mathcal{B}$. The crude rate ratio comparing population $\mathcal{A}$ to population $\mathcal{B}$ is defined to be

$$CRR = \frac{R_a}{R_b}.$$

We are also interested in the age-specific rate ratios R_{ak}/R_{bk}. As shown in the following example, crude and age-specific ratios can sometimes lead to contradictory findings.

Example 12.1 (Hypothetical Data) Table 12.1 gives hypothetical data for two populations in which there are only two age groups: young and old. The crude death rates are $R_a = .003$ and $R_b = .005$, and so the crude rate ratio is $CRR = .003/.005 = .6$. However, for both the young and old age groups, the age-specific

TABLE 12.1 Crude and Age-Specific Death Rates for Populations $\mathcal{A}$ and $\mathcal{B}$

Age group	Population $\mathcal{A}$			Population $\mathcal{B}$			Rate ratio
	D_{ak}	N_{ak}	R_{ak}	D_{bk}	N_{bk}	R_{bk}	
Young	18	9,000	.002	2	2,000	.001	2
Old	12	1,000	.012	48	8,000	.006	2
Crude	30	10,000	.003	50	10,000	.005	.6

rate ratios are $R_{ak}/R_{bk} = 2$. Depending on whether we rely on the crude rate ratio or the age-specific rate ratios, we are led to different conclusions about the mortality risk in population $\mathcal{A}$ compared to population $\mathcal{B}$.

The paradoxical findings in Example 12.1 arise because the two populations have such different age distributions. For each population, the death rate in the older age group is six times that in the younger age group. However, most of population $\mathcal{B}$ is old and so overall there are more deaths in this population than in population $\mathcal{A}$. As a result the crude death rate is larger in population $\mathcal{B}$ than in population $\mathcal{A}$. Since the deaths rates vary according to age group, and the age distributions are different, it is appropriate to view age as a confounder of the association between "place of residence" and mortality risk. We seek a method of comparing overall death rates across populations which controls for the confounding effects of age.

An approach to the problem is suggested by the form of (12.1) where the age-specific death rates in the population and the age distribution of the population appear separately. Specifically, we replace the actual age distribution of the population with the age distribution of a reference population, which we refer to as the standard population. In Chapter 13 we discuss briefly the issue of how to select an appropriate standard population for a given application. Suppose that S is such a standard population where N_{sk} is the number of individuals in the kth age group and $N_s = \sum_{k=0}^{K} N_{sk}$. The directly standardized death rate for population $\mathcal{A}$ is defined to be

$$R_{a(s)} = \sum_{k=0}^{K} \left(\frac{N_{sk}}{N_s} \right) R_{ak}. \tag{12.2}$$

$R_{a(s)}$ is a weighted average of age-specific death rates, where the weights are given by the age distribution of the standard population. From (1.8) an estimate of $\mathrm{var}(R_{a(s)})$ is

$$\widehat{\mathrm{var}}(R_{a(s)}) = \sum_{k=0}^{K} \left(\frac{N_{sk}}{N_s} \right)^2 \widehat{\mathrm{var}}(R_{ak})$$

$$= \sum_{k=0}^{K} \left(\frac{N_{sk}}{N_s} \right)^2 \frac{D_{ak}}{(N_{ak})^2}. \tag{12.3}$$

The estimate given by (12.3) may be unreliable when the number of deaths in each age group is small. Dobson et al. (1991) give a method of estimating the variance which is suited to such circumstances.

For direct standardization, all that we need to know about the standard population is its age distribution. So it is not necessary to actually specify the number of individuals in each age group. Indeed, the standard population need not exist as a real population and may simply be a particular choice of weights. It is usual to regard $R_{a(s)}$ as the crude death rate for population $\mathcal{A}$ that would have been observed if the age distribution in population $\mathcal{A}$ had been the same as that in the standard population. In at least one instance the directly standardized death rate has a clear interpretation. Suppose that we take population $\mathcal{A}$ to be the standard population. Then (12.2) and

(12.3) simplify to $R_{a(a)} = R_a$ and $\widehat{\text{var}}(R_{a(a)}) = D_a/N_a^2$. So directly standardizing a population to itself results in the crude estimates for that population.

The standardized rate ratio for population $\mathcal{A}$ compared to population $\mathcal{B}$ is defined to be the ratio of standardized death rates,

$$SRR = \frac{R_{a(s)}}{R_{b(s)}} = \frac{\sum_{k=0}^{K} R_{ak} N_{sk}}{\sum_{k=0}^{K} R_{bk} N_{sk}}. \tag{12.4}$$

We can think of SRR as an age-adjusted counterpart to CRR. In most applications the age-specific rate ratios exhibit considerable heterogeneity. Suppose, for purposes of illustration, that there is homogeneity with $R_{ak}/R_{bk} = \psi$ for all k, for some constant ψ. It follows immediately from (12.4) that $SRR = \psi$. Note that this result is independent of the choice of standard population. In Example 12.1 the age-specific rate ratios are both equal to $\psi = 2$ and so $SRR = 2$. An estimate of var(log SRR) is

$$\widehat{\text{var}}(\log SRR) = \frac{\widehat{\text{var}}(R_{a(s)})}{(R_{a(s)})^2} + \frac{\widehat{\text{var}}(R_{b(s)})}{(R_{b(s)})^2}$$

and a $(1 - \alpha) \times 100\%$ confidence interval for SSR is obtained by exponentiating

$$[\log \underline{SRR}, \log \overline{SRR}] = \log(SRR) \pm z_{\alpha/2}\sqrt{\widehat{\text{var}}(\log SRR)}$$

(Rothman and Greenland, 1998, p. 263). Other methods of estimating a confidence interval for SRR are available (Flanders, 1984).

Direct standardization is most often used to compare a population with itself over time, or to compare several distinct populations at a given time point. The following example illustrates how direct standardization can be used to analyze data from a large cohort study. In practice, a cohort study is more likely to be analyzed using the methods of the next section. However, for illustrative purposes the example is provided.

Example 12.2 (Schizophrenia) Table 12.2(a) gives data from a cohort study of mortality in 2122 males who received treatment for schizophrenia in the province of Alberta, Canada at some time during 1976–1985 (Newman and Bland, 1991). Subjects were identified through clinic records and followed until the end of 1985 using record linkage to the Statistics Canada Mortality Database, a national vital statistics registry. For the present analysis the endpoint was taken to be death from any cause. This study is an example of what is termed a retrospective cohort study because subjects were identified using archival records and followed forward as a cohort to a recent time point. Also given in Table 12.2(a) are the numbers of deaths and census counts for Alberta males in 1981. The 1981 population was chosen as the standard population since 1981 is the midpoint of the period of follow-up. In Tables 12.2(a) and 12.2(b) we use conventional demographic notation for age groups. For example, 10–19 stands for the age group [10.0, 20.0).

In this example we regard the census counts as estimates of person-years in a stationary population. As can be seen from Table 12.2(a), the distribution of person-

TABLE 12.2(a)   Death and Census Data: Schizophrenia Cohort and Alberta, Males, 1981

	Cohort			Alberta		
		Person-years			Population	
Age group	Deaths	N	(%)	Deaths	N	(%)
10–19	2	285.1	2.3	267	201,825	21.1
20–29	55	4,179.1	33.9	421	263,175	27.5
30–39	32	3,291.2	26.7	306	176,140	18.4
40–49	21	1,994.7	16.2	431	114,715	12.0
50–59	27	1,498.9	12.2	836	93,315	9.7
60–69	19	763.5	6.2	1,364	60,835	6.4
70–79	25	254.4	2.1	1,861	34,250	3.6
80+	9	46.7	0.4	1,797	12,990	1.4
Total	190	12,313.5	100	7,283	957,245	100

TABLE 12.2(b)   Death Rates and Rate Ratios: Schizophrenia Cohort and Alberta, Males, 1981

	Rate $\times 10^3$		
Age group	Cohort	Alberta	Rate ratio
10–19	7.02	1.32	5.30
20–29	13.16	1.60	8.23
30–39	9.72	1.74	5.60
40–49	10.53	3.76	2.80
50–59	18.01	8.96	2.01
60–69	24.88	22.42	1.11
70–79	98.28	54.34	1.81
80+	192.93	138.34	1.39
Crude	15.43	7.61	2.03

years is different in the cohort compared to the Alberta population. However, other than the youngest age group, the differences are not great. Table 12.2(b) gives the age-specific death rates for the two study groups. We observe that the age-specific rate ratios show considerable heterogeneity, with values ranging from 1.11 to 8.23. For the remainder of the example we denote the cohort by $\mathcal{A}$, the Alberta population by $\mathcal{B}$, and let the Alberta population be the standard population $\mathcal{S}$, that is, $\mathcal{S} = \mathcal{B}$. The crude death rates are $R_a = 15.43 \times 10^{-3}$ (per year) and $R_b = R_s = 7.61 \times 10^{-3}$, and so the crude rate ratio is $CRR = 15.43/7.61 = 2.03$. By comparison, $R_{a(s)} = 17.62 \times 10^{-3}$ and $R_{b(s)} = R_b$, and so the standardized rate ratio is $SRR = 17.62/7.61 = 2.32$. Due to the similarity of the person-years distributions noted above, the crude and standardized rate ratios are close in value. From

$$\widehat{\text{var}}(\log SRR) = \frac{(.00173)^2}{(.0176)^2} + \frac{(.0000892)^2}{(.00761)^2} = (.0992)^2$$

the 95% confidence interval for the standardized rate ratio is [1.91, 2.81].

12.3 STANDARDIZED MORTALITY RATIO

Direct standardization is frequently used to compare national populations and, less often, to compare a cohort to a standard population. To apply direct standardization in the latter setting it is necessary to have estimates of the age-specific death rates in the cohort, as was illustrated in Example 12.2. In practice, even when the total number of deaths in the cohort is reasonably large, there may be few, or even no, deaths in some of the age groups. In the latter situation the method of direct standardization effectively drops those age groups from the analysis, thereby wasting information. We now describe indirect standardization, a method that is based on the total number of deaths in the cohort.

Define age groups for the cohort using the notation of the preceding section, except that now let x_{K+1} represent the age at which follow-up ends. Let D_a denote the total number of deaths in the cohort and let N_{ak} be the number of person-years experienced by the cohort during the kth age group. As before, S denotes the standard population. For direct standardization, all that we need to know about the standard population is its age distribution. By contrast, for indirect standardization we require the age-specific death rates. For the kth age group, denote the age-specific death rate in the standard population by $R_{sk} = D_{sk}/N_{sk}$. The "expected" number of deaths in the cohort is defined to be

$$E_a = \sum_{k=0}^{K} R_{sk} N_{ak}$$

(Væth, 2000). It is sometimes said that E_a is the number of deaths that would have been observed in the cohort if the age-specific death rates in the cohort had been equal to the age-specific death rates in the standard population. However, this interpretation is incorrect (Berry, 1983). Assume for the sake of discussion that $R_{ak} > R_{sk}$ for all k. If R_{ak} had been equal to R_{sk}, the observed number of deaths in the kth age group would not have equaled $R_{sk} N_{ak}$. This is because a reduction in mortality in the cohort would have led to an increase in the number of person-years. So N_{ak} underestimates the number of person-years that the cohort would have experienced in the kth age group, and consequently $R_{sk} N_{ak}$ underestimates the number of deaths that would have been observed.

The standardized mortality ratio is defined to be the ratio of "observed" to "expected" numbers of deaths,

$$SMR_a = \frac{D_a}{E_a} = \frac{\sum_{k=0}^{K} R_{ak} N_{ak}}{\sum_{k=0}^{K} R_{sk} N_{ak}}. \tag{12.5}$$

Reversing the roles of $\mathcal{A}$ and $\mathcal{S}$ in (12.2) gives $R_{s(a)} = \sum_{k=0}^{K}(N_{ak}/N_a)R_{sk} = E_a/N_a$. Since $R_{a(a)} = D_a/N_a$, it follows that $SMR_a = R_{a(a)}/R_{s(a)}$. This shows that the standardized mortality ratio is a special case of the standardized rate ratio (Miettinen, 1972b). It is important to appreciate that, for the standardized mortality ratio, the weights come from the cohort, not the so-called standard population (Miettinen, 1972a). For this reason we sometimes use the notation $SMR_{s(a)}$ instead of SMR_a. The indirectly standardized death rate is defined to be $SMR_a \times R_s$, but in practice it is usually the standardized mortality ratio which is of primary interest.

When analyzing cohort data using the above methods, it is tempting to compare standardized mortality ratios across subcohorts. For example, in the study outlined in Example 12.2 a cohort of female patients with schizophrenia was also followed. Denote the male and female cohorts by $\mathcal{A}$ and $\mathcal{B}$, respectively, and consider $SMR_{s(a)}$ and $SMR_{s(b)}$ for some choice of standard population. The notation makes it clear that, strictly speaking, it is inappropriate to compare the standardized mortality ratios because they are based on different weighting schemes. To further illustrate the problems that can arise when standardized mortality ratios are compared inappropriately, suppose that $R_{ak}/R_{bk} = \psi$ for all k, for some constant ψ. Using (12.5) it is readily demonstrated that

$$\frac{SMR_{s(a)}}{SMR_{s(b)}} = \psi \frac{\left(\sum_{k=0}^{K} R_{bk} N_{ak}\right) \Big/ \left(\sum_{k=0}^{K} R_{sk} N_{ak}\right)}{\left(\sum_{k=0}^{K} R_{bk} N_{bk}\right) \Big/ \left(\sum_{k=0}^{K} R_{sk} N_{bk}\right)} \tag{12.6}$$

and so $SMR_{s(a)}/SMR_{s(b)}$ does not necessarily equal the common age-specific rate ratio ψ. This is in contrast to the corresponding result for the standardized rate ratio. When the cohorts have the same person-years distributions, a condition that is often approximately satisfied in practice, then $SMR_{s(a)}/SMR_{s(b)} = \psi$.

In most applications the age-specific death rates in the standard population and the person-years distribution in the cohort are known with considerable precision, at least compared to the number of deaths in the cohort. For this reason it is appropriate to treat the expected number of deaths in the cohort as a constant. From this perspective, both $SMR_a = D_a/E_a$ and $R_a = D_a/N_a$ are formally equal to the quotient of a Poisson random variable and a constant. As a result, the methods of Section 10.1 can be adapted to the analysis of standardized mortality ratios (Breslow and Day, 1985; Breslow and Day, 1987, Chapter 2; Clayton and Hills, 1993, §15.6). For example, from (10.2), an estimate of var(SMR_a) is

$$\widehat{\text{var}}(SMR_a) = \frac{D_a}{(E_a)^2} = \frac{SMR_a}{E_a}. \tag{12.7}$$

From (10.11), a $(1 - \alpha) \times 100\%$ confidence interval for SMR_a is

$$[\underline{SMR_a}, \overline{SMR_a}] = SMR_a \left(1 \pm \frac{z_{\alpha/2}}{\sqrt{D_a}}\right).$$

The hypothesis of no mortality difference between the cohort and the standard population can be tested using (10.12):

$$X^2 = \frac{(D_a - E_a)^2}{E_a} \qquad (\text{df} = 1).$$

When $E_a < 5$, exact methods should be used. From (10.7) and (10.8), an exact test of the hypothesis of no mortality difference is based on the tail probabilities

$$\exp(-E_a) \sum_{x=0}^{D_a} \frac{(E_a)^x}{x!}$$

and

$$1 - \exp(-E_a) \sum_{x=0}^{D_a-1} \frac{(E_a)^x}{x!}.$$

Corresponding to (10.9) and (10.10), an exact $(1 - \alpha) \times 100\%$ confidence interval for SMR_a is obtained by solving the equations

$$\frac{\alpha}{2} = 1 - \exp(-\underline{SMR}_a \times E_a) \sum_{x=0}^{D_a-1} \frac{(\underline{SMR}_a \times E_a)^x}{x!}$$

and

$$\frac{\alpha}{2} = \exp(-\overline{SMR}_a \times E_a) \sum_{x=0}^{D_a} \frac{(\overline{SMR}_a \times E_a)^x}{x!}$$

for $\underline{SMR}_a$ and $\overline{SMR}_a$.

In what follows we consider the cohort and standard population to be the "exposed" and "unexposed" cohorts of Section 10.3, respectively. With age as the stratifying variable, the standardized mortality ratio (12.5) is seen to be a type of standardized hazard ratio (10.38). The Mantel–Haenszel estimate of the hazard ratio (10.37) is $\widehat{HR}_{\text{mh}} = R_\bullet / S_\bullet$. Let $N_k = N_{ak} + N_{sk}$ so that, in the notation of this chapter, $R_\bullet = \sum_{k=0}^{K}(D_{ak}N_{sk})/N_k$ and $S_\bullet = \sum_{k=0}^{K}(D_{sk}N_{ak})/N_k$. Assume that N_{ak} is small compared to N_k for all k, as is usually the case in practice. Then N_k is approximately equal to N_{sk}, and so, to an approximation, $R_\bullet = \sum_{k=0}^{K} D_{ak} = D_a$ and $S_\bullet = \sum_{k=0}^{K}(D_{sk}/N_{sk})N_{ak} = E_a$. Hence $\widehat{HR}_{\text{mh}} = SMR_a$. Other Poisson methods discussed in Section 10.3 are readily adapted to the present setting. As part of the analysis it is important that homogeneity be assessed. In most applications where the standardized mortality ratio is likely to be used, there will be considerable heterogeneity, as illustrated by the rate ratios in Table 12.2(b).

Example 12.3 (Schizophrenia)   For the schizophrenia cohort, $D_a = 190$ and $E_a = 71.10$, and so $SMR_a = 2.67$. From $\widehat{\text{var}}(SMR_a) = 190/(71.11)^2 = .038$, the 95% confidence interval for the standardized mortality ratio is [2.29, 3.05]. Based on the methods of Chapter 10, $\widehat{HR}_{mh} = 2.67$, the test of homogeneity is $X_h^2 = 111.3$ ($p < .001$), and the test for linear trend is $X_t^2 = 82.49$ ($p < .001$). The absence of homogeneity is not surprising because the sample sizes are so large that even a small amount of heterogeneity can be detected. Restricting the above methods to a single age group, the standardized mortality ratio becomes an age-specific rate ratio. For the 10–19 age group, $D_a = 2$ and $E_a = .376$, and so the asymptotic approach is not suitable. Based on exact methods, a 95% confidence interval for the age-specific rate ratio is [.644, 19.21] and, based on the doubling method, the p-value for the exact test of the hypothesis of no mortality difference is $p = .11$.

12.4   AGE–PERIOD–COHORT ANALYSIS

Virtually all causes of death vary by age, and so the analysis of time trends in death rates often begins with an assessment of age-specific rates. Table 12.3 gives age-specific death rates for all causes of death in the Canadian female population for selected age groups and selected years. These data were taken from official Statistics Canada publications. In Figure 12.1(a) the rates for each age-group are graphed as a function of time. Each curve corresponds to a column of Table 12.3. The layered appearance of the curves is consistent with the well known fact that all-cause mortality increases with age. In Figure 12.1(b) the rates for each year (period) are graphed as a function of age. In this case, each curve corresponds to a row of Table 12.3. The fan-shaped appearance of the curves suggests that mortality decreased over successive time periods.

Under the assumption that there has been no net demographic change in the population due to in- and out-migration, each of the diagonals in Table 12.3 can be given a cohort interpretation. For example, those individuals in the 30 to 34-year age group in 1950 who survived for a decade became the 40 to 44-year age group in 1960, and so on. In Figure 12.1(c), rates are again graphed as a function of age, but now each curve corresponds to a diagonal of Table 12.3 (not all of which have been graphed).

TABLE 12.3   Age-Specific Death Rates for All Causes (per 100,000): Canada, Females

Year	30–34	40–44	50–54	60–64	70–74
1950	1.4	3.2	6.6	16.1	42.8
1960	0.9	2.1	5.3	13.4	35.1
1970	0.9	2.1	4.9	11.2	29.4
1980	0.7	1.6	4.2	9.8	24.9
1990	0.6	1.2	3.4	8.4	21.5

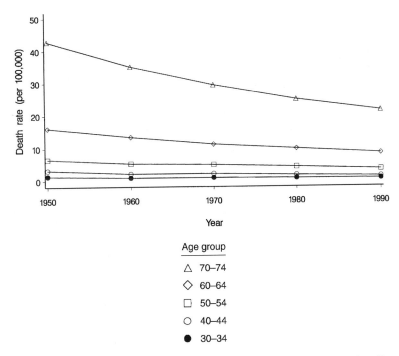

FIGURE 12.1(a)   Age-specific death rates (per 100,000) by age group: Canada, females, all causes

Each curve is labeled according to the year in which the cohort was in the 30 to 34-year age group. The appearance of the curves suggests that mortality decreased across successive cohorts.

The findings in Figures 12.1(a)–12.1(c) described above are referred to as age, period, and cohort effects, respectively. The existence of an age effect is unquestionable, but the issue is not so obvious for period and cohort effects. By definition, a period effect has an impact on an entire population at a given time point. For example, a period effect might result from an outbreak of a new strain of influenza. By contrast, a cohort effect exerts its influence on particular members of the population who then carry the consequences forward in time. For example, a cohort effect might be observed following a public health program designed to reduce adolescent smoking. The preceding examples are fairly straightforward, but even here it may be difficult to separate period from cohort effects. For instance, the smoking cessation campaign might result in an abrupt, but short-lived, decrease in the number of smokers in the general population, and so there may be a period as well as a cohort effect. These observations raise questions about the correct interpretation of Figures 12.1(a)–12.1(c), especially with respect to the relative contributions of period and cohort effects.

It might be hoped that the problem could be resolved with the aid of a multivariate model that has terms for age, period, and cohort effects (Clayton and Schifflers, 1987;

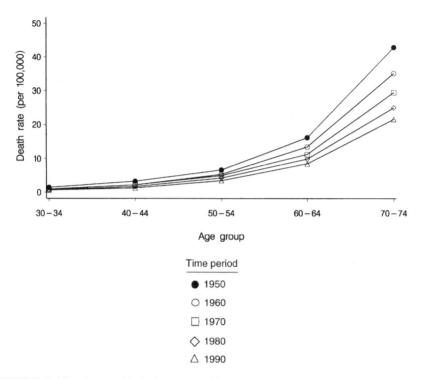

FIGURE 12.1(b)   Age-specific death rates (per 100,000) by time period: Canada, females, all causes

Holford, 1991, 1998). Following Section 11.1.3, let $r(x, t)$ be the hazard function in the population for all-cause mortality, where x is age and t is time. Let $x = 0$ and $t = 0$ correspond to a convenient baseline. The curves in Figures 12.1(b) and 12.1(c) have a more or less exponential shape and so it is reasonable to define $r(x, t) = \exp(\mu + \alpha x + \pi t)$, where μ, α, and π are constants. Corresponding to Table 12.3, assume that time is measured in decades. For a given age x, the ratio of the hazard at time $t + 1$ to that at time t is $r(x, t + 1)/r(x, t) = e^{\pi}$, which we interpret as a period effect. That is, with each successive decade the age-specific hazard increases by a factor e^{π}, where the increase is the same at every age x. By definition, an individual alive at age x and time t was born at time $y = t - x$. Substituting $t = x + y$ in $r(x, t)$ we can express the hazard function in terms of x and y, that is, $r(x, y) = \exp[\mu + (\alpha + \pi)x + \pi y]$. For a given age x, the ratio of the hazard for the cohort born at time $y + 1$ to that born at time y is $r(x, y + 1)/r(x, y) = e^{\pi}$. In this case we interpret e^{π} as a cohort effect. These calculations show that when period and cohort effects are considered together, a period effect can manifest as a cohort effect, and conversely. This problem cannot be resolved by considering more complicated mathematical models. The difficulty lies in the fact that, due to the identity $y = t - x$, age, period, and cohort effects are inextricably intertwined, a phenomenon referred to as the identifiability problem.

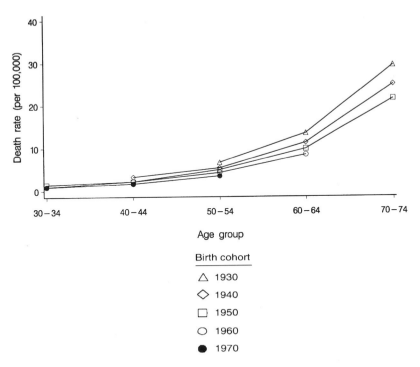

FIGURE 12.1(c) Age-specific death rates (per 100,000) by birth cohort: Canada, females, all causes

In order to separate out age, period, and cohort effects—that is, solve the identifiability problem—it is necessary to incorporate an additional equation into the age–period–cohort model. The choice of equation rests on substantive knowledge and is therefore not a statistical issue. For example, if we know from other sources that the log-age effect is twice the log-period effect, we could substitute the equation $\alpha = 2\pi$ into the model. However, it is unusual for such information to be available and so, unfortunately, age–period–cohort models cannot be relied upon to tease apart the three time-related effects on mortality.

CHAPTER 13

Life Tables

Standardization is a computationally convenient method of analyzing mortality data, but there is the drawback of having to select a standard population. Suppose that it is desired to assess temporal changes in annual death rates in a given population over the past few decades, up to and including the current year. In this instance a suitable choice of standard population would be the current census population. The resulting standardized death rates could then be interpreted as the (crude) death rates that would have been observed in the past if earlier age distributions had been the same as that in the current population. A useful feature of this choice of standard population is that the standardized death rate for the current year is the actual crude death rate. When the aim is to compare regions of a country, an appropriate standard would be the national census population. In practice it is usually not difficult to identify a suitable standard population for a given study. However, problems can arise when the analysis needs to be updated—for example, when more recent data become available. In this case, the earlier choice of standard population may no longer be suitable and, as a result, the entire collection of standardized rates will need to be recalculated.

In this chapter we present the life table approach to analyzing cross-sectional mortality and morbidity data. An attractive feature of life table methods is that they do not require a standard population. In addition, life table methods produce a number of summary indices (in particular, the life expectancy at birth) which have considerable intuitive appeal. These advantages come at the expense of increased computational complexity and, more importantly, the need to make strong assumptions about future trends in mortality and morbidity.

Life tables have a long history in actuarial science and other areas where demographic projections are made. There is a close connection between life tables and survival analysis in that both theories are expressed in terms of the follow-up of a cohort and both rely on such concepts as the survival function and hazard function. A practical difference is that the methods of survival analysis are usually applied to data collected by following a relatively small cohort over a short period of time, whereas life table methods are generally used to analyze cross-sectional data from a large population. In this chapter we discuss a number of types of life tables, including the ordinary, multiple decrement, and cause-deleted life tables. The ordinary life table is concerned with deaths from all causes and is the one routinely published by govern-

ment agencies. References for life table theory are Keyfitz (1977), Elandt-Johnson
and Johnson (1980), and Chiang (1984).

13.1 ORDINARY LIFE TABLE

Following Section 11.1.3, we define the all-causes hazard function for the population as follows: $r(x, t)$ is the instantaneous probability per unit time that an individual
alive at age x and at time t will die in the next instant. In the demographic and actuarial literature, this hazard function is sometimes referred to as the force of mortality.
Consider the (birth) cohort of individuals born into the population at some fixed time
t_0. When this cohort reaches age x, the time will be $t_0 + x$ and so the hazard for the
cohort will be $r_d(x) = r(x, t_0 + x)$. We refer to $r_d(x)$ as the diagonal hazard function.
Here we assume that in- and out-migration in the population have no net effect on
mortality in the cohort. This approach is similar to the analysis of death rates "along
the diagonal" that was presented in Section 12.4. The diagonal approach to demographic rates is of conceptual interest but has a practical limitation: In order to follow
a cohort into old age it is necessary to base the analysis on a group of individuals that
was born many decades ago.

An alternative is to use data collected at a recent time point. The cross-sectional
hazard function at time t_0 is defined to be $r_c(x) = r(x, t_0)$. Recall from Section
11.1.3 that the stationary population assumption requires that features of the population be independent of time. We can give $r_c(x)$ an interpretation as the hazard function of a cohort by assuming that the population is stationary with respect to mortality
after time t_0. With this assumption, $r(x, t_0) = r(x, t_0 + x)$ and so $r_c(x) = r_d(x)$;
that is, the cross-sectional and diagonal hazard functions are equal. In this case we
denote the common value of the hazard function by $r(x)$. The obvious problem with
the cross-sectional approach is that we are forced to assume that the population will
be stationary over the life span of the birth cohort, a period that may be in excess
of 100 years. Over the short term, as in a case-control study, the stationary population assumption may be justified, but this is no longer true when the time frame
is decades long. Nevertheless, most life table methods are based on cross-sectional
data and rely on the stationary population assumption. The usual approach in life
table analysis is to perform calculations under the stationary population assumption
and then interpret the results in the context of projected mortality trends. This can
be accomplished either using qualitative arguments or based on a formal sensitivity
analysis in which a range of future mortality scenarios are modeled. For example, if
mortality is predicted to decline, life expectancies based on cross-sectional data will
need to be adjusted upward.

Having made the stationary population assumption, we now equate the (birth)
cohort born in the population at time t_0 to a hypothetical cohort defined to have the
hazard function $r(x)$. This hypothetical cohort is the foundation of the ordinary life
table (OLT) and we refer to it as the OLT cohort. Since the mortality experience of
any cohort is governed entirely by its hazard function, the OLT cohort is equivalent to
the population cohort in terms of mortality. We use the OLT cohort as a convenient

vehicle for describing the projected mortality experience of the population cohort under the stationary population assumption. Since the hazard function $r(x)$ is defined cross-sectionally, the OLT approach allows us to express observations about cross-sectional mortality in longitudinal terms. However, it must be emphasized that this is simply a convenient form of expression and is only meaningful to the extent that the stationary population assumption is valid.

Following the notation of Section 12.2, partition the life span into $J + 1$ age groups: $[x_0, x_1), [x_1, x_2), \ldots, [x_j, x_{j+1}), \ldots, [x_{J-1}, x_J), [x_J, x_{J+1}]$. As before, $x_0 = 0$ and x_{J+1} is the upper limit of the life span. We refer to $[x_j, x_{j+1})$ as the jth age group and denote its length by $n_j = x_{j+1} - x_j$. In some applications, 1-year age groups are used, resulting in what is termed a complete ordinary life table. When the age groups are wider, the ordinary life table is said to be abridged. In the examples to follow we consider the partition based on ages 0, 1, 5, 10, $\ldots$, 80, 85, x_{19}, where the age groups between 5 and 85 are all 5 years long and x_{19} is left unspecified. The age groups resulting from this partition can be written as follows: <1, 1–4, 5–9, 10–14, $\ldots$, 80–84, 85+. It must be emphasized that we consider age to be a continuous variable and so, for example, $x_3 = 10$ is to be interpreted as $x_3 = 10.0$. Recall from Chapter 12 that the notation 10–14 is an alternate way of writing [10.0,15.0). Consequently the width of the age group 10–14 is $15.0 - 10.0 = 5.0$, not $14 - 10 = 4$.

Table 13.1 gives a description of the main OLT functions. Observe that $r(t)$ is now considered to be the hazard function for the OLT cohort. The number of individuals in the OLT birth cohort, referred to as the radix, is denoted by $l(0)$. The magnitude of $l(0)$ is unrelated to the number of births in the population. Usually $l(0)$ is defined to be some large number such as 100,000, but this is arbitrary. $S(x)$ equals the probability that a member of the OLT cohort will survive to age x. It follows that the expected number of survivors to age x is $l(x) = l(0)S(x)$. For brevity we drop the reference to "expected" in the terminology for $l(x)$ and the other life table functions denoting counts. Based on the results in Appendix F and what follows in this section, it can be shown that all the OLT functions can be expressed in terms of $r(x)$ and $l(0)$.

TABLE 13.1   Ordinary Life Table Functions ($j = 0, 1, \ldots, J$)

Function	Description
$r(x)$	Hazard function at age x
$S(x)$	Probability of surviving to age x
q_j	Conditional probability of dying in $[x_j, x_{j+1})$
p_j	Conditional probability of surviving $[x_j, x_{j+1})$
$l(x)$	Number of survivors to age x
d_j	Number of deaths in $[x_j, x_{j+1})$
L_j	Number of person-years during $[x_j, x_{j+1})$
$T(x)$	Number of person-years after age x
$e(x)$	Life expectancy at age x

For a given calendar year, let D_j denote the number of deaths in the population in the jth age group and let N_j denote the number of individuals in the midyear population ($j = 0, 1, \ldots, J$). The corresponding annual death rate in the population is defined to be $R_j = D_j/N_j$. As in Chapter 12, we drop the caret $\frown$ from notation and, for example, write R_j instead of $\hat{R}_j$. From Table 13.1, d_j is the number of deaths in the OLT cohort in the jth age group and L_j is the corresponding number of person-years. Therefore the OLT death rate for this age group is d_j/L_j. Since $r(x)$ is a continuous function, it is not amenable to calculations based on routinely collected data. For purposes of estimating OLT functions we equate the population and OLT age-specific death rates. That is, for each age group we define

$$R_j = \frac{D_j}{N_j} = \frac{d_j}{L_j}. \tag{13.1}$$

Identity (13.1) is the fundamental step in the construction of the ordinary life table from cross-sectional data. In order to proceed it is necessary to make an assumption about the functional form of $l(x)$ on each of the age groups, except for the last. The OLT functions for the last age group exhibit special features that apply regardless of the functional form of $l(x)$. We consider two possible functional forms for $l(x)$, namely, exponential and linear. Assuming that $l(x)$ is exponential on each age group (except for the last), we have from Section 10.1.2 that

$$q_j = 1 - \exp(-n_j R_j) \tag{13.2}$$

($j = 0, 1, \ldots, J - 1$). Now suppose that $l(x)$ is linear on each age group. Using the results of Appendix F, it can be shown that in general, L_j equals the area under the graph of $l(x)$ between x_j and x_{j+1}. Since $l(x_{j+1}) = l(x_j) - d_j = l(x_j) - q_j l(x_j) = (1 - q_j)l(x_j)$, it follows that

$$\begin{aligned} L_j &= \frac{[l(x_j) + l(x_{j+1})]n_j}{2} \\ &= \frac{[l(x_j) + (1 - q_j)l(x_j)]n_j}{2} \\ &= \frac{l(x_j)(2 - q_j)n_j}{2}. \end{aligned} \tag{13.3}$$

From (13.1), (13.3), and $d_j = q_j l(x_j)$, we find that

$$R_j = \frac{q_j l(x_j)}{[l(x_j)(2 - q_j)n_j]/2} = \frac{2q_j}{(2 - q_j)n_j}$$

which can be solved for q_j to give

$$q_j = \frac{n_j R_j}{1 + (n_j R_j)/2} \tag{13.4}$$

TABLE 13.2 Steps in the Construction of an Ordinary
Life Table ($j = 0, 1, \ldots, J$)

Step	Ordinary Life Table Function
1	$q_j = \begin{cases} \text{(13.2) or (13.4)} & j \neq J \\ 1 & j = J \end{cases}$
2	$p_j = 1 - q_j$
3	$l(x_j) = l(0)\, p_0\, p_1 \cdots p_{j-1}$
4	$d_j = q_j\, l(x_j)$
5	$L_j = \dfrac{d_j}{R_j}$
6	$T(x_j) = \displaystyle\sum_{i=j}^{J} L_i$
7	$e(x_j) = \dfrac{T(x_j)}{l(x_j)}$

($j = 0, 1, \ldots, J - 1$). For the last age group, $q_J = 1$ regardless of the functional form of $l(x)$ since there are no survivors past age x_{J+1}. Table 13.2 summarizes the steps involved in constructing an ordinary life table. The derivation of $l(x_j)$ in step 3 uses an argument identical to that leading to (9.4) and (9.13). In step 5, the identity $L_j = d_j/R_j$ follows directly from (13.1). Since $T(x_j)$ is the number of person-years that will be lived after age x_j, $e(x_j) = T(x_j)/l(x_j)$ is the average number of years that will be lived by those who survive to age x_j; that is, $e(x_j)$ is the life expectancy at age x_j. For the last age group, $l(x_J) = d_J$ and $T_J = L_J = d_J/R_J$. It follows that $e(x_J) = 1/R_J$ regardless of the functional form of $l(x)$.

When $n_j R_j$ is small, which is usually the case, (13.2) and (13.4) are both approximately equal to $n_j R_j$. So it generally makes little practical difference whether the exponential or linear assumption is used in the construction of the ordinary life table. This also explains why the common mistake of treating R_j as a probability rarely leads to serious difficulties.

The OLT cohort will experience $l(0)$ deaths because follow-up continues until all members of the cohort are dead. The corresponding number of person-years is $T(0)$. Therefore the crude death rate for the OLT cohort is $l(0)/T(0) = 1/e(0)$. Since $d_j = l(x_j) - l(x_{j+1})$ and $l(x_{J+1}) = 0$, we have

$$l(x_j) = \sum_{i=j}^{J} [l(x_j) - l(x_{j+1})] = \sum_{i=j}^{J} d_i. \tag{13.5}$$

This is a formal statement of the fact that all members of the OLT cohort who survive to age x_j will eventually die. Since $d_j = R_j L_j$ it follows from (13.5) that $l(0) = \sum_{j=0}^{J} R_j L_j$. By definition, $T(0) = \sum_{j=0}^{J} L_j$ and so the crude death rate for the OLT cohort is

$$R_{\text{olt}} = \frac{\sum_{j=0}^{J} L_j R_j}{\sum_{j=0}^{J} L_j}.$$

This is a directly standardized death rate for the cross-sectional population, where standardization is according to the person-years distribution in the OLT cohort.

The ordinary life table can be used to calculate a variety of mortality indices. As an example, for ages $x' < x''$, the probability of surviving to x'', given survival to x', is $l(x'')/l(x')$. Perhaps the most informative mortality index available from the ordinary life table is $e(0)$, the life expectancy at birth. The virtue of $e(0)$ as a mortality index is that it summarizes the survival experience of the OLT cohort over the entire life span. Also of interest is $l(65)/l(0)$, the probability at birth of surviving to age 65. Evidently the choice of age 65 in this definition is arbitrary.

Example 13.1 (Ordinary Life Table: Canada, Males, 1991) Table 13.3 gives the numbers of deaths from all causes, (malignant) neoplasms (140–208), circulatory diseases (390–459), and injuries (E800–E999), as well as the census population for Canadian males in 1991. The numbers in parentheses are the rubrics according to the ninth revision of the International Classification of Diseases published by the World Health Organization. The data were obtained from official Statistics Canada

TABLE 13.3 Death and Census Counts: Canada, Males, 1991

x_j	D_j	D_j^k Neoplasms	D_j^k Circulatory	D_j^k Injuries	N_j
0	1,432	5	20	38	201,600
1	298	29	18	106	774,165
5	197	36	5	91	978,220
10	253	32	10	145	962,925
15	913	50	22	704	958,405
20	1,256	65	33	1,018	985,220
25	1,502	114	59	1,002	1,182,575
30	1,683	152	121	930	1,237,685
35	1,849	230	240	806	1,133,670
40	2,248	462	475	665	1,042,180
45	2,904	846	785	550	824,200
50	3,712	1,311	1,227	444	663,285
55	5,765	2,321	1,955	404	608,085
60	9,073	3,661	3,284	411	571,940
65	12,553	4,786	4,827	367	492,505
70	14,144	4,810	5,753	337	358,950
75	16,081	4,653	6,936	357	252,530
80	14,004	3,414	6,179	330	140,130
85	15,557	2,692	7,176	441	86,305

TABLE 13.4   Ordinary Life Table: Canada, Males, 1991

x_j	q_j	p_j	$l(x_j)$	d_j	L_j	$T(x_j)$	$e(x_j)$
0	.00708	.99292	100,000	708	99,646	7,433,920	74.34
1	.00154	.99846	99,292	153	396,863	7,334,273	73.87
5	.00101	.99899	99,139	100	495,448	6,937,410	69.98
10	.00131	.99869	99,040	130	494,873	6,441,963	65.04
15	.00475	.99525	98,910	470	493,373	5,947,089	60.13
20	.00635	.99365	98,440	625	490,634	5,453,716	55.40
25	.00633	.99367	97,814	619	487,523	4,963,082	50.74
30	.00678	.99322	97,195	659	484,328	4,475,559	46.05
35	.00812	.99188	96,536	784	480,722	3,991,231	41.34
40	.01073	.98927	95,752	1,027	476,194	3,510,509	36.66
45	.01746	.98254	94,725	1,654	469,490	3,034,315	32.03
50	.02760	.97240	93,071	2,568	458,934	2,564,825	27.56
55	.04631	.95369	90,503	4,191	442,036	2,105,891	23.27
60	.07629	.92371	86,312	6,585	415,097	1,663,856	19.28
65	.11981	.88019	79,727	9,552	374,755	1,248,759	15.66
70	.17935	.82065	70,175	12,586	319,411	874,004	12.45
75	.27467	.72533	57,589	15,818	248,401	554,593	9.63
80	.39979	.60021	41,771	16,700	167,106	306,193	7.33
85	1.0000	0	25,071	25,071	139,087	139,087	5.55

publications. Table 13.4 gives the abridged ordinary life table based on the linear assumption. Under the stationary population assumption, a male born in Canada in 1991 has a life expectancy at birth of 74.34 years, and 79.73% of the birth cohort will survive to age 65.

The ordinary life table in the above example is based on the assumption that the Canadian male population will be stationary over the next 100 years or so. Based on historical evidence this assumption will almost certainly prove to be false. If past trends continue, there will be improvements in survival and so the predicted life expectancy, $e(0) = 74.34$, is a conservative estimate. A sensitivity analysis provides insight into the potential impact of declining death rates. Suppose that by the time the population birth cohort reaches the jth age group the death rate will have decreased from R_j in 1991 to $\phi_j R_j$, where $0 \le \phi_j \le 1$. To examine the mortality implications, we construct an ordinary life table as above, but with $\phi_j R_j$ in place of R_j. Exploring different death rate scenarios provides a range of possibilities for the survival experience of the population cohort. For example, suppose that the ϕ_j are all equal with common value ϕ. For Canadian males in 1991, with $\phi = .9, .8,$ and .7, the life expectancies at birth are 75.65, 77.15 and 78.89, respectively.

Example 13.2 (Schizophrenia)   Table 12.2(b) gives age-specific death rates (by 10-year age groups starting at age 10) for the schizophrenia cohort discussed in Example 12.2. It is possible to construct an ordinary life table starting at age 10 using these death rates, but due to the small numbers of deaths the OLT functions would not be reliable. An alternative is to set the ϕ_j defined above equal to the rate ratios in Table 12.2(b). Based on the ordinary life table for Canadian males in 1981, the life expectancy at age 15 is 58.06. After scaling the 1981 death rates using the ϕ_j, the life expectancy decreases to 44.75. The interpretation is that schizophrenia developing at age 15 reduces life expectancy (at age 15) by 13.31 years.

13.2 MULTIPLE DECREMENT LIFE TABLE

The ordinary life table provides a method of analyzing mortality for all causes of death combined, but gives no information on the contributions of specific causes of death to overall mortality. A multiple decrement life table (MDLT) describes the mortality experience of the group of individuals in the OLT cohort who are "due to die" of a particular cause of death. This approach makes it possible to examine specific causes of death in relation to overall mortality. The multiple decrement life table is an example of a competing risks model.

Suppose that all the causes of death have been grouped into K mutually exclusive "causes" ($k = 1, 2, \ldots, K$). Following Section 8.4 we define the crude hazard function for cause k as follows: $r^k(x, t)$ is the instantaneous probability per unit time that an individual alive at age x and at calendar time t will die of cause k in the next instant "in the presence of other causes of death." Since the causes of death are mutually exclusive and exhaustive, it follows from (8.8) that

$$r(x, t) = \sum_{k=1}^{K} r^k(x, t). \tag{13.6}$$

We now assume that the population is stationary for each of the causes of death. Arguing as in the preceding section it can be shown that, for each cause of death, the cause-specific cross-sectional and diagonal hazard functions are equal. Denoting the common cause-specific hazard function by $r^k(x)$, it follows from (13.6) that

$$r(x) = \sum_{k=1}^{K} r^k(x). \tag{13.7}$$

Since each member of the OLT cohort must die of one of the causes of death, we can, in theory, divide the OLT cohort into subcohorts consisting of individuals due to die of each of the causes. The multiple decrement life table for cause k describes the mortality experience of the subcohort of the OLT cohort due to die of cause k. For brevity we refer to this group of individuals as the MDLT cohort (for cause k).

Table 13.5 gives a description of the main MDLT functions. Other than q_j^k the functions have an interpretation analogous to their counterparts in Table 13.1. For example, $l^k(x)$ is the number of individuals in the MDLT cohort surviving to age x, and $e^k(x)$ is their life expectancy. The unique feature of $q_j^k = d_j^k / l(x_j)$ is that the denominator is $l(x_j)$ rather than $l^k(x_j)$. So q_j^k depends on the survival experience of the entire cohort and is therefore termed a crude conditional probability.

TABLE 13.5   Multiple Decrement Life Table Functions for Cause k $(j = 0, 1, \ldots, J)$

Function	Description
$q_j^k = \dfrac{d_j^k}{l(x_j)}$	Crude conditional probability of dying in $[x_j, x_{j+1})$
d_j^k	Number of deaths in $[x_j, x_{j+1})$
$l^k(x)$	Number of survivors to age x
L_j^k	Number of person-years during $[x_j, x_{j+1})$
$T^k(x)$	Number of person-years after age x
$e^k(x)$	Life expectancy at age x

For a given calendar year, let D_j^k denote the number of deaths in the population in the jth age group that are due to cause k $(j = 0, 1, \ldots, J)$. The crude cause-specific death rate in the population for this age group is defined to be $R_j^k = D_j^k / N_j$. We use the term crude because there are competing causes of death in the population, and this has an impact on the number of deaths that are due to cause k. In this sense, the term "cause-specific" is something of a misnomer (Clayton and Hills, 1993, §7.4). Since the causes of death are mutually exclusive and exhaustive, it follows that for each age group we have $D_j = \sum_{k=1}^{K} D_j^k$. Therefore

$$R_j = \sum_{k=1}^{K} R_j^k$$

which is the discrete counterpart to (13.7). From Table 13.5, d_j^k is the number of deaths in the MDLT cohort in the jth age group. Recall that L_j is the number of person-years in the OLT cohort for this age group. So the crude MDLT death rate for the jth age group is d_j^k / L_j. Analogous to the approach taken with the ordinary life table, we equate the population and MDLT crude death rates for each age group. That is, for each age group we define

$$R_j^k = \frac{D_j^k}{N_j} = \frac{d_j^k}{L_j}. \tag{13.8}$$

From $q_j = d_j / l(x_j)$ and $q_j^k = d_j^k / l(x_j)$, it follows that

$$\frac{q_j^k}{q_j} = \frac{d_j^k}{d_j}. \tag{13.9}$$

From (13.1) and (13.8) we have

$$\frac{d_j^k}{d_j} = \frac{D_j^k}{D_j}. \tag{13.10}$$

Combining (13.9) and (13.10) gives

$$q_j^k = \left(\frac{D_j^k}{D_j}\right) q_j$$

($j = 0, 1, \ldots, J$). As in the construction of the ordinary life table we need to make an assumption about the functional form of $l^k(x)$. As explained below, it is convenient to assume that $l^k(x)$ is linear on each age group. Therefore, for each age group other than the last, we define

$$L_j^k = \frac{[l^k(x_j) + l^k(x_{j+1})]n_j}{2} \tag{13.11}$$

($j = 0, 1, \ldots, J - 1$). Since x_{J+1} is unspecified, n_J cannot be calculated and so (13.11) does not apply to the last age group. For this age group we define

$$L_J^k = \left(\frac{D_J^k}{D_J}\right) L_J. \tag{13.12}$$

All members of the MDLT cohort for cause k who survive to age x_j will eventually die of this cause, and so, corresponding to (13.5), we have

$$l^k(x_j) = \sum_{i=j}^{J} d_i^k. \tag{13.13}$$

TABLE 13.6   Steps in the Construction of the Multiple Decrement Life Table for Cause k ($j = 0, 1, \ldots, J$)

Step	MDLT function
1	$q_j^k = \left(\dfrac{D_j^k}{D_j}\right) q_j$
2	$d_j^k = q_j^k \, l(x_j)$
3	$l^k(x_j) = \displaystyle\sum_{i=j}^{J} d_i^k$
4	$L_j^k = \begin{cases} (13.11) & j \neq J \\ (13.12) & j = J \end{cases}$
5	$T^k(x_j) = \displaystyle\sum_{i=j}^{J} L_i^k$
6	$e^k(x_j) = \dfrac{T^k(x_j)}{l^k(x_j)}$

Table 13.6 summarizes the steps involved in the construction of the multiple decrement life table for cause k.

It follows immediately from the definitions that

$$l(x_j) = \sum_{k=1}^{K} l^k(x_j)$$

and

$$L(x_j) = \sum_{k=1}^{K} L^k(x_j).$$

The latter identity is satisfied with the linear assumption but not with the exponential assumption, which explains the choice of functional form made above. These and other identities relating OLT and MDLT functions show that the collection of multiple decrement life tables, one for each cause of death, can be viewed as a stratification of the ordinary life table. Perhaps the most informative mortality index available from the multiple decrement life table for cause k is $e^k(0)$, the life expectancy at birth for an individual due to die of cause k. Also of interest is $l^k(0)/l(0)$, the probability at birth of eventually dying of cause k.

Example 13.3 (Multiple Decrement Life Table for Neoplasms: Canada, Males, 1991)   Table 13.7 gives the multiple decrement life table for neoplasms for Cana-

TABLE 13.7   Multiple Decrement Life Table for Neoplasms: Canada, Males, 1991

x_j	q_j^k	$l^k(x_j)$	d_j^k	L_j^k	$T^k(x_j)$	$e^k(x_j)$
0	.00002	27,167	2	27,166	1,990,557	73.27
1	.00015	27,164	15	108,628	1,963,392	72.28
5	.00018	27,150	18	135,702	1,854,764	68.32
10	.00017	27,131	16	135,615	1,719,062	63.36
15	.00026	27,115	26	135,510	1,583,446	58.40
20	.00033	27,089	32	135,365	1,447,937	53.45
25	.00048	27,057	47	135,166	1,312,572	48.51
30	.00061	27,010	59	134,900	1,177,406	43.59
35	.00101	26,950	98	134,508	1,042,506	38.68
40	.00220	26,853	211	133,736	907,998	33.81
45	.00509	26,642	482	132,003	774,262	29.06
50	.00975	26,160	907	128,531	642,259	24.55
55	.01864	25,253	1,687	122,045	513,728	20.34
60	.03078	23,565	2,657	111,185	391,683	16.62
65	.04568	20,908	3,642	95,438	280,498	13.42
70	.06099	17,267	4,280	75,633	185,060	10.72
75	.07948	12,986	4,577	53,490	109,428	8.43
80	.09746	8,410	4,071	31,870	55,938	6.65
85	.17304	4,338	4,338	24,068	24,068	5.55

dian males in 1991, which is based on the data in Table 13.3. Under the stationary population assumption, 27.17% of males born in Canada in 1991 will die of a neoplasm, and for an individual due to die of this cause, the life expectancy at birth is 73.27 years.

13.3 CAUSE-DELETED LIFE TABLE

Having examined overall and cause-specific mortality using ordinary and multiple decrement life tables, it is natural to inquire what would be the effect on mortality of eliminating a particular cause of death, say cause k. We denote by $r^{\bullet k}(x, t)$ the hazard function for the population under the assumption that cause k has been eliminated (deleted). In this notation, $\bullet k$ stands for "all causes except cause k." It is tempting to conclude from (13.6) that $r^{\bullet k}(x, t) = r(x, t) - r^k(x, t)$. However, this identity does not hold without making further assumptions. Following the example of Section 8.4, suppose that myocardial infarction ($k = 1$) and stroke ($k = 2$) are two of the causes of death under consideration. Since these two circulatory conditions have a number of risk factors in common, interventions designed to reduce the risk of one will concomitantly reduce the risk of the other. Therefore $r^{\bullet 1}(x, t) < r(x, t) - r^1(x, t)$ and $r^{\bullet 2}(x, t) < r(x, t) - r^2(x, t)$.

However, if the K causes of death are independent, the crude hazard functions become net hazard functions and, as a result, the identity $r^{\bullet k}(x, t) = r(x, t) - r^k(x, t)$ is satisfied. In practice it is often difficult to guarantee that causes of death are strictly independent. Grouping together conditions that affect a given body system helps to ensure that this is at least approximately true. For example, rather than consider myocardial infarction and stroke to be individual causes of death, they could be combined under the broader heading of circulatory conditions. For the remainder of this section it will be assumed that causes of death are independent. Under the stationary population assumption we define the cause-deleted hazard function to be

$$r^{\bullet k}(x) = r(x) - r^k(x).$$

Letting $D_j^{\bullet k} = D_j - D_j^k$ we define the cause-deleted death rate for the jth age group to be $R_j^{\bullet k} = D_j^{\bullet k}/N_j (j = 0, 1, 2, \ldots, J)$.

The cause-deleted life table (for cause k) is constructed using precisely the methods described above for the ordinary life table, except that $R_j^{\bullet k}$ is used in place of R_j. In general we use a superscript $\bullet k$ to designate the resulting cause-deleted life table (CDLT) functions. By definition, $l(0) = l^{\bullet k}(0)$. We can think of the CDLT cohort as the OLT cohort after cause k has been eliminated. This way of thinking about the CDLT cohort leads to a number of useful mortality indices. If cause k were to be eliminated, the life expectancy at birth in the OLT cohort would increase to $e^{\bullet k}(0)$, and the number of survivors to age 65 would increase to $l^{\bullet k}(65)$. Therefore the gain in life expectancy at birth would be $e^{\bullet k}(0) - e(0)$ and, since $l^{\bullet k}(0) = l(0)$, the increase in the probability of surviving to age 65 would be $[l^{\bullet k}(65) - l(65)]/l(0)$.

We now examine the effect of eliminating cause k on the group of individuals who are due to die of that cause, namely, the MDLT cohort (for cause k). Once cause

k has been eliminated, these individuals will die of some other cause, and at an age that is necessarily greater than what would have been the age of death from cause k. Consider the $l^{\bullet k}(65) - l(65)$ additional survivors to age 65 in the OLT cohort after eliminating cause k. Since causes of death are independent, eliminating cause k has no impact on individuals due to die of causes other than cause k. Consequently, all $l^{\bullet k}(65) - l(65)$ additional survivors to age 65 must come from the MDLT cohort. It follows that, after cause k has been eliminated, the probability that a member of the MDLT cohort will survive to age 65 increases to

$$\frac{l^k(65) + [l^{\bullet k}(65) - l(65)]}{l^k(0)} = \frac{l^k(65)}{l^k(0)} + \frac{[l^{\bullet k}(65) - l(65)]}{l^k(0)}.$$

Therefore, as a result of eliminating cause k, the probability that a member of the MDLT cohort will survive to age 65 increases by an amount

$$\frac{l^{\bullet k}(65) - l(65)}{l^k(0)}.$$

Now consider the $T^{\bullet k}(0) - T(0) = l(0)[e^{\bullet k}(0) - e(0)]$ additional person-years experienced by the OLT cohort as a result of eliminating cause k. Arguing as above, all of these person-years must be generated by individuals in the MDLT cohort. It follows that, once cause k has been eliminated, the life expectancy at birth for a member of the MDLT cohort increases to

$$\frac{T^k(0) + l(0)[e^{\bullet k}(0) - e(0)]}{l^k(0)} = e^k(0) + \frac{l(0)[e^{\bullet k}(0) - e(0)]}{l^k(0)}.$$

Therefore, after eliminating cause k, the life expectancy at birth for a member of the MDLT cohort will increase by an amount

$$\frac{l(0)[e^{\bullet k}(0) - e(0)]}{l^k(0)}$$

(Greville, 1948; Newman, 1986).

Example 13.4 Table 13.8 gives the summary mortality indices described above for circulatory diseases, neoplasms, and injuries for Canadian males in 1991. Circulatory diseases account for 40.09% of deaths compared to only 5.73% for injuries. However, the life expectancy (at birth) for those due to die of circulatory disease is 77.70 years compared to only 52.21 years for those due to die of injuries. We see that although injuries account for relatively few deaths compared to circulatory diseases, the loss in life expectancy is substantial as a result of death at a relatively young age. Note that the life expectancy of those due to die of circulatory disease is greater than the OLT life expectancy of 74.34 years. This shows that circulatory diseases usually do not result in premature mortality compared to other causes of death. Eliminating circulatory diseases as a cause of death would increase overall life expectancy by 6.06 years and would increase the life expectancy of those due to die of this cause by

TABLE 13.8   Summary Indices of Mortality: Canada, Males, 1991

Summary index	Circulatory	Neoplasms	Injuries
$\dfrac{l^k(0)}{l(0)} \times 100\%$	40.09	27.17	5.73
$e^k(0)$	77.70	73.27	52.21
$e^{\bullet k}(0) - e(0)$	6.06	3.87	1.69
$\dfrac{l^{\bullet k}(65) - l(65)}{l^k(0)} \times 100\%$	13.03	21.78	55.50
$\dfrac{l(0)\,[e^{\bullet k}(0) - e(0)]}{l^k(0)}$	15.12	14.25	29.51

15.12 years, a considerable gain in longevity. On the other hand, eliminating injuries as a cause of death would increase overall life expectancy by a less impressive 1.69 years due to the relatively small number of deaths due to this cause. However, for those due to die of an injury, the gain in life expectancy would be a substantial 29.51 years, an increase that is almost double that for circulatory diseases.

13.4   ANALYSIS OF MORBIDITY USING LIFE TABLES

The life table methods described above are concerned with mortality. However, there are many diseases—for example, arthritis and asthma—which are highly prevalent and which result in considerable morbidity, but which rarely cause death. The impact of conditions such as these on the population will be overlooked if the focus is exclusively on mortality. The public heath importance of morbidity has emerged in recent years as it has come to be realized that increasing the length of life does not necessarily translate into a corresponding increase in the quality of life. In this section we show how life table methods can be used to describe morbidity in a population.

13.4.1   Lifetime Probability of Developing a Disease

It follows from (13.8) and (13.13) that $d_j^k = R_j^k L_j$ and $l^k(0) = \sum_{j=0}^{J} R_j^k L_j$. Therefore the lifetime probability of dying of cause k is

$$\frac{l^k(0)}{l(0)} = \frac{1}{l(0)} \sum_{j=0}^{J} R_j^k L_j. \qquad (13.14)$$

We now extend (13.14) to the analysis of morbidity and derive an estimate of the lifetime probability of developing a given disease. Specifically, we construct an ordinary life table where "death" consists of either developing the disease or dying of some other cause, and we construct a multiple decrement life table where "death from

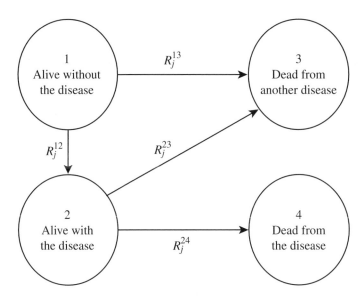

FIGURE 13.1 Live and dead states for the lifetime probability of developing a disease

cause k" is defined to be the development of the disease. The following arguments are related to those of Zdeb (1977).

Consider Figure 13.1, which depicts two live states (alive with the disease and alive without the disease) and two dead states (dead from the disease and dead from some other disease). The arrows indicate the possible transitions among the states. For the jth age group, R_j^{12}, R_j^{13}, R_j^{23}, and R_j^{24} denote the incidence and death rates in the population under the stationary population assumption. For example, R_j^{12} is the incidence rate for an individual in the population who is at risk of developing the disease. We assume that at birth all members of the OLT cohort are free of the disease (state 1). Thereafter, an individual either remains in state 1 or moves to state 2 or state 3. From state 2 it is possible to move to state 3 or state 4. Note that there is no arrow going from state 2 back to state 1, and so once the disease develops it is regarded as being present for life.

We need to estimate R_j^{12}, R_j^{13}, and R_j^{23}. For the jth age group, let D_j^{12} be the (annual) number of transitions in the population from state 1 to state 2, with analogous definitions for D_j^{13} and D_j^{23}. Let N_j^1 be the number of individuals in the midyear population in state 1, with a corresponding definition for N_j^2. Then $R_j^{12} = D_j^{12}/N_j^1$, $R_j^{13} = D_j^{13}/N_j^1$, and $R_j^{23} = D_j^{23}/N_j^2$. In what follows let k denote the disease under consideration. For the jth age group, let I_j^k be the (annual) number of incident cases in the population and let P_j^k be the number of prevalent cases at midyear. Then $D_j^{12} = I_j^k$, $N_j^1 = N_j - P_j^k$, and $N_j^2 = P_j^k$, and so

$$R_j^{12} = \frac{I_j^k}{N_j - P_j^k}$$

$R_j^{13} = D_j^{13}/(N_j - P_j^k)$, and $R_j^{23} = D_j^{23}/P_j^k$. Deaths not due to the disease under consideration can occur whether or not the disease is present, and so $D_j^{\bullet k} = D_j^{13} + D_j^{23}$. We now assume that, in each age group, individuals with and without the disease have the same death rate for other causes of death, that is, $R_j^{13} = R_j^{23}$. It follows from the preceding identities that

$$D_j^{13} = \left(\frac{N_j - P_j^k}{N_j}\right) D_j^{\bullet k}$$

and hence that

$$R_j^{13} = \frac{D_j^{\bullet k}}{N_j} = R_j^{\bullet k}.$$

We are now able to construct the ordinary and multiple decrement life tables needed to estimate the lifetime probability of developing the disease. In order to distinguish the following life table functions from those in previous sections, a superscript * is added to the notation. For the jth age group, the "overall" and "cause-specific" hazard rates are defined to be $R_j^* = R_j^{12} + R_j^{13}$ and $R_j^{k*} = R_j^{12}$, respectively, that is,

$$R_j^* = \frac{I_j^k}{N_j - P_j^k} + \frac{D_j^{\bullet k}}{N_j} \tag{13.15}$$

and

$$R_j^{k*} = \frac{I_j^k}{N_j - P_j^k}. \tag{13.16}$$

The lifetime probability of developing the disease is given by

$$\frac{l^{k*}(0)}{l^*(0)} = \frac{1}{l^*(0)} \sum_{j=0}^{J} R_j^{k*} L_j^*.$$

In practice, estimates of the number of prevalent cases are difficult to obtain. As illustrated in the example below, unless the disease is especially prevalent, little bias is introduced by ignoring prevalence and setting $P_j^k = 0$. In this case, R_j^* and R_j^{k*} simplify to

$$R_j^* = \frac{I_j^k + D_j^{\bullet k}}{N_j} \tag{13.17}$$

and

$$R_j^{k*} = \frac{I_j^k}{N_j}. \tag{13.18}$$

More elaborate multistate model are available in which individuals move among various states of health, disease, and death (Chiang, 1968, 1980; Keiding, 1991).

Example 13.5 (Breast Cancer: Canada, Females, 1991) Data on the number of deaths due to breast cancer, the number of incident cases of breast cancer, and the census population for Canadian females in 1991 were obtained from official Statistics Canada publications. Data kindly provided by the Northern Alberta Breast Cancer Registry were used to estimate the number of prevalent cases of breast cancer in Canadian females in 1991. Under the stationary population assumption and based on (13.15) and (13.16), 10.92% of females born in Canada in 1991 will develop breast cancer and 4.06% of them will die of this disease. So $4.06/10.92 = 37.18\%$ of the birth cohort who develop breast cancer will eventually succumb to this malignancy. Based on (13.17) and (13.18), the lifetime probability of developing breast cancer is 10.78%. Breast cancer is one of the more prevalent cancers due to its comparatively large incidence rate and relatively good survival. These findings suggest that it will usually be satisfactory to ignore prevalent cases when estimating the lifetime probability of developing cancer.

13.4.2 Disability-Free Life Expectancy

Let π_j be the proportion of the population in the jth age group who, at a given time point, have a particular disabling condition, and let L_j be the person-years lived by the OLT cohort as described in Section 13.1 ($j = 0, 1, \ldots, J$). In practice, π_j would usually be estimated from a population health survey. To an approximation, $\pi_j L_j$ is the number of person-years that the OLT cohort will live in a disabled state during the jth age group. Therefore the total number of person-years of disability that will be experienced by the OLT cohort after age x_j is $\sum_{i=j}^{J} \pi_i L_i$. It follows that the average number of years that a member of the OLT cohort will be free of disability after age x_j is

$$e'(x_j) = \frac{T(x_j) - \sum_{i=j}^{J} \pi_i L_i}{l(x_j)}$$

$$= e(x_j) - \frac{1}{l(x_j)} \sum_{i=j}^{J} \pi_i L_i$$

which is referred to as the disability-free life expectancy at age x_j (Sullivan, 1971; Newman, 1988). As in the preceding section, a multistate life table could be employed to obtain more sophisticated estimates; however, the data necessary for such an approach are usually unavailable.

TABLE 13.9 Age-Specific Prevalence Rates of Dementia: Canada, Males, 1994, and Ordinary Life Table Functions: Canada, Males, 1991

x_j	π_j	$l(x_j)$	L_j	T_j
65	.0087	79,727	374,755	1,248,759
70	.0293	70,175	319,411	874,004
75	.0792	57,589	248,401	554,593
80	.1960	41,771	167,106	306,193
85	.2931	25,071	139,087	139,087

Example 13.6 (Dementia: Canada, Males, 1991) Table 13.9 gives the age-specific point prevalence rates of dementia in Canadian males based on data from a national survey (CSHA Working Group, 1994). The OLT functions are taken directly from Table 13.4. The life expectancy at age 65 is $e(65) = 15.66$, and based on Table 13.9 the dementia-free life expectancy at age 65 is $e'(65) = 14.34$. It follows that, on average, $(15.66 - 14.34)/15.66 = 8.43\%$ of life after age 65 will be spent in a demented state.

CHAPTER 14

Sample Size and Power

Usually one of the first questions asked when an epidemiologic study is being designed is "What sample size do we need?" As will become clear shortly, the answer depends on a number of factors, only some of which are under the control of the investigator. It is possible to base sample size calculations on criteria related to either confidence intervals or hypothesis tests. Despite the current emphasis on confidence intervals in the analysis of epidemiologic data, most often the sample size for a study is determined using the hypothesis testing approach, and this is reflected in the material presented in this chapter. However, such sample size formulas are readily adapted to the confidence interval approach (Greenland, 1988; Bristol, 1989). The sample size formula that is used in a given study must correspond to the statistical methods that are planned for the data analysis, and the latter must in turn be appropriate to the study design. In this chapter we present sample size formulas for a number of the study designs considered in this book. Donner (1984) and Liu (2000) review methods of sample size calculation in an epidemiologic context.

All of the sample size formulas discussed below are based on asymptotic methods. When an asymptotic formula points to a small sample size, it is prudent to verify the result using exact calculations (StatXact, 1998). For simplicity we drop the caret $\hat{}$ from the notation for estimates throughout this chapter.

14.1 SAMPLE SIZE FOR A PREVALENCE STUDY

Consider a study that has the aim of estimating a binomial probability π, for example, as part of a prevalence study. The following sample size formula is derived using the confidence interval approach. Based on the explicit method of Section 3.2.1, a $(1 - \alpha) \times 100\%$ confidence interval for π is

$$\pi \pm z_{\alpha/2} \sqrt{\frac{\pi(1 - \pi)}{r}}.$$

Suppose that we want the upper and lower bounds to be at most a distance $\Delta > 0$ from π, that is,

$$z_{\alpha/2}\sqrt{\frac{\pi(1-\pi)}{r}} \leq \Delta.$$

Solving for r, the number of subjects needed for the study is

$$r = \frac{(z_{\alpha/2})^2 \pi(1-\pi)}{\Delta^2}. \tag{14.1}$$

An interesting observation is that (14.1) does not depend on the size of the underlying population. In order to apply (14.1) it is necessary to have values for α, π, and Δ. The choice of α is, in theory, at the discretion of the investigator, but by convention it is almost always taken to be .05. Having to provide a value for π is circular because the study is being conducted with the aim of estimating this quantity. Since π is unknown, a working value, which we refer to as the pre-study estimate, needs to be determined, for example, from the published literature or based on expert opinion. The estimate of π that is obtained once the study has been completed will be referred to as the post-study estimate. Obviously it is desirable to have Δ as small as possible. Since the denominator of (14.1) is Δ^2, each unit decrease in Δ leads to a substantial increase in the required sample size, as illustrated by the following example.

Example 14.1   Table 14.1 gives sample sizes based on (14.1) for selected values of π and Δ, with $\alpha = .05$. Evidently the sample size needed for a prevalence study will be extremely large if a narrow confidence interval is required.

In the following sections, sample size formulas are derived for comparative studies using the hypothesis testing approach. For each method of testing a hypothesis we consider a corresponding normal approximation with mean μ and variance σ^2, where μ is considered to be the parameter of interest. Under the null hypothesis $H_0 : \mu = \mu_0$, the variance will be denoted by σ_0^2, and under the (two-sided) alternative hypothesis $H_1 : \mu \neq \mu_1$, it will be denoted by σ_1^2. Following Section 2.1, let α be the probability of a type I error and let β be the probability of a type II error. That is, α is the probability of rejecting the null hypothesis when it is true, and β is the probability of not rejecting it when it is false. In Section 3.2.1, with $0 < \gamma < 1$, we defined z_γ to be that point which cuts off the upper γ-tail probability of the stan-

TABLE 14.1   Values of r Based on (14.1) for Selected Values of π and Δ, with $\alpha = .05$

		π	
Δ	.05	.10	.25
.01	1825	3457	7203
.02	456	864	1801
.03	203	384	800
.05	73	138	288
.10	18	35	72

dard normal distribution. Of particular relevance to the present discussion are the values $z_{.025} = 1.96$ and $z_{.20} = .842$. Most of the derivations that follow rest on the fundamental identity

$$\mu_1 - \mu_0 = z_{\alpha/2}\sigma_0 + z_\beta\sigma_1 \qquad (14.2)$$

(Lachin, 1981; Armitage and Berry, 1994, §6.6; Lachin, 2000, §3.2).

14.2 SAMPLE SIZE FOR A CLOSED COHORT STUDY

Risk Difference

Let $\mu_1 = RD$ and $\mu_0 = 0$. In the notation of Table 2.1(b), let $\pi_1 = \pi_2 + RD$ and define $\rho = r_2/r_1$ to be the ratio of unexposed to exposed subjects. It follows from (7.2) that

$$\sigma_1^2 = \frac{\pi_1(1 - \pi_1)}{r_1} + \frac{\pi_2(1 - \pi_2)}{r_2}$$

$$= \frac{1}{r_1}\left[\pi_1(1 - \pi_1) + \frac{\pi_2(1 - \pi_2)}{\rho}\right]. \qquad (14.3)$$

Under H_0,

$$\pi_0 = \frac{\pi_1 r_1 + \pi_2 r_2}{r_1 + r_2} = \frac{\pi_1 + \pi_2\rho}{1 + \rho}. \qquad (14.4)$$

Replacing π_1 and π_2 in (14.3) with π_0 gives

$$\sigma_0^2 = \left[\frac{\pi_0(1 - \pi_0)}{r_1}\right]\left(\frac{1 + \rho}{\rho}\right).$$

Substituting in (14.2) and solving for r_1, the number of exposed subjects needed for the study is

$$r_1 = \frac{\left(z_{\alpha/2}\sqrt{\pi_0(1 - \pi_0)\left(\frac{1+\rho}{\rho}\right)} + z_\beta\sqrt{\pi_1(1 - \pi_1) + \frac{\pi_2(1-\pi_2)}{\rho}}\right)^2}{(RD)^2} \qquad (14.5)$$

(Schlesselman, 1974).

From the above identities we see that r_1 is a function of α, β, ρ, π_2, and RD. Formula (14.5) is reasonable in that if any of α, β, or RD is made smaller, a larger sample size is required. In other words, to reduce either type I or type II error, or to detect a smaller risk difference, it is necessary to have more subjects. Since the tails of the standard normal distribution become progressively narrower, and since RD appears in the denominator as a squared term, reductions in α, β, and RD come at an ever increasing cost in sample size.

To compute a sample size using (14.5) it is necessary to have values for α, β, ρ, π_2, and RD. As pointed out in the preceding section, α is usually taken to be .05 and, also by convention, β is generally set equal to either .10 or .20. By definition, π_2 is the probability that someone without a history of exposure will develop the disease. By the time epidemiologic knowledge has progressed to the point where a comparative study would be undertaken, a dependable pre-study estimate of π_2 is likely to be available. Thus, to a greater or lesser extent, α, β, and π_2 are determined by factors outside the scope of the study. The situation is somewhat different for ρ and RD. When conducting a study, it may be difficult to find subjects with a history of exposure, yet relatively easy to identify individuals who have not been exposed. Instead of attempting to recruit an equal number of exposed and unexposed subjects, an alternative is to oversample the unexposed population. The choice of ρ is determined, in part, by the availability of unexposed subjects and also by the degree of efficiency introduced by oversampling, as illustrated in Example 14.2.

Usually the greatest source of difficulty encountered when calculating a sample size based on (14.5) is deciding on a pre-study estimate of RD. For good reason this value is sometimes referred to as the "difference worth detecting" (Sackett et al., 1985). When there is little prior knowledge regarding the true value of RD, it is tempting to select a pre-study estimate which ensures that even a minor difference between exposed and unexposed subjects will be detected. However, as illustrated in Example 14.2, this can result in an extremely large sample size. Given that there is relatively little latitude in the choice of α, β, and π_2 and, to a certain extent, in the choice of ρ, a sample size calculation for the risk difference largely reduces to a decision about the value of RD to use in (14.5). The usual approach is to calculate sample sizes for a range of values of RD and decide whether a sample size that is feasible will detect a "difference worth detecting."

Example 14.2   Table 14.2 gives asymptotic and exact values of r_1 for selected values of RD, with $\alpha = .05$, $\beta = .20$, $\rho = 1$, and $\pi_2 = .05$. The asymptotic sample sizes were calculated using (14.5) and exact calculations were performed using StatXact (1998). As can be seen, as RD gets smaller, the sample size needed for the study increases rapidly. Note that sample sizes based on the exact method are conservative (larger) compared to those based on the asymptotic approach.

TABLE 14.2   Asymptotic and Exact Values of r_1 Based on (14.5) and StatXact for Selected Values of RD, with $\alpha = .05$, $\beta = .20$, $\rho = 1$, and $\pi_2 = .05$

RD	Asymptotic	Exact
.01	8160	8314
.05	435	464
.10	140	151
.20	49	54
.30	27	31

TABLE 14.3 Values of r_1, r_2, and r Based on (14.5) for Selected Values of ρ, with $\alpha = .05$, $\beta = .20$, $\pi_2 = .05$, and $RD = .05$

ρ	r_1	r_2	r
1	435	435	870
2	312	624	936
3	270	810	1080
4	249	996	1245
5	236	1180	1416
10	211	2110	2321
20	198	3960	4158

Table 14.3 gives sample sizes based on (14.5) for selected values of ρ, with $\alpha = .05$, $\beta = .20$, $\pi_2 = .05$, and $RD = .05$. We can think of the far right column as the sample size needed to produce a power of 80% (see below). Consistent with the theoretical findings for case-control studies cited in Section 11.3, as ρ approaches 5 there is a progressive decrease in the number of exposed subjects needed for the study. Once ρ exceeds 5, the number of exposed subjects starts to gradually plateau, while the number of unexposed subjects continues to increase.

Risk Ratio

Let $\mu_1 = RR$ and $\mu_0 = 1$. Formula (14.5) can be adapted to the risk ratio setting by replacing π_1 with $RR\pi_2$ (Schlesselman, 1974).

Odds Ratio

Let $\mu_1 = OR$ and $\mu_0 = 1$. Formula (14.5) can be adapted to the odds ratio setting by replacing π_1 with

$$\pi_1 = \frac{OR\pi_2}{OR\pi_2 + (1 - \pi_2)}.$$

14.3 SAMPLE SIZE FOR AN OPEN COHORT STUDY

Standardized Mortality Ratio

Let D_a be Poisson with mean and variance equal to $SMR \times E_a$. From Beaumont and Breslow (1981), $\sqrt{D_a}$ is approximately normal with mean $\sqrt{SMR \times E_a}$ and variance $1/4$. Let $\mu_1 = \sqrt{SMR \times E_a}$ and $\mu_0 = \sqrt{E_a}$, and let $\sigma_1^2 = \sigma_0^2 = 1/4$. Substituting in (14.2) and solving for E_a, the expected number of deaths needed for the study is

$$E_a = \frac{\left(z_{\alpha/2} + z_\beta\right)^2}{4\left(\sqrt{SMR} - 1\right)^2}$$

(Beaumont and Breslow, 1981). Based on a crude analysis, $E_a = R_s N_a$ where R_s is the death rate in the standard population. It follows that the amount of person-time needed for the study is $N_a = E_a / R_s$.

Example 14.3 (Schizophrenia) Consider the mortality study of schizophrenia discussed in Examples 12.2 and 12.3. Suppose that at the planning stage it was desired to detect a value of *SMR* at least as small as 1.5. With $\alpha = .05$ and $\beta = .20$,

$$E_a = \frac{(1.96 + .842)^2}{4(\sqrt{1.5} - 1)^2} = 38.86.$$

The crude death for the Alberta male population in 1981 was $R_s = 7283/957,247 = 7.61 \times 10^{-3}$, and so the number of person-years needed for the study would have been $E_a / R_s = 5108$. In fact, the cohort experienced 12,314 person-years during the course of follow-up.

Hazard Ratio

Let $\mu = \log(HR)$ and $\mu_0 = 0$. Denote by φ_1 the proportion of the cohort with a history of exposure, and let $\varphi_2 = 1 - \varphi_1$. It can be shown that the total number of deaths needed for the study is

$$m = \frac{(z_{\alpha/2} + z_\beta)^2}{\varphi_1 \varphi_2 (\log HR)^2}$$

(Schoenfeld, 1983; Collett, 1994, Chapter 9). Let $S_1(t)$ and $S_2(t)$ be the survival curves for the exposed and unexposed cohorts, respectively. To estimate the number of subjects needed for the study it is necessary to take account of the method of accrual and follow-up. Suppose that subjects are accrued over a calendar time period lasting a (time units) and that the last subject recruited into the study has a maximum observation time of f. Therefore the maximum observation time for the study is $a + f$. The probability that a member of the unexposed cohort will die during follow-up is approximately

$$\pi_2 = 1 - \tfrac{1}{6}[S_2(f) + 4S_2(.5a + f) + S_2(a + f)].$$

Observe that when $a = 0$, $\pi_2 = 1 - S_2(f)$. Letting π_1 denote the corresponding probability for the exposed cohort, it follows from (8.7) that $(1 - \pi_1) = (1 - \pi_2)^{HR}$ and hence that

$$\pi_1 = 1 - (1 - \pi_2)^{HR}.$$

Therefore the probability that a member of the cohort will die during follow-up is approximately

$$\varphi_1 \pi_1 + \varphi_2 \pi_2 = \varphi_1 \left[1 - (1 - \pi_2)^{HR}\right] + \varphi_2 \pi_2.$$

So the number of subjects needed for the study is

$$\frac{m}{\varphi_1\left[1-(1-\pi_2)^{HR}\right]+\varphi_2\pi_2}.$$

Interestingly, the above formula can be used to estimate the sample size needed for either a Mantel–Haenszel or exponential analysis (George and Desu, 1974).

Example 14.4 (Breast Cancer)   Consider the breast cancer cohort in Examples 9.1 and 9.2, where subjects were enrolled in the study during 1985 and followed to the end of 1989. In this example, $a = 12$ (months) and $f = 48$. From Table 9.1 the proportion of subjects with low receptor level (exposed) is $\varphi_1 = 50/199 = .251$, and so $\varphi_2 = .749$. Suppose that when the study was being designed it was desired to detect a value of HR at least as small as 2. With $\alpha = .05$ and $\beta = .20$,

$$m = \frac{(1.96 + .842)^2}{.251(.749)(\log 2)^2} = 87.$$

As it turns out, there were only 49 deaths in the cohort. Based on the data in Table 9.1, the Kaplan–Meier estimates are $S_2(48) = S_2(54) = .836$ and $S_2(60) = .778$, and so $\pi_2 = .174$. If this estimate had been available at the planning stage, the estimate of the proportion of the cohort expected to die would have been $.251\left[1 - (1 - .174)^2\right] + .749(.174) = .210$. Therefore the total number of subjects needed for the study would have been $87/.210 = 414$. The actual sample size was 199.

14.4   SAMPLE SIZE FOR AN INCIDENCE CASE-CONTROL STUDY

Unmatched Case-Control Study
In the notation of Table 11.1, let m_1 and $m_2 = \rho m_1$ be the number of cases and controls in an incidence case-control study. As in Section 11.1, let ϕ_1 denote the probability that a case has a history of exposure and let ϕ_2 denote the corresponding probability for a control. From (4.2) and (14.4), we have

$$\phi_1 = \frac{OR\phi_2}{OR\phi_2 + (1 - \phi_2)} \tag{14.6}$$

and

$$\phi_0 = \frac{\phi_1 + \phi_2\rho}{1 + \rho}. \tag{14.7}$$

As shown in Section 11.1.3, the odds ratio for an incidence case-control study is the same whether we consider the row or column marginal totals fixed. Arguing as above for the risk difference, the sample size needed for an incidence case-control study is

$$m_1 = \frac{\left(z_{\alpha/2}\sqrt{\phi_0(1-\phi_0)\left(\frac{1+\rho}{\rho}\right)} + z_\beta\sqrt{\phi_1(1-\phi_1) + \frac{\phi_2(1-\phi_2)}{\rho}}\right)^2}{(\phi_1 - \phi_2)^2} \tag{14.8}$$

(Schlesselman, 1981, p. 150).

Example 14.5   Table 14.4 gives asymptotic and exact values of m_1 based on (14.8) and StatXact (1998) for selected values of OR, with $\alpha = .05$, $\beta = .20$, $\rho = 1$, and $\phi_2 = .05$. As can be seen, the exact method is conservative compared to (14.8).

Matched-Pairs Case-Control Study

In the notation of Section 11.2.3, consider the binomial distribution with parameters (Π, r), where $\Pi = OR/(OR+1)$ and $r = f_{(1,0)} + f_{(0,1)}$. Let $\mu_1 = \Pi$ and $\mu_0 = 1/2$, so that $\sigma_1^2 = [\Pi(1-\Pi)]/r$ and $\sigma_0^2 = 1/(4r)$. Substituting in (14.2) and solving for r, the number of discordant pairs needed for the study is

$$\begin{aligned} r &= \frac{\left[(z_{\alpha/2}/2) + z_\beta\sqrt{\Pi(1-\Pi)}\right]^2}{(\Pi - 1/2)^2} \\ &= \frac{\left[z_{\alpha/2}(OR+1) + 2z_\beta\sqrt{OR}\right]^2}{(OR-1)^2}. \end{aligned} \tag{14.9}$$

Table 14.5(a) gives the probability that a matched pair has a particular type of configuration under the assumption that, in the population, the matching variables are not associated with exposure. This means that, in effect, matched pairs are formed at random. Therefore the probability of a pair being discordant is

$$\begin{aligned} \varphi_0 &= \phi_1(1-\phi_2) + \phi_2(1-\phi_1) \\ &= \frac{\phi_2(1-\phi_2)(OR+1)}{OR\phi_2 + (1-\phi_2)} \end{aligned} \tag{14.10}$$

where the second equality follows from (14.6). Consequently the number of matched pairs needed for the study is $J_0 = r/\varphi_0$ (Schlesselman, 1982, p. 161).

TABLE 14.4   Asymptotic and Exact Values of m_1 Based on (14.8) and StatXact for Selected Values of OR, with $\alpha = .05$, $\beta = .20$, $\rho = 1$, and $\phi_2 = .05$

OR	Asymptotic	Exact
2	516	550
3	177	192
4	100	109
5	69	76
10	27	31

TABLE 14.5(a) Probabilities of Configurations: Matched-Pairs Case-Control Study

Case	Control exposed	unexposed	
exposed	$\phi_1\phi_2$	$\phi_1(1-\phi_2)$	ϕ_1
unexposed	$(1-\phi_1)\phi_2$	$(1-\phi_1)(1-\phi_2)$	$1-\phi_1$
	ϕ_2	$1-\phi_2$	1

TABLE 14.5(b) Probabilities of Configurations: Matched-Pairs Case-Control Study

Case	Control exposed	unexposed	
exposed	ϕ_{11}	ϕ_{12}	ϕ_1^*
unexposed	ϕ_{21}	ϕ_{22}	$1-\phi_1^*$
	ϕ_2^*	$1-\phi_2^*$	1

In most applications the assumption that the matching variables are unrelated to exposure is untenable. Table 14.5(b) gives the notation corresponding to Table 14.5(a) in this more realistic setting. We seek an estimate of $\varphi = \phi_{12} + \phi_{21}$, the probability that a matched pair will be discordant. In Table 14.5(b), the probability that a sampled case has a history of exposure is denoted by $\phi_1^* (= \phi_{11} + \phi_{12})$, and the corresponding probability for a sampled control is $\phi_2^* (= \phi_{11} + \phi_{21})$. A superscript * is used to distinguish the sample probabilities in Table 14.5(b) from the corresponding population probabilities, ϕ_1 and ϕ_2, in Table 14.5(a). In most case-control studies, the cases are (or can be thought of as) a simple random sample of cases arising in the population, and so we generally have $\phi_1^* = \phi_1$. Aside from the situation where the matching variables are unrelated to exposure, the identity $\phi_2^* = \phi_2$ does not hold, as we now illustrate.

Consider a case-control study investigating hypertension (high blood pressure) as a risk factor for stroke, where age is the matching variable. The process of matching on age ensures that the case and control samples have the same age distribution. The risk of stroke increases with age, and so the average age of the case and control samples will be greater than the average age of the nonstroke population from which the controls were selected. The risk of hypertension also increases with age, and consequently sampled controls are more likely to have hypertension than individuals in the nonstroke population, that is, $\phi_2^* > \phi_2$. As a consequence, matched pairs in the study will more often be concordant with respect to the presence or absence of hypertension than pairs formed at random in the population. Therefore $\varphi < \varphi_0$ and so, in most applications, J_0 will be an underestimate of the number of matched pairs needed for the study.

TABLE 14.5(c)   Probabilities of Configurations: Matched-Pairs Case-Control Study

Case	Control exposed	Control unexposed	
	exposed	unexposed	
exposed	$\phi_1^* \phi_2^*$	$\phi_1^*(1 - \phi_2^*)$	ϕ_1^*
unexposed	$(1 - \phi_1^*)\phi_2^*$	$(1 - \phi_1^*)(1 - \phi_2^*)$	$1 - \phi_1^*$
	ϕ_2^*	$1 - \phi_2^*$	1

From Table 14.5(b) define the parameter

$$\vartheta = \frac{\phi_{11}\phi_{22}}{\phi_{12}\phi_{21}}.$$

It can be shown that $\vartheta = 1$ if and only if the matching variables are not associated with exposure in the case and control samples; equivalently, if and only if Table 14.5(b) can be expressed as Table 14.5(c). For example, $\vartheta = 1$ implies

$$\phi_{11} = \frac{\phi_{12}\phi_{21}}{\phi_{22}} = \frac{(\phi_1^* - \phi_{11})(\phi_2^* - \phi_{11})}{1 - \phi_1^* - \phi_2^* + \phi_{11}}$$

which can be solved for ϕ_{11} to give $\phi_{11} = \phi_1^* \phi_2^*$. As can seen from the denominator of ϑ, as φ increases (decreases) there is a corresponding decrease (increase) in ϑ. The preceding observations provide a rationale for viewing ϑ as a measure of concordance (of exposure) in the case and control samples. For $\vartheta \neq 1$, a "corrected" version of φ_0 which accounts for this concordance is

$$\varphi = \varphi_0 \left(\frac{\sqrt{1 + 4(\vartheta - 1)\phi_1(1 - \phi_1)} - 1}{2(\vartheta - 1)\phi_1(1 - \phi_1)} \right) \tag{14.11}$$

(Fliess and Levin, 1988), where ϕ_1 and φ_0 are given by (14.6) and (14.10). So the number of matched pairs needed for the case-control study is

$$J_1 = \frac{r}{\varphi}.$$

From (14.6) and (14.9)–(14.11), we observe that J_1 is a function α, β, ϕ_2, OR, and ϑ. In practice, pre-study estimates of ϕ_2 and OR can usually be specified, but the same cannot be said for ϑ. From Table 11.8, an estimate of ϑ is

$$\widehat{\vartheta} = \frac{f_{(1,1)}f_{(0,0)}}{f_{(1,0)}f_{(0,1)}}.$$

However, if matched-pairs data such as that in Table 11.8 are available, φ can be estimated directly using $\widehat{\varphi} = [f_{(1,0)} + f_{(0,1)}]/J$. When pre-study information on ϑ

is limited, Fleiss and Levin (1988) recommend using a relatively large value of ϑ in (14.11), such as 2.5. Figures 14.1(a) and 14.1(b) show graphs of J_0/J_1 as a function of ϑ, for $\phi_2 = .05$ and .5, and $OR = 2$ and 5, with $\vartheta > 1$. As can be seen, J_0/J_1 is a steadily decreasing function of ϑ and so J_0 may seriously underestimate J_1 when ϑ differs substantially from 1.

Lachin (1992) gives an alternative formulation of (14.11) and shows that, for matched-pairs designs, the methods of Fleiss and Levin (1988) and Dupont (1988) produce similar results. Dupont (1988) gives a sample size formula for studies with $(1 : M)$ matching but, except for the matched-pairs case, the calculations are rather involved. Following Schlesselman (1982, p. 168), for $(1 : M)$ matching, an approximate sample size formula for the number of cases is

$$J_M = \left(\frac{M+1}{2M} \right) J_1.$$

Dupont (1988) shows that J_M can greatly overestimate the sample size when the probability is small that a sampled control has a history of exposure.

14.5 CONTROLLING FOR CONFOUNDING

Except for the matched-pairs design, the above sample size formulas do not take confounding into account. Methods of sample size estimation which make allowance for stratification and other forms of confounder control have been described. For example, see Gail (1973), Muñoz and Rosner (1984), Wilson and Gordon (1986), Woolson et al. (1986), Self and Mauritsen (1988), Lubin and Gail (1990), and Self et al. (1992). EGRET SIZ (1997) is a software package that performs asymptotic sample size calculations with adjustment for confounders.

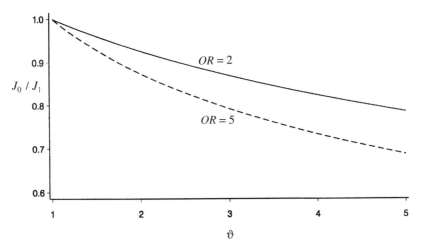

FIGURE 14.1(a) J_0/J_1 as a function of ϑ for selected values of OR, with $\phi_2 = .05$

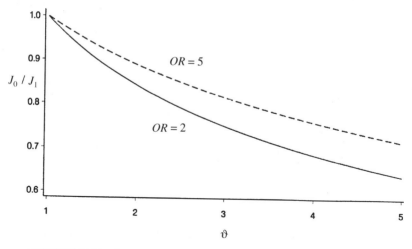

FIGURE 14.1(b) J_0/J_1 as a function of ϑ for selected values of OR, with $\phi_2 = .5$

Example 14.6 (Estrogen–Endometrial Cancer)   Consider the matched-pairs study of Example 11.3. Suppose that at the planning stage it was desired to detect a value of OR at least as small as 3. With $\alpha = .05$ and $\beta = .20$,

$$r = \frac{\left[1.96(3+1) + 2(.842)\sqrt{3}\right]^2}{(3-1)^2} = 29.$$

The actual number of discordant pairs was 50. Based on the data in Table 11.10, $\phi_2^* = 19/183 = .104$ and $\vartheta = 4.82$. Suppose for the sake of illustration that $\phi_2 = .05$. From the above formulas, $\varphi_0 = .173$ and $\varphi = .129$. So the total number of matched pairs needed for the study would have been $J_1 = 29/.129 = 224$. The actual number of matched pairs was 183.

14.6   POWER

In practice it is not unusual for the maximum possible sample size for a proposed study to be determined by factors outside the control of the investigator. For example, financial considerations or the availability of subjects may place a ceiling on the number of subjects that can be enrolled, or the study may involve the analysis of existing data. In situations such as this, the question arises as to whether the given sample size is large enough to detect "a difference worth detecting."

By definition, β is the probability of not rejecting the null hypothesis when it is false. Consequently $1 - \beta$, which is termed the power, is the probability of rejecting the null hypothesis when it is false. In other words, the power is the probability of detecting "a difference worth detecting" when it is there to be detected. Strictly speaking, power is a property of statistical tests; however, due to the close connection between power and sample size, it is usual to speak of the power of a study. When designing a study, it is generally desired to have a power of at least 80%. The sample size formulas considered above are identities involving certain variables. For example, (14.8) expresses a mathematical relationship between the six quantities m_1, α, β, ρ, ϕ_2, and OR. Once any five of them have been specified, the sixth is automatically determined. Suppose that a case-control study is planned and the number of cases is fixed at m_1. We now derive a formula for the power of the study by solving (14.8) for $1 - \beta$.

Let Z be standard normal and define $\Phi(z) = P(Z \leq z)$. From the definition of z_γ, $P(Z \geq z_\gamma) = \gamma$ and so $\Phi(z_\gamma) = P(Z \leq z_\gamma) = 1 - P(Z \geq z_\gamma) = 1 - \gamma$. For example, $\Phi(1.96) = 1 - .025 = .975$. We need to take the square root of both sides of (14.8). In practice, α and β are always less than .5, and so $z_{\alpha/2}$ and z_β are greater than 0. Therefore the square root of the numerator of (14.8) is the term in large parentheses. However, the sign of $\phi_1 - \phi_2$ depends on whether OR is greater than or less than 1. The square root of the denominator can be written as $|\phi_1 - \phi_2|$, the absolute value of $\phi_1 - \phi_2$. Solving (14.8) for z_β gives

$$z_\beta = \frac{\left(\sqrt{m_1}\,|\phi_1 - \phi_2|\right) - z_{\alpha/2}\sqrt{\phi_0(1 - \phi_0)\left(\frac{1+\rho}{\rho}\right)}}{\sqrt{\phi_1(1 - \phi_1) + \frac{\phi_2(1-\phi_2)}{\rho}}}. \qquad (14.12)$$

Applying $\Phi(z)$ to both sides of (14.12) yields

$$1 - \beta = \Phi\left(\frac{\left(\sqrt{m_1}\,|\phi_1 - \phi_2|\right) - z_{\alpha/2}\sqrt{\phi_0(1 - \phi_0)\left(\frac{1+\rho}{\rho}\right)}}{\sqrt{\phi_1(1 - \phi_1) + \frac{\phi_2(1-\phi_2)}{\rho}}}\right) \qquad (14.13)$$

which is the power formula corresponding to (14.8).

Example 14.7   Suppose that a case-control study is to be conducted with 100 cases and 200 controls ($\rho = 2$). As usual, let $\alpha = .05$ and suppose that the pre-study estimates are $\phi_2 = .10$ and $OR = 2$. From (14.13), the power is 52%; that is, there is a 52% probability that the null hypothesis $H_0 : OR = 1$ will be rejected when the true value of the odds ratio is 2. Figure 14.2 shows the graph of $1 - \beta$ as a function of OR. The power is below 80% unless OR is less than .16 or greater than 2.58.

Example 14.8   (Oral Contraceptives–Myocardial Infarction: 35 to 44-Year Age Group)   Consider the 35 to 44-year age group in Table 11.5. Suppose that when the study was being planned, the pre-study estimates of ρ and ϕ_2 were precisely equal to what turned out to be the post-study estimates, that is, $\rho = 727/108 = 6.73$ and $\phi_2 = 35/727 = .048$. Also suppose that, from the published literature, the pre-study estimate of the odds ratio was taken to be $OR = 3$. Based on (14.8), with $\alpha = .05$ and $\beta = .20$, 89 cases would have been needed for the study. In fact, the actual number of cases in the study was 108.

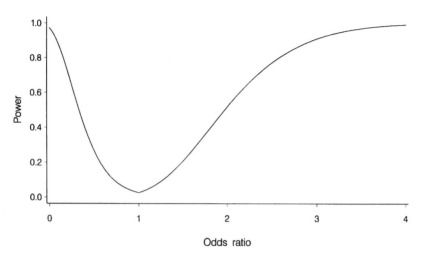

FIGURE 14.2   Power function for Example 14.7

From Table 11.5, $\widehat{OR}_u = 2.02$ and $X^2_{mh} = 3.64$ ($p = .06$). Note that the post-study estimate of the odds ratio is smaller than the above (hypothetical) pre-study estimate. If we adhere rigidly to $\alpha = .05$, the null hypothesis (that oral contraceptive use is not a risk factor for myocardial infarction) is not rejected. When a null hypothesis is not rejected, the *"post hoc"* power is sometimes considered. The *post hoc* power is calculated from (14.13) using post-study, as opposed to pre-study, estimates. Based on Table 11.5 the *post hoc* power is 48.4%. With this information, it might be concluded that the study had insufficient power to detect an odds ratio as small as that which was observed. It has been argued by Greenland (1988) and Goodman and Berlin (1994) that, although power calculations are entirely appropriate when a study is being designed, *post hoc* power has no place in the interpretation of study findings. The reason is that (pre-study) power is concerned with the probability of rejecting the null hypothesis. Once the study has been completed and the null hypothesis has either been rejected or not, it makes no sense to talk about the probability of an event that has already occurred. However, post-study estimates can, and should, be used to make decisions about future studies—in particular, the sample size. Based on Table 11.5, to achieve a power of 80%, a future case-control study of the 35 to 44-year age group would need 269 cases.

CHAPTER 15

Logistic Regression and Cox Regression

In previous chapters we considered a range of statistical methods for analyzing data from cohort and case-control studies, many of which have served data analysts well for decades. However, these classical techniques have limitations. In particular, in order to control for confounding using these methods it is necessary for exposure variables to be categorical (discrete). This is satisfactory, even desirable, at the early stages of data analysis. By categorizing a continuous exposure variable, category-specific parameter estimates can be examined and functional relationships between exposure and disease uncovered, as illustrated below. However, the inability to model variables in continuous form means that a risk relationship that could be summarized concisely in terms of a continuous variable must be expressed as a series of category-specific parameter estimates. When a continuous variable is categorized, there is often a loss of information, which can lead to statistical inefficiency. Nonregression methods based on stratification are particularly prone to this problem because tables with too many zero cells are effectively dropped from the analysis.

In this chapter we present an overview of two of the most important regression techniques in epidemiology: logistic regression and Cox regression. Logistic regression extends odds ratio methods to the regression setting, and Cox regression does the same for hazard ratio methods. Linear regression, analysis of variance, repeated measures analysis of variance, and other multivariate methods designed for continuous outcome (dependent) variables also have a place in the analysis of epidemiologic data. The feature of logistic regression and Cox regression which makes them so useful in epidemiology is that they are concerned with dichotomous outcomes and can accommodate both continuous and categorical predictor (independent) variables. There are many books that have excellent discussions of logistic regression and Cox regression. For further reading on logistic regression the reader is referred to Breslow and Day (1980), Kleinbaum et al. (1982), Cox and Snell (1989), Hosmer and Lemeshow (1989), Collett (1991), and Kleinbaum (1994). References for Cox regression are included among the citations at the beginning of Chapter 8.

There are other regression techniques designed for categorical outcomes which are useful in epidemiology. Loglinear analysis provides a method of analyzing multidimensional contingency tables (Bishop et al., 1975; Feinberg, 1981). Poisson regression, which is closely related to loglinear analysis, can be used to analyze censored survival data that is grouped in the sense of Section 9.3 (Frome 1983; Frome and Checkoway, 1985; Breslow and Day, 1987; Seber, 2000). Parametric survival regression models, such as those defined using the Weibull and exponential distributions, offer an alternative to Cox regression, as described below (Kalbfleisch and Prentice, 1980; Lawless, 1982; Cox and Oakes, 1984; Lee, 1992; Collett, 1994; Hosmer and Lemeshow, 1999). Methods are available for the analysis of longitudinal data in which repeated measurements are taken on each individual and where the outcome variable is dichotomous (Lindsey, 1993; Diggle et al., 1994).

15.1 LOGISTIC REGRESSION

Consider a closed cohort study but, unlike earlier discussions, assume that the exposure variable is continuous. For example, in the breast cancer study considered in Example 5.1, we treated receptor level as a dichotomous variable but in fact it is measured by the laboratory on a continuous scale. Disease processes often exhibit dose–response relationships of the type depicted in Figure 15.1, where the horizontal axis is exposure (dose) and the vertical axis is the probability of disease (response). The sigmoidal shape means that the probability of disease is low until a certain exposure threshold is reached, after which the risk increases rapidly until all but the most resilient subjects have become ill as a result of exposure. There are a number of functions that have this sigmoidal shape and that have proved useful in modeling dose–response relationships (Cox and Snell, 1989, §1.5). One of these is the logistic function,

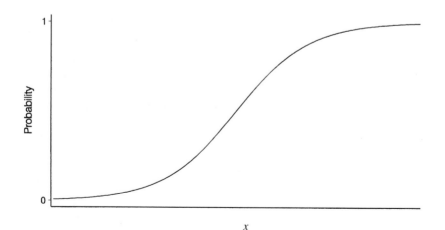

FIGURE 15.1 Logistic curve

$$\pi_x = \frac{\exp(\alpha + \beta x)}{1 + \exp(\alpha + \beta x)}. \tag{15.1}$$

In (15.1), x is exposure, π_x is the probability of disease at exposure x, and α and β are unknown constants that must be estimated from study data. Since the exponential function is strictly positive, it follows that $0 < \pi_x < 1$. When $\beta > 0$, π_x goes to 0 as x goes to $-\infty$, and π_x goes to 1 as x goes to $+\infty$; when $\beta < 0$ the limits are reversed. For each x, define the odds to be $\omega_x = \pi_x/(1 - \pi_x)$. Then (15.1) can be written in the form of a logistic regression model,

$$\log(\omega_x) = \log\left(\frac{\pi_x}{1 - \pi_x}\right) = \alpha + \beta x \tag{15.2}$$

where the log-odds is expressed as a linear function of exposure. In keeping with the terminology of linear regression, we refer to α as the intercept and to βx as a main effect. For a given subject, let x be the (observed) value of the exposure variable and let π_x be the (unknown) probability of developing the disease at exposure x. Random variation can be incorporated into (15.2) by introducing the binomial random variable δ with parameters $(\pi_x, 1)$, where $\delta = 1$ if the subject develops the disease at exposure x, and $\delta = 0$ otherwise. Models (15.1) and (15.2) are readily adapted to the situation where exposure is dichotomous by defining x to be a dummy variable, where $x = 1$ if the subject is exposed, and $x = 0$ otherwise. Note that, unlike δ, x is not a random variable.

With x assumed to be dichotomous, the observed counts and expected values for a closed cohort study can be displayed as in Tables 15.1(a) and 15.1(b), which are seen to correspond to Tables 2.1(a) and 2.1(b), respectively. In this notation, a_{11} is the number of exposed subjects who develop the disease, a_{10} is the number of unexposed subjects who develop the disease, and so on.

TABLE 15.1(a) Observed Counts: Closed Cohort Study

	$x = 1$	$x = 0$
$\delta = 1$	a_{11}	a_{10}
$\delta = 0$	a_{01}	a_{00}
	r_1	r_0

TABLE 15.1(b) Expected Values: Closed Cohort Study

	$x = 1$	$x = 0$
$\delta = 1$	$\pi_1 r_1$	$\pi_0 r_0$
$\delta = 0$	$(1 - \pi_1)r_1$	$(1 - \pi_0)r_0$
	r_1	r_0

Model (15.2) has two parameters, the maximum possible for the 2×2 case, and is therefore said to be saturated. From (15.2), we have

$$\log(\omega_0) = \log\left(\frac{\pi_0}{1 - \pi_0}\right) = \alpha$$

and

$$\log(\omega_1) = \log\left(\frac{\pi_1}{1 - \pi_1}\right) = \alpha + \beta.$$

It follows that $\log(OR) = \log(\omega_1) - \log(\omega_0) = \beta$ and so $OR = \exp(\beta)$. When performing a regression analysis, logistic or otherwise, one of the aims is usually to find the least complicated model that fits the data and at the same time accounts for random error. The intercept α is (usually) required and so the only simplification possible in (15.2) is to have $\beta = 0$, in which case $OR = 1$. So $\beta = 0$ corresponds to the model of no association between exposure and disease.

Now suppose there are two dichotomous exposure variables, x and y, and denote the corresponding probability of disease by π_{xy} ($x = 0, 1$; $y = 0, 1$). The expected values are given in Table 15.2.

For each x and y, the odds is defined to be $\omega_{xy} = \pi_{xy}/(1 - \pi_{xy})$ and the odds ratio for stratum y is defined to be

$$OR_y = \frac{\omega_{1y}}{\omega_{0y}} = \frac{\pi_{1y}(1 - \pi_{0y})}{\pi_{0y}(1 - \pi_{1y})}.$$

For the case of two dichotomous exposure variables the most general logistic regression model is

$$\log(\omega_{xy}) = \log\left(\frac{\pi_{xy}}{1 - \pi_{xy}}\right) = \alpha + \beta x + \gamma y + \varphi xy \qquad (15.3)$$

where α is the intercept, βx and γy are main effects, and φxy is an interaction term. This model has four parameters, the maximum possible for the $2 \times 2 \times 2$ case, and thus is saturated. For $y = 0$, $\log(\omega_{10}) = \alpha + \beta$ and $\log(\omega_{00}) = \alpha$, and so $\log(OR_0) = \beta$. Similarly, for $y = 1$, $\log(\omega_{11}) = \alpha + \beta + \gamma + \varphi$ and $\log(\omega_{01}) = \alpha + \gamma$, and so

TABLE 15.2 Expected Values: Closed Cohort Study

	\multicolumn{2}{c}{$y = 1$}			\multicolumn{2}{c}{$y = 0$}	
	$x = 1$	$x = 0$		$x = 1$	$x = 0$
$\delta = 1$	$\pi_{11} r_{11}$	$\pi_{01} r_{01}$		$\pi_{10} r_{10}$	$\pi_{00} r_{00}$
$\delta = 0$	$(1 - \pi_{11}) r_{11}$	$(1 - \pi_{01}) r_{01}$		$(1 - \pi_{10}) r_{10}$	$(1 - \pi_{00}) r_{00}$
	r_{11}	r_{01}		r_{10}	r_{00}

$\log(OR_1) = \beta + \varphi$. Therefore the parameter φ is the amount by which the log-odds ratios, $\log(OR_0)$ and $\log(OR_1)$, differ across strata determined by y. This means that when $\varphi = 0$ the odds ratios for the association between by x and δ are homogeneous across strata determined by y. In this case we interpret $\exp(\beta)$ as the common value of the stratum-specific odds ratio for the association between x and δ, after adjusting for y.

Reworking Table 15.2, we can stratify the same expected values according to x. The logistic regression model (15.3) remains the same, but now when $\varphi = 0$ we interpret $\exp(\gamma)$ as the common value of the stratum-specific odds ratio for the association between y and δ, after adjusting for x. This demonstrates that the variables in a logistic regression model have equal status from a statistical point of view. Unlike the stratified methods described in Chapter 5, where a distinction was made between the "exposure" variable and the "stratifying" variable, in logistic regression analysis there are only "variables." Consequently, in a logistic regression model—and other regression models for that matter—variables are adjusted for each other simultaneously. Of course, in a given analysis, particular emphasis will usually be placed on certain variables and they will be given an appropriate interpretation as risk factors, confounders, effect modifiers, and so on.

The general logistic regression model is

$$\log\left(\frac{\pi}{1-\pi}\right) = \alpha + \sum_{i=1}^{n}\beta_i x_i \qquad (15.4)$$

where π is a function of the x_i, and each x_i is either a continuous or dummy variable.

The above discussion was presented in terms of a closed cohort study. We showed in Section 11.1.3 that odds ratio methods for closed cohort studies can be adapted to the analysis of incidence case-control data. This raises the question of whether logistic regression can also be utilized in this way. Outwardly it seems that there is the same problem of interpretation that was encountered in Section 11.1.1; that is, we are in a position to estimate the odds of exposure but are really interested in estimating the odds of disease. It is a remarkable fact that, analogous to the results of Section 11.1.3, logistic regression can be used to analyze data from a case-control study by proceeding as if the data had been collected using a closed cohort design (Anderson, 1972; Breslow and Powers, 1978; Prentice and Pyke, 1979; Breslow and Day, 1980, §6.3). The β_i from the logistic regression model are interpreted as log-odds ratios relating exposure to disease. However, due to the case-control design, α has no epidemiologic meaning.

With logistic regression it is possible to perform the types of analyses described in Chapters 5, including point estimation, interval estimation, and testing for association, homogeneity, and linear trend. Furthermore, unconditional, conditional, and exact methods of logistic regression are available (Breslow and Day, 1980; Kleinbaum et al., 1982; Hirji et al., 1987; Mehta and Patel, 1995; LogXact, 1999). An important feature of logistic regression is that both categorical and continuous independent variables can appear in the same model.

Example 15.1 (Stage–Receptor Level–Breast Cancer)   We illustrate the power and flexibility of logistic regression by reanalyzing the breast cancer data considered in Chapter 5. The following analysis is based on asymptotic unconditional methods and was performed using EGRET (1999). Since both receptor level and stage are important predictors of breast cancer survival, we start with the model that has these two variables as main effects,

$$\log(\omega) = \alpha + \beta x + \gamma_1 y_1 + \gamma_2 y_2 \tag{15.5}$$

where

$$x = \begin{cases} 1 & \text{low receptor level} \\ 0 & \text{high receptor level} \end{cases}$$

$$y_1 = \begin{cases} 1 & \text{stage II} \\ 0 & \text{stage I or III} \end{cases} \qquad y_2 = \begin{cases} 1 & \text{stage III} \\ 0 & \text{stage I or II} \end{cases} \; .$$

Note that in order to specify stage, which has three categories, two dummy variables are required. In particular, for a subject in stage I, we have $y_1 = 0$ and $y_2 = 0$. Since (15.5) does not contain an interaction term for receptor level and stage, the model assumes homogeneity. Table 15.3 gives the estimates based on model (15.5), where θ is a generic symbol for a model parameter. By definition, for each variable, $\widehat{OR} = \log(\hat{\theta})$, and the 95% confidence interval for *OR* is obtained by exponentiating $\hat{\theta} \pm 1.96\sqrt{\widehat{\text{var}}(\hat{\theta})}$. The variable descriptions and estimates in each row correspond to the presence of a characteristic compared to its absence. For example, the row for "low receptor level" gives the estimates comparing subjects with low receptor level ($x = 1$) to those with high receptor level ($x = 0$).

It is of interest to compare the point and interval estimates for receptor level in Table 15.3 to the corresponding estimates in Example 5.1. Since the present example and Example 5.1 are both based on unconditional maximum likelihood methods, and whereas both analyses incorporate adjustment for stage, the estimates for receptor level are necessarily identical. The point and interval estimates for stage in Table 15.3 are similar to those in Table 5.11 where the MH–RBG method is used.

Based on model (15.5), for a subject with variables x, y_1, and y_2, the estimated probability of dying of breast cancer is

TABLE 15.3   Logistic Regression Output for Model (15.5): Breast Cancer

Variable	$\hat{\theta}$	$\sqrt{\widehat{\text{var}}(\hat{\theta})}$	$\widehat{OR}$	$\underline{OR}$	$\overline{OR}$
Intercept	−2.37	.420	.09	.04	.21
Low receptor level	.92	.395	2.51	1.16	5.44
Stage II	1.13	.466	3.11	1.25	7.75
Stage III	2.94	.586	18.84	5.98	59.37

$$\hat{\pi} = \frac{\exp(-2.37 + .92x + 1.13y_1 + 2.94y_2)}{1 + \exp(-2.37 + .92x + 1.13y_1 + 2.94y_2)}. \tag{15.6}$$

The estimate of $\hat{\pi}$ for a given subject is obtained by substituting in (15.6) the values of x, y_1, and y_2 for that individual. As an illustration, for a woman with low receptor level ($x = 1$) and stage II disease ($y_1 = 1$, $y_2 = 0$), the estimated probability of dying during the 5-year period of follow-up is

$$\hat{\pi} = \frac{\exp(-2.37 + .92 + 1.13)}{1 + \exp(-2.37 + .92 + 1.13)} = .421.$$

The model that extends (15.5) by including an interaction between receptor level and stage is

$$\log(\omega) = \alpha + \beta x + \gamma_1 y_1 + \gamma_2 y_2 + \varphi_1 x y_1 + \varphi_2 x y_2. \tag{15.7}$$

Observe that two additional parameters are needed because stage has three categories. Model (15.7) has six parameters, the maximum possible for the $2 \times 2 \times 3$ case, and thus is saturated. The Wald and likelihood ratio tests based on the logistic model are precisely the Wald and likelihood ratio tests given in Example 5.1 (provided the logistic model uses the null variance estimate for the Wald test). So there is considerable evidence for homogeneity, that is, $\varphi_1 = \varphi_2 = 0$. Having decided at the outset that main effects for receptor level and stage are needed, the absence of interaction means that (15.5) is the "final" logistic regression model. The fitted counts based on (15.5) are precisely those given in Table 5.5.

Receptor level is measured in the laboratory on a continuous scale, a feature that is not exploited in the preceding analysis. Prior to performing logistic regression with a continuous independent variable, it is necessary to ensure that the variable is linearly related to the log-odds of disease. If the variable does not exhibit linearity, it may be possible to transform the data so that the linear condition is met. Using receptor level as an example, we present a method of deciding whether a variable should be retained in its original form or transformed. In what follows, when receptor level is considered to be a continuous variable it will be denoted by x'.

The values of x' range from 0 to 2621, with a distribution which is highly skewed (median $= 37$). We begin by creating five categories based on quintiles of x', as shown in Table 15.4. For the ith category we denote the midpoint by x_i', the number

TABLE 15.4 Data for Receptor Level as a Categorical Variable

Category	x_i'	$\log(x_i')$	a_i	b_i	$\hat{\omega}_i$	$\log(\hat{\omega}_i)$
0–6	3.0	1.10	19	19	1	0
7–27	17.0	2.83	16	22	.73	−.32
28–55	41.5	3.73	11	28	.39	−.93
56–159	107.5	4.68	5	33	.15	−1.89
160–2621	1390.5	7.24	3	36	.08	−2.48

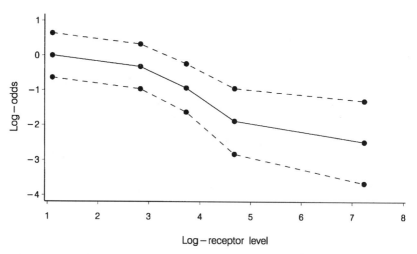

FIGURE 15.2   Log-odds (and 95% confidence intervals) as a function of log-receptor level, based on Table 15.4

of deaths from breast cancer by a_i, the number of survivors by b_i, and the estimated odds by $\hat{\omega}_i = a_i/b_i$ ($i = 1, 2, \ldots, 5$). Note that $\hat{\omega}_i$ and $\log(\hat{\omega}_i)$ decrease as $\log(x_i')$ increases. To check for linearity of the untransformed variable, we graph $\log(\hat{\omega}_i)$ against x_i'. The resulting curve (not shown) has an approximately exponential shape, indicating that a transformation is necessary and that the logarithmic transformation may be helpful. Figure 15.2 shows the graph of $\log(\hat{\omega}_i)$ against $\log(x_i')$, along with the 95% confidence intervals calculated using (3.13). Allowing for random error, it appears that the log-odds of dying of breast cancer is roughly a linear function of log-receptor level (treated as a categorical variable). Evidently more complicated functional forms could be explored.

The above finding provides a rationale for considering the model with the log-odds of dying of breast cancer as a linear function of the continuous variable $\log(x')$. An issue that is not addressed by the preceding analysis is whether the linear assumption is still valid when other variables are included in the logistic regression model along with $\log(x')$—in particular, stage of disease. This question can be addressed by performing the preceding graphical analysis for each stage separately. An alternative approach is described below.

The logistic regression model with $\log(x')$ and stage as main effects is

$$\log(\omega) = \alpha + \beta \log(x') + \gamma_1 y_1 + \gamma_2 y_2. \qquad (15.8)$$

Table 15.5 gives the estimates based on model (15.8). We interpret $\widehat{OR} = .74$ as the amount by which the odds of dying of breast cancer is increased (decreased) by each unit decrease (increase) in $\log(x')$. For example, consider two members of the cohort who will be referred to as subject 1 and subject 2. Denote their receptor levels by x_1' and x_2', and their odds of disease by ω_1 and ω_2, respectively, and sup-

TABLE 15.5 Logistic Regression Output for Model (15.8): Breast Cancer

Variable	$\hat{\theta}$	$\sqrt{\widehat{\text{var}}(\hat{\theta})}$	$\widehat{OR}$	$\underline{OR}$	$\overline{OR}$
Intercept	−1.14	.49	.32	.12	.84
$\log(x')$	−.31	.09	.74	.61	.89
Stage II	1.14	.47	3.11	1.23	7.86
Stage III	2.94	.59	18.89	5.93	60.20

pose they are both at the same stage of disease. From (15.8), we have $\log(\omega_1/\omega_2) = \log(\omega_1) - \log(\omega_2) = (-.31)[\log(x_1') - \log(x_2')]$, and so the odds that subject 1 will die of breast cancer equals the odds for subject 2 multiplied by $(.74)^{\log(x_1') - \log(x_2')}$. Although this result is technically correct, it illustrates that when continuous variables are included in a logistic regression model, it may be difficult to give the model an intuitive interpretation, especially when transformations are involved. This can be problematic when the results of a regression analysis need to be explained to those with a limited background in statistical methods. Based on model (15.8), the estimated probability of dying of breast cancer for a subject with variables x', y_1, and y_2 is

$$\hat{\pi} = \frac{\exp(-1.14 - .31\log(x') + 1.14y_1 + 2.94y_2)}{1 + \exp(-1.14 - .31\log(x') + 1.14y_1 + 2.94y_2)}. \tag{15.9}$$

It is of interest to examine the logistic regression model in which receptor level is treated as a discrete variable with the five categories given in Table 15.4. The model with main effects for receptor level and stage is

$$\log(\omega) = \alpha + \beta_1 x_1 + \beta_2 x_2 + \beta_3 x_3 + \beta_4 x_4 + \gamma_1 y_1 + \gamma_2 y_2. \tag{15.10}$$

In model (15.10), category 0–6 is taken to be the reference category and the x_i are dummy variables for the remaining categories. Table 15.6 gives the estimates based on model (15.10). We can use the four log-odds ratio estimates for receptor level to examine the linear assumption. Figure 15.3 shows the graph of $\log(\widehat{OR}_i)$ against

TABLE 15.6 Logistic Regression Output for Model (15.10): Breast Cancer

Variable	$\hat{\theta}$	$\sqrt{\widehat{\text{var}}(\hat{\theta})}$	$\widehat{OR}$	$\underline{OR}$	$\overline{OR}$
Intercept	−1.68	.56	.19	.06	.57
2nd receptor level	.39	.53	1.47	.52	4.18
3rd receptor level	−.44	.55	.65	.22	1.91
4th receptor level	−1.40	.63	.25	.07	.86
5th receptor level	−2.09	.73	.12	.03	.52
Stage II	1.25	.49	3.49	1.34	9.07
Stage III	3.13	.63	22.96	6.69	78.83

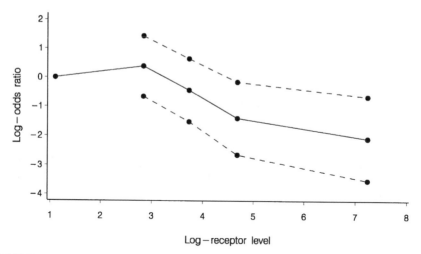

FIGURE 15.3 Log-odds ratio (and 95% confidence intervals) as a function of log-receptor level, based on Model 15.10

$\log(x_i')$, along with the 95% confidence intervals ($i = 1, 2, \ldots, 5$); $\log(\widehat{OR}_1) = 0$ has been included as a baseline value. The curve is not as straight as the one in Figure 15.2, but the linear assumption still seems acceptable, at least for purposes of exploring the data. Since model (15.10) includes a term for stage, we can think of Figure 15.3 as "adjusted" for the confounding effects of this variable, a feature that is absent from Figure 15.2.

Table 15.7 gives the observed and fitted counts of breast cancer deaths based on models (15.8) and (15.10). To estimate the fitted counts for a given receptor level-stage category, we view each subject as a cohort having a sample size of 1. The expected value for this cohort equals the probability of dying of breast cancer as determined by either model (15.8) or (15.10). For model (15.8) the probabilities are estimated using (15.9). For model (15.10) they are estimated using a formula similar to (15.6), but based on the results in Table 15.6. In both cases the fitted count for a given category is obtained by summing the estimated expected values over

TABLE 15.7 Observed and Fitted Counts of Breast Cancer Deaths Based on Models (15.8) and (15.10): Breast Cancer

Receptor level	Stage I			Stage II			Stage III		
	Obs	(15.8)	(15.10)	Obs	(15.8)	(15.10)	Obs	(15.8)	(15.10)
1	2	1.61	.95	6	9.11	7.51	11	10.77	10.55
2	2	1.91	3.24	10	5.84	9.31	4	2.90	3.45
3	2	1.61	1.84	5	3.97	4.75	4	3.93	4.41
4	1	1.14	.67	3	4.10	2.79	1	1.79	1.55
5	0	.73	.32	2	2.98	1.64	1	1.61	1.04

all subjects in the category. For model (15.10), everyone in a given category has the same probability of dying and so the fitted count for a category is simply the estimated probability for that category multiplied by the number of subjects in the category. As can be seen from Table 15.7, both models fit the data moderately well. Interestingly, despite the fact that model (15.8) uses only one parameter to account for receptor level, it appears to fit the data as well as model (15.10), which uses four parameters for this purpose. This may be due to the fact that, by categorizing receptor level, a certain amount of information has been lost and this is reflected in the fit of model (15.10).

15.2 COX REGRESSION

Consider an open cohort study in which exposure is dichotomous, and denote the hazard functions for the exposed and unexposed cohorts by $h_1(t)$ and $h_0(t)$, respectively. We refer to $h_0(t)$ as the baseline hazard function. As in the preceding section, let x be a dummy variable indicating whether a subject is exposed ($x = 1$) or unexposed ($x = 0$). We assume that the proportional hazards assumption is satisfied, that is, $h_1(t)/h_0(t) = \exp(\beta)$, where $HR = \exp(\beta)$. With these definitions, we can write $h_x(t)$ in the form of a proportional hazards regression model,

$$h_x(t) = h_0(t) \exp(\beta x)$$

which can be expressed as

$$\log[h_x(t)] = \log[h_0(t)] + \beta x. \tag{15.11}$$

The general model is

$$h(t) = h_0(t) \exp\left(\sum_{i=1}^{n} \beta_i x_i\right)$$

or, equivalently,

$$\log[h(t)] = \log[h_0(t)] + \sum_{i=1}^{n} \beta_i x_i.$$

There are parallels between logistic regression and proportional hazards regression. An important difference is that, unlike logistic regression where the intercept is merely a constant, in proportional hazards regression, $\log[h_0(t)]$ may have a complicated functional form. There are two possibilities for handling this problem. The first, referred to as the parametric approach, is to specify a functional form for $h_0(t)$. Usually this requires substantive knowledge of the disease under consideration as well as an examination of the data to determine whether the parametric assumption is reasonable. The second option uses conditional arguments to eliminate $h_0(t)$, in much the same way as a nuisance parameter is eliminated in conditional logistic regression.

TABLE 15.8 Cox Regression Output: Breast Cancer

Variable	$\hat{\theta}$	$\sqrt{\widehat{\text{var}}(\hat{\theta})}$	$\widehat{HR}$	$\underline{HR}$	$\overline{HR}$
Low receptor level	.91	.29	2.48	1.40	4.41
Stage II	1.02	.46	2.78	1.13	6.84
Stage III	2.39	.47	10.87	4.30	27.47

Perhaps the most frequently used parametric survival model in epidemiology defines the baseline hazard function in terms of the Weibull distribution, that is, $h_0(t) = \alpha\lambda(\lambda t)^{\alpha-1}$ (Section 10.1.2). The result is the Weibull regression model,

$$h(t) = \alpha\lambda(\lambda t)^{\alpha-1} \exp\left(\sum_{i=1}^{n} \beta_i x_i\right). \tag{15.12}$$

When $\alpha = 1$, (15.12) simplifies to the exponential regression model. The applicability of the Weibull and exponential models is limited by the strong assumption that must be made regarding the functional form of the baseline hazard function.

Undoubtedly the most widely used regression method for analyzing censored survival data in epidemiology is the Cox regression model. This area of survival analysis was pioneered by Cox (1972), who emphasized the importance of the proportional hazards assumption and described a method of parameter estimation and hypothesis testing based on conditional methods. The advantage of this approach is that $h_0(t)$ is treated as a nuisance function and thereby eliminated from the likelihood. Consequently it is not necessary to make any assumptions about the functional form of $h_0(t)$. There is a close connection between the Cox regression model and the analysis of censored survival data based on conditional odds ratio methods as presented in Section 9.2 (Prentice and Breslow, 1978). In particular, for a single dichotomous exposure variable, and assuming there is only one death at each death time, the hazard ratio estimate based on the Cox regression model is identical to $\widehat{OR}_c$, and the score test of association based on the Cox regression model is identical to the logrank test X^2_{mh} (Cox, 1972).

Example 15.2 (Stage–Receptor Level–Breast Cancer) We illustrate Cox regression with an analysis of the breast cancer survival data considered in Section 9.2. The following analysis was performed using EGRET (1999). Table 15.8 gives estimates for the model with main effects for receptor level (as a dichotomous variable) and stage. The hazard ratio estimates are quite close to the adjusted estimates in Tables 9.10 and 9.11.

APPENDIX A

Odds Ratio Inequality

In the notation of Section 2.4.5, let $\xi_{1j} = p_{1j}(1 - \pi_{1j})$ and $\xi_{2j} = p_{1j}(1 - \pi_{2j})$. Then (2.15) can be written as

$$OR = \theta \frac{\left(\sum_{j=1}^{J} \xi_{1j}\omega_{2j}\right) \bigg/ \left(\sum_{j=1}^{J} \xi_{1j}\right)}{\left(\sum_{j=1}^{J} \xi_{2j}\omega_{2j}\right) \bigg/ \left(\sum_{j=1}^{J} \xi_{2j}\right)}. \tag{A.1}$$

From $\pi_{1j} = \omega_{1j}/(1+\omega_{1j}) = \theta\omega_{2j}/(1+\theta\omega_{2j})$ and $\pi_{2j} = \omega_{2j}/(1+\omega_{2j})$, it follows that

$$\xi_{1j} = \frac{p_{1j}}{1 + \theta\omega_{2j}}$$

$$\xi_{2j} = \frac{p_{1j}}{1 + \omega_{2j}}$$

and

$$\frac{\xi_{2j}}{\xi_{1j}} = \frac{1 + \theta\omega_{2j}}{1 + \omega_{2j}}. \tag{A.2}$$

Assume that $\theta > 1$. Using (A.2) it is readily demonstrated that

$$1 < \frac{\xi_{2j}}{\xi_{1j}} < \theta \tag{A.3}$$

and hence

$$\sum_{j=1}^{J} \xi_{1j} < \sum_{j=1}^{J} \xi_{2j}. \tag{A.4}$$

Now consider the case where the p_{1j} are all equal. In what follows, ω denotes a continuous variable. For given $\theta > 0$ and $\tau > 0$, define

$$\xi_1(\omega) = \frac{1}{1 + \theta\omega}$$

and

$$\xi_2(\omega) = \frac{1}{1 + \omega}$$

for ω in $[0, \tau]$. The probability functions corresponding to $\xi_1(\omega)$ and $\xi_2(\omega)$ are

$$f_1(\omega) = \frac{\theta}{\log(1 + \theta\tau)(1 + \theta\omega)}$$

and

$$f_2(\omega) = \frac{1}{\log(1 + \tau)(1 + \omega)}$$

and the survival functions are

$$S_1(\omega) = 1 - \frac{\log(1 + \theta\omega)}{\log(1 + \theta\tau)}$$

and

$$S_2(\omega) = 1 - \frac{\log(1 + \omega)}{\log(1 + \tau)}$$

respectively. It can be shown that $S_1(\omega) \leq S_2(\omega)$, for all ω, and so $f_2(\omega)$ is distributed to the right compared to $f_1(\omega)$. This is illustrated in Figures A.1(a) and A.1(b), which show the probability functions and survival functions for $\theta = 5$ and $\tau = 10$. The interval $[0, 10]$ is typical of the range of values of ω that might be observed in practice. Returning to the discrete case, we assume, without loss of generality, that the ω_{2j} are in increasing order. It follows from the preceding observations that

$$\frac{\sum_{j=1}^{J} \xi_{1j}\omega_{2j}}{\sum_{j=1}^{J} \xi_{1j}} < \frac{\sum_{j=1}^{J} \xi_{2j}\omega_{2j}}{\sum_{j=1}^{J} \xi_{2j}}$$

and so, from (A.1), that $OR < \theta$. From (A.1), (A.3), and (A.4), we have

$$OR > \theta \frac{\left(\sum_{j=1}^{J} (\xi_{2j}/\theta)\omega_{2j}\right) \Big/ \left(\sum_{j=1}^{J} \xi_{1j}\right)}{\left(\sum_{j=1}^{J} \xi_{2j}\omega_{2j}\right) \Big/ \left(\sum_{j=1}^{J} \xi_{2j}\right)} = \frac{\sum_{j=1}^{J} \xi_{2j}}{\sum_{j=1}^{J} \xi_{1j}} > 1.$$

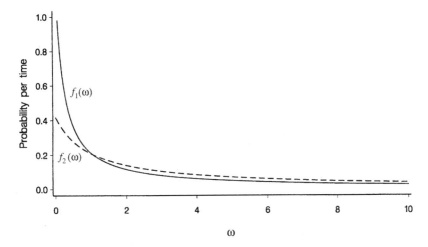

FIGURE A.1(a) Probability functions with $\theta = 5$ and $\tau = 10$

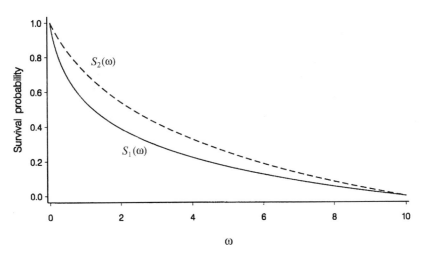

FIGURE A.1(b) Survival functions with $\theta = 5$ and $\tau = 10$

TABLE A.1(a)   Values of $\xi_{1j}/\xi_{1\bullet}$ and $\xi_{2j}/\xi_{2\bullet}$ for
Selected Values of p_{1j} and π_{2j}, with $\theta = 5$ $(OR = 3.27)$

j	p_{1j}	π_{2j}	$\xi_{1j}/\xi_{1\bullet}$	$\xi_{2j}/\xi_{2\bullet}$
1	.1	.1	.266	.176
2	.2	.2	.368	.314
3	.3	.4	.287	.353
4	.4	.8	.079	.157

TABLE A.1(b)   Values of $\xi_{1j}/\xi_{1\bullet}$ and $\xi_{2j}/\xi_{2\bullet}$ for
Selected Values of p_{1j} and π_{2j}, with $\theta = 5$ $(OR = 3.60)$

j	p_{1j}	π_{2j}	$\xi_{1j}/\xi_{1\bullet}$	$\xi_{2j}/\xi_{2\bullet}$
1	.4	.1	.583	.486
2	.3	.2	.302	.324
3	.2	.4	.105	.162
4	.1	.8	.011	.027

In summary, if $\theta > 1$ then $1 < OR < \theta$. An analogous argument shows that if $\theta < 1$ then $\theta < OR < 1$. Tables A.1(a) and A.1(b) give examples of the weights, $\xi_{1j}/\xi_{1\bullet}$ and $\xi_{2j}/\xi_{2\bullet}$, entering into (A.1), with $\theta = 5$. In these examples, the original definitions of ξ_{1j} and ξ_{2j} have been used—that is, values for the p_{1j} have been included. In both examples, the $\xi_{2j}/\xi_{2\bullet}$ are distributed to the right compared to the $\xi_{1j}/\xi_{1\bullet}$, and the inequality $1 < OR < \theta$ is satisfied.

Maximum Likelihood Theory

B.1 UNCONDITIONAL MAXIMUM LIKELIHOOD

A general reference for likelihood theory is Cox and Hinkley (1974).

B.1.1 Unconditional Likelihood and Newton–Raphson Algorithm

Let X be a random variable with probability function $f(x; \Theta)$, where

$$\Theta = (\theta_1, \ldots, \theta_p, \ldots, \theta_P)^{\mathrm{T}}$$

is a vector of parameters and T denotes matrix transposition. For simplicity of notation we sometimes omit the superscript T when no confusion can result. For a sample $X_1, X_2, \ldots, X_J$ from $f(x; \Theta)$, the unconditional likelihood is

$$L(\Theta) = \prod_{j=1}^{J} f(x_j, \Theta)$$

and the log-likelihood is $l = \log(L)$. Define the score vector, Hessian matrix, and Fisher information matrix to be

$$\mathbf{U} = (U_1, \ldots, U_P) = \left(\frac{\partial l}{\partial \theta_1}, \ldots, \frac{\partial l}{\partial \theta_p}, \ldots, \frac{\partial l}{\partial \theta_P} \right)^{\mathrm{T}}$$

$$\mathbf{H} = \left(\frac{\partial^2 l}{\partial \theta_p \partial \theta_q} \right)_{(P \times P)}$$

and

$$\mathbf{I} = \left(E \left[\frac{-\partial^2 l}{\partial \theta_p \partial \theta_q} \right] \right)_{(P \times P)}$$

respectively. By definition, the unconditional maximum likelihood estimate of Θ, denoted by $\hat{\Theta}$, is that value of Θ that maximizes $L(\Theta)$. From maximum likelihood theory, $\hat{\Theta}$ is asymptotically normal with mean $E(\hat{\Theta}) = \Theta$ and variance–covariance matrix $\text{var}(\hat{\Theta}) = \mathbf{I}(\Theta)^{-1}$, the latter having the estimate

$$\widehat{\text{var}}(\hat{\Theta}) = \mathbf{I}(\hat{\Theta})^{-1}. \tag{B.1}$$

The Newton–Raphson algorithm is an iterative method of calculating an estimate of Θ. The nth iteration is given by

$$\hat{\Theta}^{(n+1)} = \hat{\Theta}^{(n)} - (\hat{\mathbf{H}}^{(n)})^{-1}\hat{\mathbf{U}}^{(n)} \tag{B.2}$$

where $\hat{\mathbf{H}}^{(n)}$ and $\hat{\mathbf{U}}^{(n)}$ denote $\mathbf{H}(\hat{\Theta}^{(n)})$ and $\mathbf{U}(\hat{\Theta}^{(n)})$. The process begins with an initial estimate $\hat{\Theta}^{(1)}$ and iteration continues until the desired accuracy is achieved. In some situations $\mathbf{I} = -E(\mathbf{H})$ is a less complicated expression than $\mathbf{H}$, and so using

$$\hat{\Theta}^{(n+1)} = \hat{\Theta}^{(n)} + (\hat{\mathbf{I}}^{(n)})^{-1}\hat{\mathbf{U}}^{(n)}$$

in place of (B.2) may be computationally more convenient. When there are constraints on the parameter values, it must be verified that the estimates satisfy these conditions; otherwise, alternate methods of maximization must be employed.

Let $\Theta = (\Phi, \Psi)$ be a partition of Θ and denote the corresponding partitions of $\mathbf{U}$, $\mathbf{I}$, and $\mathbf{I}^{-1}$ by $\mathbf{U} = (\mathbf{U}_\Phi, \mathbf{U}_\Psi)$,

$$\mathbf{I} = \begin{pmatrix} \mathbf{I}_{\Phi\Phi} & \mathbf{I}_{\Phi\Psi} \\ \mathbf{I}_{\Psi\Phi} & \mathbf{I}_{\Psi\Psi} \end{pmatrix}$$

and

$$\mathbf{I}^{-1} = \begin{pmatrix} \mathbf{I}^{\Phi\Phi} & \mathbf{I}^{\Phi\Psi} \\ \mathbf{I}^{\Psi\Phi} & \mathbf{I}^{\Psi\Psi} \end{pmatrix}.$$

Since $\mathbf{I}$ and $\mathbf{I}^{-1}$ are symmetric matrices, $\mathbf{I}_{\Psi\Phi} = (\mathbf{I}_{\Phi\Psi})^{\mathrm{T}}$ and $\mathbf{I}^{\Psi\Phi} = (\mathbf{I}^{\Phi\Psi})^{\mathrm{T}}$. A generalization of (B.1) is

$$\widehat{\text{var}}(\hat{\Phi}) = \mathbf{I}^{\Phi\Phi}(\hat{\Theta}) = \left[\mathbf{I}_{\Phi\Phi}(\hat{\Theta}) - \mathbf{I}_{\Phi\Psi}(\hat{\Theta})\mathbf{I}_{\Psi\Psi}(\hat{\Theta})^{-1}\mathbf{I}_{\Psi\Phi}(\hat{\Theta})\right]^{-1} \tag{B.3}$$

where the last equality follows from (B.16).

B.1.2 Wald, Score, and Likelihood Ratio Tests

The hypothesis $H_0 : \Theta = \Theta_0$ can be tested using the following statistics, each of which is asymptotically chi-square:

Wald: $\qquad X_{\mathrm{w}}^2 = (\hat{\Theta} - \Theta_0)^{\mathrm{T}}\mathbf{I}(\hat{\Theta})(\hat{\Theta} - \Theta_0)$

Score: $\qquad X_{\mathrm{s}}^2 = \mathbf{U}(\Theta_0)^{\mathrm{T}}\mathbf{I}(\Theta_0)^{-1}\mathbf{U}(\Theta_0)$

Likelihood ratio: $X_{lr}^2 = 2\log\left[\dfrac{L(\hat{\Theta})}{L(\Theta_0)}\right] = 2\left[l(\hat{\Theta}) - l(\Theta_0)\right].$

In each of the above tests, the degrees of freedom equals dim(Θ), the dimension of Θ. An asymptotically equivalent alternative to the Wald test is obtained by replacing $\mathbf{I}(\hat{\Theta})$ with $\mathbf{I}(\Theta_0)$.

Suppose we wish to test the hypothesis $H_0 : \Phi = \Phi_0$, where no conditions are placed on Ψ. Denote the maximum likelihood estimate of Θ by $\hat{\Theta} = (\hat{\Phi}, \hat{\Psi})$, let $\tilde{\Psi}$ be the value of Ψ which maximizes $L(\Phi_0, \Psi)$, and let $\tilde{\Theta} = \left(\Phi_0, \tilde{\Psi}\right)$. The hypothesis $H_0 : \Phi = \Phi_0$ can be tested using the following statistics, each of which is asymptotically chi-square:

Wald: $X_w^2 = (\hat{\Phi} - \Phi_0)^{\mathrm{T}}\left[\mathbf{I}^{\Phi\Phi}(\hat{\Theta})\right]^{-1}(\hat{\Phi} - \Phi_0)$

Score: $X_s^2 = \mathbf{U}_\Phi(\tilde{\Theta})^{\mathrm{T}}\mathbf{I}^{\Phi\Phi}(\tilde{\Theta})\mathbf{U}_\Phi(\tilde{\Theta})$

Likelihood ratio: $X_{lr}^2 = 2\log\left[\dfrac{L(\hat{\Theta})}{L(\tilde{\Theta})}\right] = 2\left[l(\hat{\Theta}) - l(\tilde{\Theta})\right].$ (B.4)

In each of the above tests the degrees of freedom equals dim(Φ), the dimension of Φ. An asymptotically equivalent alternative to the Wald test is obtained by replacing $\mathbf{I}^{\Phi\Phi}(\hat{\Theta})$ with $\mathbf{I}^{\Phi\Phi}(\tilde{\Theta})$. For further results on the score test in the context of epidemiology, see Day and Byar (1979), Gart (1985), Gart and Nam (1988, 1990), and Lachin (2000, Chapter 6).

B.2 BINOMIAL DISTRIBUTION

In this section we develop asymptotic unconditional methods for $J(2 \times 2)$ tables based on the binomial distribution.

B.2.1 Unconditional Likelihood

Dropping terms that do not involve the parameters, the unconditional likelihood is

$$L = \prod_{j=1}^{J} \pi_{1j}^{a_{1j}}(1 - \pi_{1j})^{b_{1j}}\pi_{2j}^{a_{2j}}(1 - \pi_{2j})^{b_{2j}}$$

and the log-likelihood is

$$l = \sum_{j=1}^{J}[a_{1j}\log(\pi_{1j}) + b_{1j}\log(1 - \pi_{1j}) + a_{2j}\log(\pi_{2j}) + b_{2j}\log(1 - \pi_{2j})].$$

Let $\pi_{1j} = \pi_{1j}(\theta_j, \pi_{2j})$; that is, assume that π_{1j} is a function of θ_j and π_{2j}, where θ_j is the parameter of interest in the jth stratum.

B.2.2 Variance Estimate

Assume that the θ_j are homogeneous across strata with common value θ. Let the vector of parameters be $\Theta = (\Phi, \Psi)$, where $\Phi = (\theta)$ and $\Psi = (\pi_{21}, \ldots, \pi_{2j}, \ldots, \pi_{2J})$. Then π_{1j} can be written as $\pi_{1j} = \pi_{1j}(\theta, \pi_{2j})$. The elements of the score vector are

$$
\frac{\partial l}{\partial \theta} = \sum_{j=1}^{J} \left[\frac{a_{1j}}{\pi_{1j}} - \frac{b_{1j}}{1 - \pi_{1j}} \right] \left(\frac{\partial \pi_{1j}}{\partial \theta} \right)
$$

$$
= \sum_{j=1}^{J} \frac{a_{1j} - \pi_{1j} r_{1j}}{\pi_{1j}(1 - \pi_{1j})} \left(\frac{\partial \pi_{1j}}{\partial \theta} \right) \tag{B.5}
$$

and

$$
\frac{\partial l}{\partial \pi_{2j}} = \left[\frac{a_{1j}}{\pi_{1j}} - \frac{b_{1j}}{1 - \pi_{1j}} \right] \left(\frac{\partial \pi_{1j}}{\partial \pi_{2j}} \right) + \left[\frac{a_{2j}}{\pi_{2j}} - \frac{b_{2j}}{1 - \pi_{2j}} \right]
$$

$$
= \frac{a_{1j} - \pi_{1j} r_{1j}}{\pi_{1j}(1 - \pi_{1j})} \left(\frac{\partial \pi_{1j}}{\partial \pi_{2j}} \right) + \frac{a_{2j} - \pi_{2j} r_{2j}}{\pi_{2j}(1 - \pi_{2j})}.
$$

The elements of the Hessian matrix are

$$
\frac{\partial^2 l}{\partial \theta^2} = \sum_{j=1}^{J} \left\{ \frac{a_{1j} - \pi_{1j} r_{1j}}{\pi_{1j}(1 - \pi_{1j})} \left(\frac{\partial^2 \pi_{1j}}{\partial \theta^2} \right) - \left[\frac{a_{1j}}{\pi_{1j}^2} + \frac{b_{1j}}{(1 - \pi_{1j})^2} \right] \left(\frac{\partial \pi_{1j}}{\partial \theta} \right)^2 \right\}
$$

$$
\frac{\partial^2 l}{\partial \theta \partial \pi_{2j}} = \frac{a_{1j} - \pi_{1j} r_{1j}}{\pi_{1j}(1 - \pi_{1j})} \left(\frac{\partial^2 \pi_{1j}}{\partial \theta \partial \pi_{2j}} \right) - \left[\frac{a_{1j}}{\pi_{1j}^2} + \frac{b_{1j}}{(1 - \pi_{1j})^2} \right] \left(\frac{\partial \pi_{1j}}{\partial \pi_{2j}} \right) \left(\frac{\partial \pi_{1j}}{\partial \theta} \right)
$$

$$
\frac{\partial^2 l}{\partial \pi_{2j}^2} = \frac{a_{1j} - \pi_{1j} r_{1j}}{\pi_{1j}(1 - \pi_{1j})} \left(\frac{\partial^2 \pi_{1j}}{\partial \pi_{2j}^2} \right) - \left[\frac{a_{1j}}{\pi_{1j}^2} + \frac{b_{1j}}{(1 - \pi_{1j})^2} \right] \left(\frac{\partial \pi_{1j}}{\partial \pi_{2j}} \right)^2
$$

$$
- \left[\frac{a_{2j}}{\pi_{2j}^2} + \frac{b_{2j}}{(1 - \pi_{2j})^2} \right]
$$

and

$$
\frac{\partial^2 l}{\partial \pi_{2j} \partial \pi_{2k}} = 0 \qquad (j \neq k).
$$

Define

$$x_j = \frac{r_{1j}}{\pi_{1j}(1 - \pi_{1j})} \left(\frac{\partial \pi_{1j}}{\partial \theta} \right)^2$$

$$y_j = \frac{r_{1j}}{\pi_{1j}(1 - \pi_{1j})} \left(\frac{\partial \pi_{1j}}{\partial \pi_{2j}} \right) \left(\frac{\partial \pi_{1j}}{\partial \theta} \right)$$

$$z_j = \frac{r_{1j}}{\pi_{1j}(1 - \pi_{1j})} \left(\frac{\partial \pi_{1j}}{\partial \pi_{2j}} \right)^2 + \frac{r_{2j}}{\pi_{2j}(1 - \pi_{2j})}$$

and

$$v_j = x_j - \frac{y_j^2}{z_j} = \frac{r_{1j}r_{2j} \left(\frac{\partial \pi_{1j}}{\partial \theta} \right)^2}{\pi_{2j}(1 - \pi_{2j})r_{1j} \left(\frac{\partial \pi_{1j}}{\partial \pi_{2j}} \right)^2 + \pi_{1j}(1 - \pi_{1j})r_{2j}}. \tag{B.6}$$

From

$$E \left[\frac{a_{1j} - \pi_{1j}r_{1j}}{\pi_{1j}(1 - \pi_{1j})} \right] = 0 \tag{B.7}$$

$$E \left[\frac{a_{1j}}{\pi_{1j}^2} + \frac{b_{1j}}{(1 - \pi_{1j})^2} \right] = \frac{r_{1j}}{\pi_{1j}(1 - \pi_{1j})} \tag{B.8}$$

and

$$E \left[\frac{a_{2j}}{\pi_{2j}^2} + \frac{b_{2j}}{(1 - \pi_{2j})^2} \right] = \frac{r_{2j}}{\pi_{2j}(1 - \pi_{2j})} \tag{B.9}$$

it follows that the elements of the Fisher information matrix are

$$E \left(\frac{-\partial^2 l}{\partial \theta^2} \right) = \sum_{j=1}^{J} \frac{r_{1j}}{\pi_{1j}(1 - \pi_{1j})} \left(\frac{\partial \pi_{1j}}{\partial \theta} \right)^2 = \sum_{j=1}^{J} x_j$$

$$E \left(\frac{-\partial^2 l}{\partial \theta \partial \pi_{2j}} \right) = \frac{r_{1j}}{\pi_{1j}(1 - \pi_{1j})} \left(\frac{\partial \pi_{1j}}{\partial \pi_{2j}} \right) \left(\frac{\partial \pi_{1j}}{\partial \theta} \right) = y_j$$

and

$$E \left(\frac{-\partial^2 l}{\partial \pi_{2j}^2} \right) = \frac{r_{1j}}{\pi_{1j}(1 - \pi_{1j})} \left(\frac{\partial \pi_{1j}}{\partial \pi_{2j}} \right)^2 + \frac{r_{2j}}{\pi_{2j}(1 - \pi_{2j})} = z_j.$$

With $\Phi = (\theta)$ and $\Psi = (\pi_{21}, \ldots, \pi_{2j}, \ldots, \pi_{2J})$, we have

$$\mathbf{I}_{\Phi\Phi}(\hat{\Theta}) = \left(\sum_{j=1}^{J} \hat{x}_j\right)$$

$$\mathbf{I}_{\Phi\Psi}(\hat{\Theta}) = (\hat{y}_1, \ldots, \hat{y}_J)$$

and

$$\mathbf{I}_{\Psi\Psi}(\hat{\Theta}) = \mathrm{diag}(\hat{z}_1, \ldots, \hat{z}_J).$$

It follows from (B.3) that

$$\widehat{\mathrm{var}}(\hat{\theta}) = \mathbf{I}^{\Phi\Phi}(\hat{\Theta}) = \left(\sum_{j=1}^{J} \hat{v}_j\right)^{-1}. \tag{B.10}$$

As an example, let $\theta = \log(OR)$. In the notation of Chapter 5, with v_j and the partial derivatives given in Tables B.1(a) and B.1(b), we have

TABLE B.1(a) Values for the Binomial Distribution Under Homogeneity

θ	π_{1j}	$\dfrac{\partial \pi_{1j}}{\partial \theta}$	$\dfrac{\partial \pi_{1j}}{\partial \pi_{2j}}$
$\log(OR)$	$\dfrac{e^{\theta}\pi_{2j}}{e^{\theta}\pi_{2j} + (1 - \pi_{2j})}$	$\pi_{1j}(1 - \pi_{1j})$	$\dfrac{\pi_{1j}(1 - \pi_{1j})}{\pi_{2j}(1 - \pi_{2j})}$
$\log(RR)$	$e^{\theta}\pi_{2j}$	π_{1j}	$\dfrac{\pi_{1j}}{\pi_{2j}}$
RD	$\pi_{2j} + \theta$	1	1

TABLE B.1(b) Values for the Binomial Distribution Under Homogeneity

θ	v_j
$\log(OR)$	$\left[\dfrac{1}{\pi_{1j}r_{1j}} + \dfrac{1}{(1 - \pi_{1j})r_{1j}} + \dfrac{1}{\pi_{2j}r_{2j}} + \dfrac{1}{(1 - \pi_{2j})r_{2j}}\right]^{-1}$
$\log(RR)$	$\left[\dfrac{1 - \pi_{1j}}{\pi_{1j}r_{1j}} + \dfrac{1 - \pi_{2j}}{\pi_{2j}r_{2j}}\right]^{-1}$
RD	$\left[\dfrac{\pi_{1j}(1 - \pi_{1j})}{r_{1j}} + \dfrac{\pi_{2j}(1 - \pi_{2j})}{r_{2j}}\right]^{-1}$

$$\widehat{\text{var}}(\log \widehat{OR}_u) = \left(\sum_{j=1}^{J} \left[\frac{1}{\hat{\pi}_{1j} r_{1j}} + \frac{1}{(1-\hat{\pi}_{1j}) r_{1j}} + \frac{1}{\hat{\pi}_{2j} r_{2j}} + \frac{1}{(1-\hat{\pi}_{2j}) r_{2j}} \right]^{-1} \right)^{-1}.$$

B.2.3 Tests of Association

Let the vector of parameters be $\Theta = (\Phi, \Psi)$, where $\Phi = (\theta)$ and $\Psi = (\pi_{21}, \ldots, \pi_{2j}, \ldots, \pi_{2J})$. Under the hypothesis of no association $H_0 : \theta = \theta_0$, we have $\tilde{\Theta} = (\Phi_0, \tilde{\Psi})$, where $\Phi_0 = (\theta_0)$ and $\tilde{\Psi} = (\tilde{\pi}_{21}, \ldots, \tilde{\pi}_{2j}, \ldots, \tilde{\pi}_{2J})$. It follows from (B.5) and (B.10) that

$$\mathbf{U}_\Phi(\tilde{\Theta}) = \sum_{j=1}^{J} \frac{(a_{1j} - \tilde{a}_{1j})}{\tilde{\pi}_{1j}(1-\tilde{\pi}_{1j})} \left(\frac{\partial \tilde{\pi}_{1j}}{\partial \theta} \right)$$

and

$$\mathbf{I}^{\Phi\Phi}(\tilde{\Theta}) = \left(\sum_{j=1}^{J} \tilde{v}_j \right)^{-1}.$$

From (B.4), the Wald, score, and likelihood ratio tests of association are

$$X_w^2 = (\hat{\theta} - \theta_0)^2 \left(\sum_{j=1}^{J} \tilde{v}_j \right)$$

$$X_s^2 = \left[\sum_{j=1}^{J} \frac{(a_{1j} - \tilde{a}_{1j})}{\tilde{\pi}_{1j}(1-\tilde{\pi}_{1j})} \left(\frac{\partial \tilde{\pi}_{1j}}{\partial \theta} \right) \right]^2 \bigg/ \left(\sum_{j=1}^{J} \tilde{v}_j \right)$$

and

$$X_{lr}^2 = 2 \sum_{j=1}^{J} \left[a_{1j} \log \left(\frac{a_{1j}}{\tilde{a}_{1j}} \right) + a_{2j} \log \left(\frac{a_{2j}}{\tilde{a}_{2j}} \right) + b_{1j} \log \left(\frac{b_{1j}}{\tilde{b}_{1j}} \right) + b_{2j} \log \left(\frac{b_{2j}}{\tilde{b}_{2j}} \right) \right].$$

B.2.4 Score Test for Linear Trend

Let $\theta_j = \alpha + \beta s_j$, where s_j is the exposure level for the jth stratum and α and β are constants ($j = 1, 2, \ldots, J$). Then π_{1j} can be written as $\pi_{1j} = \pi_{1j}(\alpha, \beta, \pi_{2j})$. Let the vector of parameters be $\Theta = (\Phi, \Psi)$, where $\Phi = (\beta)$ and $\Psi = (\alpha, \pi_{21}, \ldots, \pi_{2j}, \ldots, \pi_{2J})$. The elements of the score vector are

$$\frac{\partial l}{\partial \beta} = \sum_{j=1}^{J} \frac{s_j(a_{1j} - \pi_{1j} r_{1j})}{\pi_{1j}(1-\pi_{1j})} \left(\frac{\partial \pi_{1j}}{\partial \theta_j} \right) \tag{B.11}$$

$$\frac{\partial l}{\partial \alpha} = \sum_{j=1}^{J} \frac{a_{1j} - \pi_{1j} r_{1j}}{\pi_{1j}(1-\pi_{1j})} \left(\frac{\partial \pi_{1j}}{\partial \theta_j} \right)$$

and

$$\frac{\partial l}{\partial \pi_{2j}} = \frac{a_{1j} - \pi_{1j} r_{1j}}{\pi_{1j}(1 - \pi_{1j})} \left(\frac{\partial \pi_{1j}}{\partial \pi_{2j}} \right) + \frac{a_{2j} - \pi_{2j} r_{2j}}{\pi_{2j}(1 - \pi_{2j})}.$$

The elements of the Hessian matrix are

$$\frac{\partial^2 l}{\partial \beta^2} = \sum_{j=1}^{J} s_j^2 \left\{ \frac{a_{1j} - \pi_{1j} r_{1j}}{\pi_{1j}(1 - \pi_{1j})} \left(\frac{\partial^2 \pi_{1j}}{\partial \theta_j^2} \right) - \left[\frac{a_{1j}}{\pi_{1j}^2} + \frac{b_{1j}}{(1 - \pi_{1j})^2} \right] \left(\frac{\partial \pi_{1j}}{\partial \theta_j} \right)^2 \right\}$$

$$\frac{\partial^2 l}{\partial \alpha^2} = \sum_{j=1}^{J} \left\{ \frac{a_{1j} - \pi_{1j} r_{1j}}{\pi_{1j}(1 - \pi_{1j})} \left(\frac{\partial^2 \pi_{1j}}{\partial \theta_j^2} \right) - \left[\frac{a_{1j}}{\pi_{1j}^2} + \frac{b_{1j}}{(1 - \pi_{1j})^2} \right] \left(\frac{\partial \pi_{1j}}{\partial \theta_j} \right)^2 \right\}$$

$$\frac{\partial^2 l}{\partial \beta \partial \alpha} = \sum_{j=1}^{J} s_j \left\{ \frac{a_{1j} - \pi_{1j} r_{1j}}{\pi_{1j}(1 - \pi_{1j})} \left(\frac{\partial^2 \pi_{1j}}{\partial \theta_j^2} \right) - \left[\frac{a_{1j}}{\pi_{1j}^2} + \frac{b_{1j}}{(1 - \pi_{1j})^2} \right] \left(\frac{\partial \pi_{1j}}{\partial \theta_j} \right)^2 \right\}$$

$$\frac{\partial^2 l}{\partial \beta \partial \pi_{2j}} = s_j \left\{ \frac{a_{1j} - \pi_{1j} r_{1j}}{\pi_{1j}(1 - \pi_{1j})} \left(\frac{\partial^2 \pi_{1j}}{\partial \theta_j \partial \pi_{2j}} \right) \right.$$
$$\left. - \left[\frac{a_{1j}}{\pi_{1j}^2} + \frac{b_{1j}}{(1 - \pi_{1j})^2} \right] \left(\frac{\partial \pi_{1j}}{\partial \pi_{2j}} \right) \left(\frac{\partial \pi_{1j}}{\partial \theta_j} \right) \right\}$$

$$\frac{\partial^2 l}{\partial \alpha \partial \pi_{2j}} = \frac{a_{1j} - \pi_{1j} r_{1j}}{\pi_{1j}(1 - \pi_{1j})} \left(\frac{\partial^2 \pi_{1j}}{\partial \theta_j \partial \pi_{2j}} \right) - \left[\frac{a_{1j}}{\pi_{1j}^2} + \frac{b_{1j}}{(1 - \pi_{1j})^2} \right] \left(\frac{\partial \pi_{1j}}{\partial \pi_{2j}} \right) \left(\frac{\partial \pi_{1j}}{\partial \theta_j} \right)$$

$$\frac{\partial^2 l}{\partial \pi_{2j}^2} = \frac{a_{1j} - \pi_{1j} r_{1j}}{\pi_{1j}(1 - \pi_{1j})} \left(\frac{\partial^2 \pi_{1j}}{\partial \pi_{2j}^2} \right) - \left[\frac{a_{1j}}{\pi_{1j}^2} + \frac{b_{1j}}{(1 - \pi_{1j})^2} \right] \left(\frac{\partial \pi_{1j}}{\partial \pi_{2j}} \right)^2$$
$$- \left[\frac{a_{2j}}{\pi_{2j}^2} + \frac{b_{2j}}{(1 - \pi_{2j})^2} \right]$$

and

$$\frac{\partial^2 l}{\partial \pi_{2j} \partial \pi_{2k}} = 0 \qquad (j \neq k).$$

Define

$$x_j = \frac{r_{1j}}{\pi_{1j}(1 - \pi_{1j})} \left(\frac{\partial \pi_{1j}}{\partial \theta_j} \right)^2$$

$$y_j = \frac{r_{1j}}{\pi_{1j}(1 - \pi_{1j})} \left(\frac{\partial \pi_{1j}}{\partial \pi_{2j}} \right) \left(\frac{\partial \pi_{1j}}{\partial \theta_j} \right)$$

$$z_j = \frac{r_{1j}}{\pi_{1j}(1 - \pi_{1j})} \left(\frac{\partial \pi_{1j}}{\partial \pi_{2j}}\right)^2 + \frac{r_{2j}}{\pi_{2j}(1 - \pi_{2j})}$$

and

$$v_j = x_j - \frac{y_j^2}{z_j} = \frac{r_{1j} r_{2j} \left(\frac{\partial \pi_{1j}}{\partial \theta_j}\right)^2}{\pi_{2j}(1 - \pi_{2j}) r_{1j} \left(\frac{\partial \pi_{1j}}{\partial \pi_{2j}}\right)^2 + \pi_{1j}(1 - \pi_{1j}) r_{2j}}. \tag{B.12}$$

(B.12) is the same as (B.6) except that θ_j appears instead of θ. From (B.7)–(B.9), the elements of the Fisher information matrix $\mathbf{I}$ are

$$E\left(\frac{-\partial^2 l}{\partial \beta^2}\right) = \sum_{j=1}^{J} s_j^2 \left\{ \frac{r_{1j}}{\pi_{1j}(1 - \pi_{1j})} \left(\frac{\partial \pi_{1j}}{\partial \theta_j}\right)^2 \right\} = \sum_{j=1}^{J} s_j^2 x_j$$

$$E\left(\frac{-\partial^2 l}{\partial \alpha^2}\right) = \sum_{j=1}^{J} \frac{r_{1j}}{\pi_{1j}(1 - \pi_{1j})} \left(\frac{\partial \pi_{1j}}{\partial \theta_j}\right)^2 = \sum_{j=1}^{J} x_j$$

$$E\left(\frac{-\partial^2 l}{\partial \beta \partial \alpha}\right) = \sum_{j=1}^{J} s_j \left\{ \frac{r_{1j}}{\pi_{1j}(1 - \pi_{1j})} \left(\frac{\partial \pi_{1j}}{\partial \theta_j}\right)^2 \right\} = \sum_{j=1}^{J} s_j x_j$$

$$E\left(\frac{-\partial^2 l}{\partial \beta \partial \pi_{2j}}\right) = s_j \left\{ \frac{r_{1j}}{\pi_{1j}(1 - \pi_{1j})} \left(\frac{\partial \pi_{1j}}{\partial \pi_{2j}}\right) \left(\frac{\partial \pi_{1j}}{\partial \theta_j}\right) \right\} = s_j y_j$$

$$E\left(\frac{-\partial^2 l}{\partial \alpha \partial \pi_{2j}}\right) = \frac{r_{1j}}{\pi_{1j}(1 - \pi_{1j})} \left(\frac{\partial \pi_{1j}}{\partial \pi_{2j}}\right) \left(\frac{\partial \pi_{1j}}{\partial \theta_j}\right) = y_j$$

$$E\left(\frac{-\partial^2 l}{\partial \pi_{2j}^2}\right) = \frac{r_{1j}}{\pi_{1j}(1 - \pi_{1j})} \left(\frac{\partial \pi_{1j}}{\partial \pi_{2j}}\right)^2 + \frac{r_{2j}}{\pi_{2j}(1 - \pi_{2j})} = z_j$$

and

$$E\left(\frac{-\partial^2 l}{\partial \pi_{2j} \partial \pi_{2k}}\right) = 0 \qquad (j \neq k).$$

Under the hypothesis of no linear trend $H_0 : \beta = 0$, we have $\tilde{\Theta} = (\Phi_0, \tilde{\Psi})$ where $\Phi_0 = (0)$ and $\tilde{\Psi} = (\tilde{\alpha}, \tilde{\pi}_{21}, \ldots, \tilde{\pi}_{2j}, \ldots, \tilde{\pi}_{2J})$. However, $\beta = 0$ is equivalent to $\theta_j = \alpha$ for all j, which is the same as saying that the θ_j are homogeneous. Under homogeneity, $\partial \pi_{1j}/\partial \theta_j$ is the same $\partial \pi_{1j}/\partial \theta$ and so (B.6) and (B.12) are identical. Partition $\mathbf{I}$ so that the subscripts 1, 2, and 3 correspond to the vectors (β), (α) and $(\pi_{21}, \ldots, \pi_{2j}, \ldots, \pi_{2J})$, respectively. It is readily verified that

$$\mathbf{I}_{11} - \mathbf{I}_{13} \mathbf{I}_{33}^{-1} \mathbf{I}_{31} = \sum_{j=1}^{J} s_j^2 v_j$$

$$\mathbf{I}_{12} - \mathbf{I}_{13}\mathbf{I}_{33}^{-1}\mathbf{I}_{32} = \sum_{j=1}^{J} s_j v_j$$

$$\mathbf{I}_{22} - \mathbf{I}_{23}\mathbf{I}_{33}^{-1}\mathbf{I}_{32} = \sum_{j=1}^{J} v_j$$

and

$$\mathbf{I}_{21} - \mathbf{I}_{23}\mathbf{I}_{33}^{-1}\mathbf{I}_{31} = \sum_{j=1}^{J} s_j v_j.$$

It follows from (B.17) that

$$\mathbf{I}^{11} = \left[\sum_{j=1}^{J} s_j^2 v_j - \left(\sum_{j=1}^{J} s_j v_j \right)^2 \bigg/ \sum_{j=1}^{J} v_j \right]^{-1}. \tag{B.13}$$

From (B.11) and (B.13), we have

$$\mathbf{U}_\Phi(\tilde{\Theta}) = \sum_{j=1}^{J} \frac{s_j(a_{1j} - \tilde{a}_{1j})}{\tilde{\pi}_{1j}(1 - \tilde{\pi}_{1j})} \left(\frac{\partial \tilde{\pi}_{1j}}{\partial \theta} \right)$$

and

$$\mathbf{I}^{\Phi\Phi}(\tilde{\Theta}) = \left[\sum_{j=1}^{J} s_j^2 \tilde{v}_j - \left(\sum_{j=1}^{J} s_j \tilde{v}_j \right)^2 \bigg/ \sum_{j=1}^{J} \tilde{v}_j \right]^{-1}.$$

From (B.4) the score test of $H_0 : \beta = 0$ is

$$X_t^2 = \left[\sum_{j=1}^{J} \frac{s_j(a_{1j} - \tilde{a}_{1j})}{\tilde{\pi}_{1j}(1 - \tilde{\pi}_{1j})} \left(\frac{\partial \tilde{\pi}_{1j}}{\partial \theta} \right) \right]^2 \bigg/ \left[\sum_{j=1}^{J} s_j^2 \tilde{v}_j - \left(\sum_{j=1}^{J} s_j \tilde{v}_j \right)^2 \bigg/ \sum_{j=1}^{J} \tilde{v}_j \right].$$
$$\tag{B.14}$$

Let $\psi_j = g(\theta_j) = g(\alpha + \beta s_j)$, where $g(\cdot)$ is monotonic and has first and second derivatives. Reparameterizing the likelihood in terms of ψ_j and working through the above argument, it can be shown that the score test is once again (B.14). The key to this result is the observation that, under the hypothesis $H_0 : \beta = 0$, we have $\tilde{\theta}_j = \tilde{\alpha}$. So $\tilde{\theta}_j$ is independent of j and hence so is $dg(\tilde{\theta}_j)/d\theta_j$.

B.3 POISSON DISTRIBUTION

The derivation of formulas for the Poisson distribution is similar to the binomial case.

B.3.1 Unconditional Likelihood

Dropping terms that do not involve the parameters, the unconditional likelihood is

$$L = \prod_{j=1}^{J} \exp(-\lambda_{1j} n_{1j}) \lambda_{1j}^{d_{1j}} \exp(-\lambda_{2j} n_{2j}) \lambda_{2j}^{d_{2j}}$$

and the log-likelihood is

$$l = \sum_{j=1}^{J} [-\lambda_{1j} n_{1j} + d_{1j} \log(\lambda_{1j}) - \lambda_{2j} n_{2j} + d_{2j} \log(\lambda_{2j})].$$

Let $\lambda_{1j} = \lambda_{1j}(\theta_j, \lambda_{2j})$.

B.3.2 Variance Estimate

Assume that the θ_j are homogeneous across strata with common value θ. Let the vector of parameters be $\Theta = (\Phi, \Psi)$, where $\Phi = (\theta)$ and $\Psi = (\lambda_{21}, \ldots, \lambda_{2j}, \ldots, \lambda_{2J})$. Then λ_{1j} can be written as $\lambda_{1j} = \lambda_{1j}(\theta, \lambda_{2j})$. The elements of the score vector are

$$\frac{\partial l}{\partial \theta} = \sum_{j=1}^{J} \frac{d_{1j} - \lambda_{1j} n_{1j}}{\lambda_{1j}} \left(\frac{\partial \lambda_{1j}}{\partial \theta} \right) \qquad (B.15)$$

and

$$\frac{\partial l}{\partial \lambda_{2j}} = \frac{d_{1j} - \lambda_{1j} n_{1j}}{\lambda_{1j}} \left(\frac{\partial \lambda_{1j}}{\partial \lambda_{2j}} \right) + \frac{d_{2j} - \lambda_{2j} n_{2j}}{\lambda_{2j}}.$$

The elements of the Hessian matrix are

$$\frac{\partial^2 l}{\partial \theta^2} = \sum_{j=1}^{J} \left[\frac{d_{1j} - \lambda_{1j} n_{1j}}{\lambda_{1j}} \left(\frac{\partial^2 \lambda_{1j}}{\partial \theta^2} \right) - \frac{d_{1j}}{\lambda_{1j}^2} \left(\frac{\partial \lambda_{1j}}{\partial \theta} \right)^2 \right]$$

$$\frac{\partial^2 l}{\partial \theta \partial \lambda_{2j}} = \frac{d_{1j} - \lambda_{1j} n_{1j}}{\lambda_{1j}} \left(\frac{\partial^2 \lambda_{1j}}{\partial \theta \partial \lambda_{2j}} \right) - \frac{d_{1j}}{\lambda_{1j}^2} \left(\frac{\partial \lambda_{1j}}{\partial \lambda_{2j}} \right) \left(\frac{\partial \lambda_{1j}}{\partial \theta} \right)$$

$$\frac{\partial^2 l}{\partial \lambda_{2j}^2} = \frac{d_{1j} - \lambda_{1j} n_{1j}}{\lambda_{1j}} \left(\frac{\partial^2 \lambda_{1j}}{\partial \lambda_{2j}^2} \right) - \frac{d_{1j}}{\lambda_{1j}^2} \left(\frac{\partial \lambda_{1j}}{\partial \lambda_{2j}} \right)^2 - \frac{d_{2j}}{\lambda_{2j}^2}$$

and

$$\frac{\partial^2 l}{\partial \lambda_{2j} \partial \lambda_{2k}} = 0 \qquad (j \neq k).$$

From

$$E\left(\frac{d_{1j} - \lambda_{1j}n_{1j}}{\lambda_{1j}}\right) = 0$$

$$E\left(\frac{d_{1j}}{\lambda_{1j}^2}\right) = \frac{n_{1j}}{\lambda_{1j}}$$

and

$$E\left(\frac{d_{2j}}{\lambda_{2j}^2}\right) = \frac{n_{2j}}{\lambda_{2j}}$$

it follows that the elements of the Fisher information matrix are

$$E\left(\frac{-\partial^2 l}{\partial\theta^2}\right) = \sum_{j=1}^{J} \frac{n_{1j}}{\lambda_{1j}}\left(\frac{\partial\lambda_{1j}}{\partial\theta}\right)^2$$

$$E\left(\frac{-\partial^2 l}{\partial\theta\partial\lambda_{2j}}\right) = \frac{n_{1j}}{\lambda_{1j}}\left(\frac{\partial\lambda_{1j}}{\partial\lambda_{2j}}\right)\left(\frac{\partial\lambda_{1j}}{\partial\theta}\right)$$

and

$$E\left(\frac{-\partial^2 l}{\partial\lambda_{2j}^2}\right) = \frac{n_{1j}}{\lambda_{1j}}\left(\frac{\partial\lambda_{1j}}{\partial\lambda_{2j}}\right)^2 + \frac{n_{2j}}{\lambda_{2j}}.$$

Let

$$v_j = \frac{n_{1j}n_{2j}\left(\frac{\partial\lambda_{1j}}{\partial\theta}\right)^2}{\lambda_{2j}n_{1j}\left(\frac{\partial\lambda_{1j}}{\partial\lambda_{2j}}\right)^2 + \lambda_{1j}n_{2j}}.$$

With $\Phi = (\theta)$ and $\Psi = (\lambda_{21}, \ldots, \lambda_{2j}, \ldots, \lambda_{2J})$, it follows from (B.3) that

$$\widehat{\mathrm{var}}(\hat{\theta}) = \mathbf{I}^{\Phi\Phi}(\hat{\Theta}) = \left(\sum_{j=1}^{J} \hat{v}_j\right)^{-1}.$$

As an example, let $\theta = \log(HR)$. In the notation of Chapter 10, with v_j and the partial derivatives given in Tables B.2(a) and B.2(b), we have

$$\widehat{\mathrm{var}}(\log \widehat{HR}) = \left(\sum_{j=1}^{J}\left[\frac{1}{\hat{\lambda}_{1j}n_{1j}} + \frac{1}{\hat{\lambda}_{2j}n_{2j}}\right]^{-1}\right)^{-1}.$$

In Tables B.2(a) and B.2(b), the hazard difference is defined to be $HD = \lambda_{1j} - \lambda_{2j}$.

TABLE B.2(a) Values for the Poisson Distribution Under Homogeneity

θ	λ_{1j}	$\dfrac{\partial \lambda_{1j}}{\partial \theta}$	$\dfrac{\partial \lambda_{1j}}{\partial \lambda_{2j}}$
$\log(HR)$	$e^{\theta} \lambda_{2j}$	λ_{1j}	$\dfrac{\lambda_{1j}}{\lambda_{2j}}$
HD	$\lambda_{2j} + \theta$	1	1

TABLE B.2(b) Values of v_j for the Poisson Distribution

θ	v_j
$\log(HR)$	$\left(\dfrac{1}{\lambda_{1j} n_{1j}} + \dfrac{1}{\lambda_{2j} n_{2j}} \right)^{-1}$
HD	$\left(\dfrac{\lambda_{1j}}{n_{1j}} + \dfrac{\lambda_{2j}}{n_{2j}} \right)^{-1}$

B.3.3 Score Test for Linear Trend

Let $\theta_j = \alpha + \beta s_j$, where s_j is the exposure level for the jth stratum and α and β are constants ($j = 1, 2, \ldots, J$). The score test of $H_0 : \beta = 0$ is

$$X_t^2 = \left[\sum_{j=1}^{J} \frac{s_j (d_{1j} - \tilde{d}_{1j})}{\tilde{\lambda}_{1j}} \left(\frac{\partial \tilde{\lambda}_{1j}}{\partial \theta} \right) \right]^2 \Big/ \left[\sum_{j=1}^{J} s_j^2 \tilde{v}_j - \left(\sum_{j=1}^{J} s_j \tilde{v}_j \right)^2 \Big/ \sum_{j=1}^{J} \tilde{v}_j \right].$$

B.4 MATRIX INVERSION

Let $\mathbf{A}$ be a symmetric invertible matrix having the partition

$$\mathbf{A} = \begin{pmatrix} \mathbf{A}_{11} & \mathbf{A}_{12} \\ \mathbf{A}_{21} & \mathbf{A}_{22} \end{pmatrix}$$

and denote the inverse of $\mathbf{A}$ by

$$\mathbf{A}^{-1} = \begin{pmatrix} \mathbf{A}^{11} & \mathbf{A}^{12} \\ \mathbf{A}^{21} & \mathbf{A}^{22} \end{pmatrix}.$$

The submatrices of $\mathbf{A}^{-1}$ can be calculated using the following identities (Rao, 1973, p. 33):

$$\mathbf{A}^{11} = (\mathbf{A}_{11} - \mathbf{A}_{12}\mathbf{A}_{22}^{-1}\mathbf{A}_{21})^{-1} \tag{B.16}$$

$$\mathbf{A}^{12} = -\mathbf{A}_{11}^{-1}\mathbf{A}_{12}(\mathbf{A}_{22} - \mathbf{A}_{21}\mathbf{A}_{11}^{-1}\mathbf{A}_{12})^{-1}$$

$$\mathbf{A}^{21} = -(\mathbf{A}_{22} - \mathbf{A}_{21}\mathbf{A}_{11}^{-1}\mathbf{A}_{12})^{-1}\mathbf{A}_{21}\mathbf{A}_{11}^{-1}$$

$$\mathbf{A}^{22} = (\mathbf{A}_{22} - \mathbf{A}_{21}\mathbf{A}_{11}^{-1}\mathbf{A}_{12})^{-1}.$$

Let $\mathbf{I}$ be a symmetric invertible matrix with the partition

$$\mathbf{I} = \left(\begin{array}{cc|c} \mathbf{I}_{11} & \mathbf{I}_{12} & \mathbf{I}_{13} \\ \mathbf{I}_{21} & \mathbf{I}_{22} & \mathbf{I}_{23} \\ \hline \mathbf{I}_{31} & \mathbf{I}_{32} & \mathbf{I}_{33} \end{array}\right) = \begin{pmatrix} \mathbf{A}_{11} & \mathbf{A}_{12} \\ \mathbf{A}_{21} & \mathbf{A}_{22} \end{pmatrix}.$$

In this notation, $\mathbf{I}^{11}$ is the upper left submatrix of $\mathbf{A}^{11}$. From (B.16),

$$\mathbf{A}^{11} = (\mathbf{A}_{11} - \mathbf{A}_{12}\mathbf{A}_{22}^{-1}\mathbf{A}_{21})^{-1}$$

$$= \left[\begin{pmatrix} \mathbf{I}_{11} & \mathbf{I}_{12} \\ \mathbf{I}_{21} & \mathbf{I}_{22} \end{pmatrix} - \begin{pmatrix} \mathbf{I}_{13} \\ \mathbf{I}_{23} \end{pmatrix} \mathbf{I}_{33}^{-1} \left(\mathbf{I}_{31}\mathbf{I}_{32}\right)\right]^{-1}$$

$$= \begin{pmatrix} \mathbf{I}_{11} - \mathbf{I}_{13}\,\mathbf{I}_{33}^{-1}\,\mathbf{I}_{31} & \mathbf{I}_{12} - \mathbf{I}_{13}\,\mathbf{I}_{33}^{-1}\,\mathbf{I}_{32} \\ \mathbf{I}_{21} - \mathbf{I}_{23}\,\mathbf{I}_{33}^{-1}\,\mathbf{I}_{31} & \mathbf{I}_{22} - \mathbf{I}_{23}\,\mathbf{I}_{33}^{-1}\,\mathbf{I}_{32} \end{pmatrix}^{-1}$$

and so, again from (B.16),

$$\mathbf{I}^{11} = \Big[(\mathbf{I}_{11} - \mathbf{I}_{13}\,\mathbf{I}_{33}^{-1}\,\mathbf{I}_{31})$$

$$- (\mathbf{I}_{12} - \mathbf{I}_{13}\,\mathbf{I}_{33}^{-1}\,\mathbf{I}_{32})(\mathbf{I}_{22} - \mathbf{I}_{23}\mathbf{I}_{33}^{-1}\,\mathbf{I}_{32})^{-1}(\mathbf{I}_{21} - \mathbf{I}_{23}\,\mathbf{I}_{33}^{-1}\,\mathbf{I}_{31})\Big]^{-1} \tag{B.17}$$

Hypergeometric and Conditional Poisson Distributions

C.1 HYPERGEOMETRIC

Let A_1 and A_2 be independent binomial random variables with parameters (π_1, r_1) and (π_2, r_2), respectively, and consider the conditional probability

$$P(A_1 = a_1 | A_1 + A_2 = m_1) = \frac{P(A_1 = a_1, A_1 + A_2 = m_1)}{P(A_1 + A_2 = m_1)}$$

$$= \frac{P(A_1 = a_1, A_2 = m_1 - a_1)}{P(A_1 + A_2 = m_1)}.$$

Since A_1 and A_2 are independent,

$$P(A_1 = a_1, A_2 = m_1 - a_1) = P(A_1 = a_1) P(A_2 = m_1 - a_1)$$

$$= \binom{r_1}{a_1} \pi_1^{a_1} (1 - \pi_1)^{r_1 - a_1}$$

$$\times \binom{r_2}{m_1 - a_1} \pi_2^{m_1 - a_1} (1 - \pi_2)^{r_2 - (m_1 - a_1)}$$

$$= (1 - \pi_1)^{r_1} \pi_2^{m_1} (1 - \pi_2)^{r_2 - m_1} \binom{r_1}{a_1} \binom{r_2}{m_1 - a_1} OR^{a_1}.$$

With $l = \max(0, r_1 - m_2)$ and $u = \min(r_1, m_1)$,

$$P(A_1 + A_2 = m_1) = \sum_{x=l}^{u} P(A_1 = x) P(A_2 = m_1 - x)$$

$$= (1 - \pi_1)^{r_1} \pi_2^{m_1} (1 - \pi_2)^{r_2 - m_1} \sum_{x=l}^{u} \binom{r_1}{x} \binom{r_2}{m_1 - x} OR^{x}.$$

So the probability function for the conditional distribution of A_1 is

$$P(A_1 = a_1 | OR) = \frac{1}{C}\binom{r_1}{a_1}\binom{r_2}{m_1 - a_1} OR^{a_1}$$

where

$$C = \sum_{x=l}^{u}\binom{r_1}{x}\binom{r_2}{m_1 - x} OR^{x}.$$

C.2   CONDITIONAL POISSON

Let D_1 and D_2 be independent Poisson random variables with parameters v_1 and v_2, respectively. Arguing as above,

$$P(D_1 = d_1 | D_1 + D_2 = m) = \frac{P(D_1 = d_1, D_2 = m - d_1)}{P(D_1 + D_2 = m)}$$

$$P(D_1 = d_1, D_2 = m - d_1) = P(D_1 = d_1)P(D_2 = m - d_1)$$

$$= \frac{e^{-v_1} v_1^{d_1}}{d_1!} \times \frac{e^{-v_2} v_2^{m-d_1}}{(m - d_1)!}$$

$$= \frac{e^{-(v_1+v_2)}}{m!}\binom{m}{d_1} v_1^{d_1} v_2^{m-d_1}$$

and

$$P(D_1 + D_2 = m) = \sum_{x=0}^{m} P(D_1 = x)P(D_2 = m - x)$$

$$= \frac{e^{-(v_1+v_2)}}{m!}\sum_{x=0}^{m}\binom{m}{x} v_1^{x} v_2^{m-x}$$

$$= \frac{e^{-(v_1+v_2)}}{m!}(v_1 + v_2)^{m}.$$

So the probability function for the conditional distribution of D_1 is

$$P(D_1 = d_1 | \pi) = \binom{m}{d_1}\pi^{d_1}(1 - \pi)^{m-d_1}$$

where $\pi = v_1/(v_1 + v_2)$.

C.3 HYPERGEOMETRIC VARIANCE ESTIMATE

Consider the independent hypergeometric random variables A_{1j} ($j = 1, 2, \ldots J$) and assume that the OR_j are homogeneous with $\theta = \log(OR)$. Then A_{1j} has the probability function

$$P(A_{1j} = a_{1j}|OR) = \frac{1}{C_j}\binom{r_{1j}}{a_{1j}}\binom{r_{2j}}{m_{1j} - a_{1j}}e^{\theta a_{1j}}$$

where

$$C_j = \sum_{x=l_j}^{u_j}\binom{r_{1j}}{x}\binom{r_{2j}}{m_{1j} - x}e^{\theta x}.$$

Dropping terms not involving θ, the conditional log-likelihood is

$$l(\theta) = \theta a_{1\bullet} - \sum_{j=1}^{J}\log(C_j).$$

It is readily verified that

$$E(A_{1j}|OR) = \frac{1}{C_j}\left(\frac{\partial C_j}{\partial \theta}\right)$$

and

$$E(A_{1j}^2|OR) = \frac{1}{C_j}\left(\frac{\partial^2 C_j}{\partial \theta^2}\right)$$

from which it follows that

$$\frac{\partial l(\theta)}{\partial \theta} = a_{1\bullet} - \sum_{j=1}^{J}\frac{1}{C_j}\left(\frac{\partial C_j}{\partial \theta}\right) = a_{1\bullet} - \sum_{j=1}^{J}E(A_{1j}|OR) \qquad \text{(C.1)}$$

and

$$\frac{-\partial^2 l(\theta)}{\partial \theta^2} = \sum_{j=1}^{J}\left\{\frac{1}{C_j}\left(\frac{\partial^2 C_j}{\partial \theta^2}\right) - \left[\frac{1}{C_j}\left(\frac{\partial C_j}{\partial \theta}\right)\right]^2\right\}$$

$$= \sum_{j=1}^{J}\{E(A_{1j}^2|OR) - [E(A_{1j}|OR)]^2\}$$

$$= \sum_{j=1}^{J}\text{var}(A_{1j}|OR). \qquad \text{(C.2)}$$

The last equality in (C.2) follows from the general result that for any random variable X, $\text{var}(X) = E(X^2) - [E(X)]^2$. From (C.1), the conditional maximum likelihood equation is

$$a_{1\bullet} = \sum_{j=1}^{J} E(A_{1j}|\widehat{OR}_c).$$

With $\hat{\theta} = \log(\widehat{OR}_c)$, it follows from (C.2) and Anderson (1970) that an estimate of $\text{var}(\log \widehat{OR}_c)$ is

$$\widehat{\text{var}}(\log \widehat{OR}_c) = \left(\frac{-\partial^2 l(\hat{\theta})}{\partial \theta^2}\right)^{-1} = \left[\sum_{j=1}^{J} \widehat{\text{var}}(A_{1j}|\widehat{OR}_c)\right]^{-1}.$$

C.4   CONDITIONAL POISSON VARIANCE ESTIMATE

Consider the independent binomial (conditional Poisson) random variables D_{1j} ($j = 1, 2, \ldots J$) and assume that the HR_j are homogeneous with $\theta = \log(HR)$. Recall that $v_{1j} = HR\lambda_{2j}n_{1j}$, $v_{2j} = \lambda_{2j}n_{2j}$, and $\pi_j = v_{1j}/(v_{1j} + v_{2j})$. Then D_{1j} has the probability function

$$P(D_{1j} = d_{1j}|HR) = \binom{m_j}{d_{1j}}\pi_j^{d_{1j}}(1 - \pi_j)^{m_j - d_{1j}}$$

$$= \frac{1}{G_j}n_{1j}^{d_{1j}}n_{2j}^{m_j - d_{1j}}e^{\theta d_{1j}}$$

where

$$G_j = (e^{\theta}n_{1j} + n_{2j})^{m_j}\Big/\binom{m_j}{d_{1j}}.$$

Arguing as in Section C.3, an estimate of $\text{var}(\log \widehat{HR})$ is

$$\widehat{\text{var}}(\log \widehat{HR}) = \left[\sum_{j=1}^{J} \widehat{\text{var}}(D_{1j}|\widehat{HR})\right]^{-1}.$$

APPENDIX D

Quadratic Equation for the Odds Ratio

Dropping the superscript $*$ for convenience of notation, (4.28) can be written as

$$OR = \frac{a_1(r_2 - m_1 + a_1)}{(m_1 - a_1)(r_1 - a_1)} \qquad \text{(D.1)}$$

which has the solutions

$$a_1 = \frac{-y \pm \sqrt{y^2 - 4xz}}{2x}$$

where

$$x = OR - 1$$
$$y = -[(m_1 + r_1)OR - m_1 + r_2]$$
$$z = ORm_1r_1.$$

We now show that only the negative root guarantees that $a_1 \geq 0$. Let $\pi_1 = a_1/r_1$, $\pi_2 = (m_1 - a_1)/r_2$, $\omega_1 = \pi_1/(1 - \pi_1)$, and $\omega_2 = \pi_2/(1 - \pi_2)$, so that $\omega_1 = OR\omega_2$. Then (D.1) can be written as

$$OR = \frac{\pi_1(r_2 - m_1 + \pi_1 r_1)}{(m_1 - \pi_1 r_1)(1 - \pi_1)} = \frac{\omega_1[\omega_1(r - m_1) + r_2 - m_1]}{\omega_1(m_1 - r_1) + m_1}$$

which has the solutions

$$\omega_1 = \frac{-Y \pm \sqrt{Y^2 - 4XZ}}{2X} \qquad \text{(D.2)}$$

where

$$X = r - m_1$$
$$Y = -[(m_1 - r_1)OR + m_1 - r_2]$$
$$Z = -ORm_1.$$

Since $X > 0$ and ω_1 must be nonnegative, we need to determine the root that makes the numerator of (D.2) nonnegative. It follows from $-4XZ = 4(r - m_1)ORm_1 \geq 0$ that $\sqrt{Y^2 - 4XZ} \geq |Y|$. This means that, regardless of the sign of Y, the positive root must be chosen. So

$$\omega_1 = \frac{-Y + \sqrt{Y^2 - 4XZ}}{2X}$$

and hence

$$\pi_1 = \frac{\omega_1}{1 + \omega_1} = \frac{-Y + \sqrt{Y^2 - 4XZ}}{2X - Y + \sqrt{Y^2 - 4XZ}}. \tag{D.3}$$

From $\omega_1 > 0$, it follows that $0 < \pi_1 < 1$. Multiplying numerator and denominator of (D.3) by $2X - Y - \sqrt{Y^2 - 4XZ}$ gives

$$\pi_1 = \frac{-Y + 2Z + \sqrt{Y^2 - 4XZ}}{2(X - Y + Z)}. \tag{D.4}$$

It is readily demonstrated that

$$x = \frac{-(X - Y + Z)}{r_1}$$

$$y = -Y + 2Z$$

$$z = -r_1 Z$$

from which it follows that

$$y^2 - 4xz = Y^2 - 4XZ.$$

Substituting in (D.4), and noting that $a_1 = \pi_1 r_1$, we obtain

$$a_1 = \frac{-y - \sqrt{y^2 - 4xz}}{2x}.$$

APPENDIX E

Matrix Identities and Inequalities

Except for the first inequality in Proposition E.4, the following results are based on Peto and Pike (1973) and Crowley and Breslow (1975).

E.1 IDENTITIES AND INEQUALITIES FOR J ($1 \times I$) AND J ($2 \times I$) TABLES

We begin with a series of definitions. The data layout for the jth stratum is given in Table E.1, where the r_{ij} will later represent either the numbers of persons at risk or person-time ($j = 1, 2, \ldots, J$).

Let γ_j be a constant such that $0 < \gamma_j \le m_{1j}$ and define

$$p_{ij} = \frac{r_{ij}}{r_j} \qquad e_{ij} = p_{ij} m_{1j} \qquad g_{ij} = p_{ij} \gamma_j$$

$$e_{\bullet\bullet} = \varepsilon \qquad g_{\bullet\bullet} = \gamma.$$

It follows that

$$e_{\bullet j} = m_{1j} \qquad g_{\bullet j} = \gamma_j \qquad e_{\bullet\bullet} = m_{1\bullet}$$

$$g_{ij} \le e_{ij} \qquad g_{i\bullet} \le e_{i\bullet} \qquad \gamma \le \varepsilon.$$

Let

$$P_i = \frac{1}{\gamma} \sum_{j=1}^{J} \gamma_j p_{ij} = \frac{g_{i\bullet}}{\gamma}$$

TABLE E.1 Data Layout for the jth Stratum

	Exposure category						
	1	2	$\cdots$	i	$\cdots$	I	
count	a_{1j}	a_{2j}	$\cdots$	a_{ij}	$\cdots$	a_{Ij}	m_{1j}
	r_{1j}	r_{2j}	$\cdots$	r_{ij}	$\cdots$	r_{Ij}	r_j

$$Q_i = \frac{1}{\varepsilon} \sum_{j=1}^{J} m_{1j} p_{ij} = \frac{e_{i\bullet}}{\varepsilon}$$

$$\mathbf{a}^{\mathrm{T}} = (a_{1\bullet}, \ldots, a_{i\bullet}, \ldots, a_{I\bullet})$$

$$\mathbf{e}^{\mathrm{T}} = (e_{1\bullet}, \ldots, e_{i\bullet}, \ldots, e_{I\bullet})$$

and

$$\mathbf{s}^{\mathrm{T}} = (s_1, \ldots, s_i, \ldots, s_I)$$

where the s_i are arbitrary constants and T denotes matrix transposition. Let $\mathbf{V} = (v_{ik})$, $\mathbf{U} = (u_{ik})$, and $\mathbf{W} = (w_{ik})$ be $I \times I$ matrices, where

$$v_{ik} = \sum_{j=1}^{J} \gamma_j p_{ij} (\delta_{ik} - p_{kj}) = g_{i\bullet} \delta_{ik} - \sum_{j=1}^{J} \frac{g_{ij} g_{kj}}{\gamma_j}$$

$$u_{ik} = \gamma P_i (\delta_{ik} - P_k) = g_{i\bullet} \delta_{ik} - \frac{g_{i\bullet} g_{k\bullet}}{\gamma}$$

$$w_{ik} = \varepsilon Q_i (\delta_{ik} - Q_k) = e_{i\bullet} \delta_{ik} - \frac{e_{i\bullet} e_{k\bullet}}{\varepsilon}$$

and

$$\delta_{ik} = \begin{cases} 1 & i = k \\ 0 & i \neq k. \end{cases}$$

Note that v_{ik} can be expressed as

$$v_{ik} = \begin{cases} \displaystyle\sum_{j=1}^{J} \frac{\gamma_j r_{ij} (r_j - r_{ij})}{r_j^2} & \text{if } i = k \\[2ex] -\displaystyle\sum_{j=1}^{J} \frac{\gamma_j r_{ij} r_{kj}}{r_j^2} & \text{if } i \neq k. \end{cases} \tag{E.1}$$

The columns of $\mathbf{V}$, $\mathbf{U}$, and $\mathbf{W}$ are linearly dependent and so inverses do not exist. Let $\mathbf{V}_*$ denote the matrix obtained from $\mathbf{V}$ by dropping the Ith row and Ith column, and let $\mathbf{V}_*^{-1}$ be its inverse. Define $\mathbf{U}_*$ and $\mathbf{U}_*^{-1}$ in an analogous manner. Let $\mathbf{a}_*$ and $\mathbf{e}_*$ denote the vectors obtained by dropping the Ith rows of $\mathbf{a}$ and $\mathbf{e}$, and define

$$X_{\mathrm{mh}}^2 = (\mathbf{a}_* - \mathbf{e}_*)^{\mathrm{T}} \mathbf{V}_*^{-1} (\mathbf{a}_* - \mathbf{e}_*)$$

$$X_{\mathrm{pp}}^2 = \sum_{i=1}^{I} \frac{(a_{i\bullet} - e_{i\bullet})^2}{g_{i\bullet}}$$

$$X_{\text{oe}}^2 = \sum_{i=1}^{I} \frac{(a_{i\bullet} - e_{i\bullet})^2}{e_{i\bullet}}$$

and

$$X_{\text{t}}^2 = \frac{[\mathbf{s}^{\mathsf{T}}(\mathbf{a} - \mathbf{e})]^2}{\mathbf{s}^{\mathsf{T}}\mathbf{V}\mathbf{s}}.$$

Since $g_{ij} = (\gamma_j/m_{1j})e_{ij}$ and $m_{1j} = e_{\bullet j}$, it follows that

$$
\begin{aligned}
\mathbf{s}^{\mathsf{T}}\mathbf{V}\mathbf{s} &= \sum_{i,k} s_i s_k \left(g_{i\bullet}\delta_{ik} - \sum_{j=1}^{J} \frac{g_{ij}g_{kj}}{\gamma_j} \right) \\
&= \sum_{i=1}^{I} s_i^2 g_{i\bullet} - \sum_{j=1}^{J} \frac{1}{\gamma_j} \left(\sum_{i=1}^{I} s_i g_{ij} \right)^2 \\
&= \sum_{j=1}^{J} \left[\sum_{i=1}^{I} s_i^2 g_{ij} - \frac{1}{\gamma_j} \left(\sum_{i=1}^{I} s_i g_{ij} \right)^2 \right] \\
&= \sum_{j=1}^{J} \frac{\gamma_j}{m_{1j}} \left[\sum_{i=1}^{I} s_i^2 e_{ij} - \left(\sum_{i=1}^{I} s_i e_{ij} \right)^2 \Big/ e_{\bullet j} \right].
\end{aligned}
\tag{E.2}
$$

We also have

$$
\begin{aligned}
\mathbf{s}^{\mathsf{T}}\mathbf{U}\mathbf{s} &= \sum_{i,k} s_i s_k \left(g_{i\bullet}\delta_{ik} - \frac{g_{i\bullet}g_{k\bullet}}{\gamma} \right) \\
&= \sum_{i=1}^{I} s_i^2 g_{i\bullet} - \frac{1}{\gamma} \left(\sum_{i=1}^{I} s_i g_{i\bullet} \right)^2 \\
&= \sum_{i=1}^{I} (s_i - \bar{s}_g)^2 g_{i\bullet}
\end{aligned}
\tag{E.3}
$$

and

$$
\begin{aligned}
\mathbf{s}^{\mathsf{T}}\mathbf{W}\mathbf{s} &= \sum_{i,k} s_i s_k \left(e_{i\bullet}\delta_{ik} - \frac{e_{i\bullet}e_{k\bullet}}{\varepsilon} \right) \\
&= \sum_{i=1}^{I} s_i^2 e_{i\bullet} - \frac{1}{\varepsilon} \left(\sum_{i=1}^{I} s_i e_{i\bullet} \right)^2
\end{aligned}
$$

$$= \sum_{i=1}^{I} (s_i - \bar{s}_e)^2 e_{i\bullet} \qquad \text{(E.4)}$$

where

$$\bar{s}_g = \frac{1}{\gamma} \sum_{i=1}^{I} s_i g_{i\bullet}$$

and

$$\bar{s}_e = \frac{1}{\varepsilon} \sum_{i=1}^{I} s_i e_{i\bullet}.$$

Proposition E.1. $\mathbf{U}_*^{-1} = (u_*^{ik})$, where

$$u_*^{ik} = \left(\frac{\delta_{ik}}{g_{i\bullet}} + \frac{1}{g_{I\bullet}} \right) = \frac{1}{\gamma} \left(\frac{\delta_{ik}}{P_i} + \frac{1}{P_I} \right).$$

Proof
The i, kth term of $\mathbf{U}_* \mathbf{U}_*^{-1}$ is

$$\sum_{h=1}^{I-1} u_{ih} u_*^{hk} = P_i \left[\sum_{h=1}^{I-1} (\delta_{ih} - P_h) \left(\frac{\delta_{hk}}{P_h} + \frac{1}{P_I} \right) \right]$$

$$= P_i \left[\sum_{h=1}^{I-1} \frac{(\delta_{ih} - P_h)\delta_{hk}}{P_h} + \frac{1}{P_I} \sum_{h=1}^{I-1} (\delta_{ih} - P_h) \right]$$

$$= P_i \left(\frac{\delta_{ik} - P_k}{P_k} + 1 \right) = \frac{P_i \delta_{ik}}{P_k} = \delta_{ik}.$$

Proposition E.2. $X_{\mathrm{pp}}^2 = (\mathbf{a}_* - \mathbf{e}_*)^{\mathrm{T}} \mathbf{U}_*^{-1} (\mathbf{a}_* - \mathbf{e}_*).$

Proof

$$(\mathbf{a}_* - \mathbf{e}_*)^{\mathrm{T}} \mathbf{U}_*^{-1} (\mathbf{a}_* - \mathbf{e}_*) = \sum_{i,k}^{I-1} \left[\frac{\delta_{ik}(a_{i\bullet} - e_{i\bullet})(a_{k\bullet} - e_{k\bullet})}{g_{i\bullet}} + \frac{(a_{i\bullet} - e_{i\bullet})(a_{k\bullet} - e_{k\bullet})}{g_{I\bullet}} \right]$$

$$= \sum_{i=1}^{I-1} \frac{(a_{i\bullet} - e_{i\bullet})^2}{g_{i\bullet}} + \frac{1}{g_{I\bullet}} \left[\sum_{i=1}^{I-1} (a_{i\bullet} - e_{i\bullet}) \right]^2$$

$$= \sum_{i=1}^{I} \frac{(a_{i\bullet} - e_{i\bullet})^2}{g_{i\bullet}}$$

where the last equality follows from $\sum_{i=1}^{I-1}(a_{i\bullet} - e_{i\bullet}) = -(a_{I\bullet} - e_{I\bullet})$.

Theorem E.3. $X_{oe}^2 \leq X_{pp}^2 \leq X_{mh}^2$.

Proof
Since $g_{i\bullet} \leq e_{i\bullet}$ it follows that $X_{oe}^2 \leq X_{pp}^2$. For a proof of $X_{pp}^2 \leq X_{mh}^2$, see Peto and Pike (1973).

Proposition E.4. $\mathbf{s}^T \mathbf{W} \mathbf{s} \geq \mathbf{s}^T \mathbf{U} \mathbf{s} \geq \mathbf{s}^T \mathbf{V} \mathbf{s}$.

Proof
To prove the first inequality, consider

$$\sum_{i=1}^{I} [(s_i - \bar{s}_e)^2 - (s_i - \bar{s}_g)^2] g_{i\bullet} = \sum_{i=1}^{I} [(s_i - \bar{s}_e) + (s_i - \bar{s}_g)][(s_i - \bar{s}_e) - (s_i - \bar{s}_g)] g_{i\bullet}$$

$$= \sum_{i=1}^{I} [2s_i - (\bar{s}_e + \bar{s}_g)](\bar{s}_g - \bar{s}_e) g_{i\bullet}$$

$$= (\bar{s}_g - \bar{s}_e)[2\bar{s}_g \gamma - (\bar{s}_e + \bar{s}_g)\gamma]$$

$$= \gamma(\bar{s}_g - \bar{s}_e)^2 \geq 0.$$

From $e_{i\bullet} \geq g_{i\bullet}$ it follows that

$$\sum_{i=1}^{I} (s_i - \bar{s}_e)^2 e_{i\bullet} \geq \sum_{i=1}^{I} (s_i - \bar{s}_e)^2 g_{i\bullet} \geq \sum_{i=1}^{I} (s_i - \bar{s}_g)^2 g_{i\bullet}$$

and, from (E.3) and (E.4), that $\mathbf{s}^T \mathbf{W} \mathbf{s} \geq \mathbf{s}^T \mathbf{U} \mathbf{s}$.
To prove the second inequality, multiply

$$\frac{1}{\gamma_j^2} \left(\sum_{i=1}^{I} s_i g_{ij} \right)^2 - 2 \left(\frac{1}{\gamma_j} \sum_{i=1}^{I} s_i g_{ij} \right) \left(\frac{1}{\gamma} \sum_{i=1}^{I} s_i g_{i\bullet} \right) + \frac{1}{\gamma^2} \left(\sum_{i=1}^{I} s_i g_{i\bullet} \right)^2$$

$$= \left[\sum_{i=1}^{I} s_i \left(\frac{g_{ij}}{\gamma_j} - \frac{g_{i\bullet}}{\gamma} \right) \right]^2 \geq 0$$

by γ_j and sum over j to obtain

$$\sum_{j=1}^{J} \frac{1}{\gamma_j} \left(\sum_{i=1}^{I} s_i g_{ij} \right)^2 - \frac{1}{\gamma} \left(\sum_{i=1}^{I} s_i g_{i\bullet} \right)^2 \geq 0.$$

From (E.2) and (E.3) it follows that $\mathbf{s}^T \mathbf{U} \mathbf{s} \geq \mathbf{s}^T \mathbf{V} \mathbf{s}$.

Corollary E.5. $\dfrac{[\mathbf{s}^T(\mathbf{a} - \mathbf{e})]^2}{\mathbf{s}^T \mathbf{W} \mathbf{s}} \leq \dfrac{[\mathbf{s}^T(\mathbf{a} - \mathbf{e})]^2}{\mathbf{s}^T \mathbf{U} \mathbf{s}} \leq X_t^2.$

E.2 IDENTITIES AND INEQUALITIES FOR A SINGLE TABLE

We now specialize to the case $J = 1$. Dropping the index j from the above notation, we have $g_i = (\gamma/m_1)e_i$ and $v_{ik} = u_{ik} = \gamma p_i(\delta_{ik} - p_k)$. Since $\mathbf{U} = \mathbf{V}$, it follows from Proposition E.2 that

$$\left(\frac{m_1}{\gamma} \right) X_{oe}^2 = X_{pp}^2 = X_{mh}^2 \tag{E.5}$$

From (E.2), we have

$$\mathbf{s}^T \mathbf{V} \mathbf{s} = \left(\frac{\gamma}{m_1} \right) \left[\sum_{i=1}^{I} s_i^2 e_i - \left(\sum_{i=1}^{I} s_i e_i \right)^2 \Big/ e_\bullet \right]. \tag{E.6}$$

E.3 HYPERGEOMETRIC DISTRIBUTION

For each j, consider the random vector $\mathbf{A}_j = (A_{1j}, \ldots, A_{ij}, \ldots A_{Ij})$, where the A_{ij} are binomial random variables that satisfy the constraint $\sum_{i=1}^{I} A_{ij} = m_{1j}$. Then $\mathbf{A}_j$ is multidimensional hypergeometric and, under the hypothesis of no association in each stratum, the variance–covariance matrix is $\mathbf{V} = (v_{ik})$, with γ_j given by

$$\gamma_j = \left(\frac{m_{2j}}{r_j - 1} \right) m_{1j}$$

(Breslow, 1979; Breslow and Day, 1980, p. 147). From (E.1),

$$v_{ik} = \begin{cases} \displaystyle\sum_{j=1}^{J} \frac{m_{1j} m_{2j} r_{ij}(r_j - r_{ij})}{r_j^2(r_j - 1)} & \text{if } i = k \\[4mm] -\displaystyle\sum_{j=1}^{J} \frac{m_{1j} m_{2j} r_{ij} r_{kj}}{r_j^2(r_j - 1)} & \text{if } i \neq k. \end{cases}$$

Evidently the inequality $\gamma_j \leq m_{1j}$ is satisfied. By definition, X_{mh}^2 is the Mantel–Haenszel test of association for odds ratios. From $\gamma_j / m_{1j} = m_{2j}/(r_j - 1)$, it follows from (E.2) that

$$\mathbf{s}^T \mathbf{V} \mathbf{s} = \sum_{j=1}^{J} \left(\frac{m_{2j}}{r_j - 1} \right) \left[\sum_{i=1}^{I} s_i^2 e_{ij} - \left(\sum_{i=1}^{I} s_i e_{ij} \right)^2 \bigg/ e_{\bullet j} \right]$$

which is the denominator of (5.40). From (E.3), we have

$$\mathbf{s}^T \mathbf{U} \mathbf{s} = \sum_{i=1}^{I} s_i^2 g_{i\bullet} - \left(\sum_{i=1}^{I} s_i g_{i\bullet} \right)^2 \bigg/ g_{\bullet\bullet}$$

which is the denominator of (5.41). When $J = 1$, it follows from (E.5) that

$$X_{mh}^2 = \left(\frac{r-1}{m_2} \right) \sum_{i=1}^{I} \frac{(a_i - e_i)^2}{e_i}$$

which is (4.37). From (E.6), we have

$$\mathbf{s}^T \mathbf{V} \mathbf{s} = \left(\frac{m_2}{r-1} \right) \left[\sum_{i=1}^{I} s_i^2 e_i - \left(\sum_{i=1}^{I} s_i e_i \right)^2 \bigg/ e_{\bullet} \right]$$

which is the denominator of (4.38).

E.4 CONDITIONAL POISSON DISTRIBUTION

In this section we switch to the notation of Chapter 10. For each k, consider the random vector $\mathbf{D}_k = (D_{1k}, \ldots, D_{ik}, \ldots D_{Ik})$, where the D_{ik} are Poisson random variables that satisfy the constraint $\sum_{i=1}^{I} D_{ik} = m_k$. Then $\mathbf{D}_k$ is multinomial and, under the hypothesis of no association in each stratum, the variance–covariance matrix is $\mathbf{V} = (v_{ij})$, with γ_k given by $\gamma_k = m_k$ (Breslow and Day, 1987, p. 113). From (E.1),

$$v_{ij} = \begin{cases} \displaystyle\sum_{k=1}^{K} \frac{m_k n_{ik}(n_k - n_{ik})}{n_k^2} & \text{if } i = j \\[4mm] -\displaystyle\sum_{k=1}^{K} \frac{m_k n_{ik} n_{jk}}{n_k^2} & \text{if } i \neq j. \end{cases}$$

By definition, X_{pt}^2 is the Mantel–Haenszel test of association for person-time data. Since $\gamma_k = m_k$, it follows that $g_{ik} = e_{ik}$, and so, from Theorem E.3, we have

$$X_{oe}^2 = X_{pp}^2 \leq X_{pt}^2.$$

With $\gamma_j/m_{1j} = 1$, it follows from (E.2) that

$$
\mathbf{s}^T\mathbf{V}\mathbf{s} = \sum_{k=1}^{K}\left[\sum_{i=1}^{I} s_i^2 e_{ik} - \left(\sum_{i=1}^{I} s_i e_{ik}\right)^2 \bigg/ e_{\bullet k}\right]
$$

$$
= \sum_{i=1}^{I} s_i^2 e_{i\bullet} - \sum_{k=1}^{K}\left(\sum_{i=1}^{I} s_i e_{ik}\right)^2 \bigg/ e_{\bullet k}
$$

which is the denominator of (10.40). From (E.3) and (E.4), we have

$$
\mathbf{s}^T\mathbf{U}\mathbf{s} = \mathbf{s}^T\mathbf{W}\mathbf{s} = \sum_{i=1}^{I} s_i^2 e_{i\bullet} - \left(\sum_{i=1}^{I} s_i e_{i\bullet}\right)^2 \bigg/ e_{\bullet\bullet}
$$

which is the denominator of (10.41). When $K = 1$, it follows from (E.5) that

$$
X_{oe}^2 = X_{pp}^2 = X_{pt}^2 = \sum_{i=1}^{I} \frac{(d_i - e_i)^2}{e_i}
$$

which is (10.24). From (E.2)–(E.4) and (E.6), we have

$$
\mathbf{s}^T\mathbf{U}\mathbf{s} = \mathbf{s}^T\mathbf{W}\mathbf{s} = \mathbf{s}^T\mathbf{V}\mathbf{s} = \sum_{i=1}^{I} s_i^2 e_i - \left(\sum_{i=1}^{I} s_i e_i\right)^2 \bigg/ e_{\bullet}
$$

which is the denominator of (10.25).

APPENDIX F

Survival Analysis and Life Tables

F.1 SINGLE COHORT

In the notation of Chapter 8, let T be a continuous random variable with the sample space $[0, \tau]$. Denote the probability function, survival function, and hazard function for T by $f(t)$, $S(t)$, and $h(t)$, respectively. By definition,

$$S(t) = P(T \geq t) = \int_t^\tau f(u)\,du \qquad (F.1)$$

and

$$h(t) = \lim_{\varepsilon \to 0} \frac{P(t \leq T < t + \varepsilon | T \geq t)}{\varepsilon}.$$

Differentiating both sides of (F.1) with respect to t gives

$$f(t) = -\frac{dS(t)}{dt}.$$

From

$$P(t \leq T < t + \varepsilon | T \geq t) = \frac{P(t \leq T < t + \varepsilon)}{P(T \geq t)} = \frac{S(t) - S(t + \varepsilon)}{S(t)}$$

it follows that

$$h(t) = \frac{1}{S(t)}\left[-\frac{dS(t)}{dt}\right] = \frac{f(t)}{S(t)} = -\frac{d[\log S(t)]}{dt}.$$

Integrating $h(t) = -d \log S(t)/dt$ and exponentiating gives

$$S(t) = \exp\left[-\int_0^t h(u)\,du \right]. \qquad (F.2)$$

339

Substituting $h(t)S(t) = f(t)$ in (F.1) we find that

$$S(t) = \int_t^\tau h(u)S(u)\,du. \tag{F.3}$$

Note that $h(u)\,du$ is approximately equal to the conditional probability of dying in $[u, u + du)$, given survival to u. It follows that for a cohort consisting of one subject, $h(u)S(u)\,du$ is approximately equal to the (expected) number of deaths in $[u, u + du)$. So (F.3) says that the number of survivors to time t equals the number who will die after t.

F.2   COMPARISON OF COHORTS

For $\varepsilon > 0$, define

$$Q_\varepsilon(t) = \frac{S(t - \varepsilon) - S(t + \varepsilon)}{S(t - \varepsilon)}$$

and

$$\omega_\varepsilon(t) = \frac{Q_\varepsilon(t)}{1 - Q_\varepsilon(t)}.$$

It is readily demonstrated that

$$\frac{\partial Q_\varepsilon(t)}{\partial \varepsilon} = \frac{S(t + \varepsilon)[h(t - \varepsilon) + h(t + \varepsilon)]}{S(t - \varepsilon)}$$

and

$$\frac{\partial \omega_\varepsilon(t)}{\partial \varepsilon} = \frac{S(t - \varepsilon)[h(t - \varepsilon) + h(t + \varepsilon)]}{S(t + \varepsilon)}.$$

Consider two cohorts, one of which is exposed ($i = 1$) and the other unexposed ($i = 2$). Define

$$RR_\varepsilon(t) = \frac{Q_{\varepsilon 1}(t)}{Q_{\varepsilon 2}(t)}$$

and

$$OR_\varepsilon(t) = \frac{\omega_{\varepsilon 1}(t)}{\omega_{\varepsilon 2}(t)}.$$

Using l'Hôpital's rule,

$$\lim_{\varepsilon \to 0} RR_\varepsilon(t) = \lim_{\varepsilon \to 0} OR_\varepsilon(t) = \frac{h_1(t)}{h_2(t)}.$$

Suppose that the proportional hazards assumption is satisfied, with $h_1(t) = \psi h_2(t)$, for some constant ψ. It follows from (F.2) that

$$S_1(t) = [S_2(t)]^{\psi} \tag{F.4}$$

which is equivalent to

$$\log[-\log S_1(t)] = \log[-\log S_2(t)] + \log(\psi).$$

F.3   LIFE TABLES

In the notation of Section 13.1, $l(x) = l(0)S(x)$ is the (expected) number of survivors to age x in the OLT cohort. From (F.3), the number of deaths in the jth age group is

$$d_j = \int_{x_j}^{x_{j+1}} r(u)l(u)\,du = l(0) \int_{x_j}^{x_{j+1}} r(u)S(u)\,du. \tag{F.5}$$

The number of person-years experienced by the cohort during $[u, u + du)$ is approximately equal to $l(u)\,du$, and so the number of person-years experienced by the cohort after age x is

$$T(x) = \int_{x}^{\tau} l(u)\,du = l(0) \int_{x}^{\tau} S(u)\,du. \tag{F.6}$$

Denote the survival time random variable by X. The expected survival time for a member of the cohort who has survived to age x—that is, the life expectancy at age x—is given by

$$e(x) = E(X - x | X \geq x) = \frac{1}{S(x)} \int_{x}^{\tau} (u - x)f(u)\,du$$

$$= \frac{1}{S(x)} \int_{x}^{\tau} S(u)\,du = \frac{T(x)}{l(x)}.$$

Confounding in Open Cohort and Case-Control Studies

In what follows we make repeated use of results from Section F.1 of Appendix F.

G.1 OPEN COHORT STUDIES

G.1.1 Counterfactual Definition of Confounding in Open Cohort Studies

Consider two cohorts, one of which is exposed ($i = 1$) and the other unexposed ($i = 2$), and let the period of observation be $[0, \tau]$. For the ith cohort, denote the survival function, probability function, and hazard function by $S_i(t)$, $f_i(t)$, and $h_i(t) = f_i(t)/S_i(t)$, and let $l_i(t) = l_i(0)S_i(t)$ be the (expected) number of survivors to time $t \leq \tau$. Suppose that the exposed and unexposed cohorts are stratified according to a categorical variable F. For the kth stratum of the ith cohort, denote the corresponding functions by $S_{ik}(t)$, $f_{ik}(t)$, $h_{ik}(t) = f_{ik}(t)/S_{ik}(t)$, and $l_{ik}(t) = l_{ik}(0)S_{ik}(t)$ ($k = 1, 2, \ldots, K$). It follows that

$$l_i(0)S_i(t) = \sum_{k=1}^{K} l_{ik}(0)S_{ik}(t). \tag{G.1}$$

Let $p_{ik}(t) = l_{ik}(t)/l_i(t)$ be the proportion of the ith cohort in the kth stratum at time t. Then the $p_{ik}(t)$ ($k = 1, 2, \ldots, K$) give the distribution of F in the ith cohort at this time point. For brevity, denote $p_{ik}(0)$ by p_{ik}. Note that, even if $p_{1k} = p_{2k}$, it does not necessarily follow that $p_{1k}(t) = p_{2k}(t)$ for $t > 0$. From (G.1),

$$S_i(t) = \sum_{k=1}^{K} p_{ik} S_{ik}(t)$$

and consequently

$$f_i(t) = \sum_{k=1}^{K} p_{ik} f_{ik}(t) = \sum_{k=1}^{K} p_{ik} S_{ik}(t) h_{ik}(t) \tag{G.2}$$

and

$$h_i(t) = \frac{f_i(t)}{S_i(t)} = \frac{\sum_{k=1}^{K} p_{ik} S_{ik}(t) h_{ik}(t)}{\sum_{k=1}^{K} p_{ik} S_{ik}(t)}.$$

Therefore, $h_i(t)$ is a weighted average of the $h_{ik}(t)$, where the weights are functions of t. The ratio of hazard functions at time t is defined to be $h_1(t)/h_2(t)$. Since $f_i(t) = S_i(t)h_i(t)$, it follows from (G.2) that

$$S_i(t)h_i(t) = \sum_{k=1}^{K} p_{ik} S_{ik}(t) h_{ik}(t).$$

Therefore the (expected) number of deaths in the ith cohort during $[0, \tau]$ is

$$l_i(0) \int_0^\tau S_i(t) h_i(t)\, dt = l_i(0) \sum_{k=1}^{K} p_{ik} \int_0^\tau S_{ik}(t) h_{ik}(t)\, dt$$

and the (expected) amount of person-time is

$$l_i(0) \int_0^\tau S_i(t)\, dt = l_i(0) \sum_{k=1}^{K} p_{ik} \int_0^\tau S_{ik}(t)\, dt.$$

The (crude) hazard rate for ith cohort is defined to be

$$R_i = \frac{\sum_{k=1}^{K} p_{ik} \int_0^\tau S_{ik}(t) h_{ik}(t)\, dt}{\sum_{k=1}^{K} p_{ik} \int_0^\tau S_{ik}(t)\, dt} \tag{G.3}$$

and the ratio of hazard rates is defined to be

$$\rho = \frac{R_1}{R_2}.$$

The counterfactual definition of confounding in open cohort studies parallels the definition given in Section 2.5.1 for closed cohort studies. Suppose that F is the only potential confounder. Let $p_{1k}^*(t)$ be the proportion of the counterfactual unexposed cohort in the kth stratum at time t, and let $S_{1k}^*(t)$ and $h_{1k}^*(t)$ be the corresponding survival function and hazard function for the kth stratum. From (G.3) it follows that the hazard rate in the counterfactual unexposed cohort is

$$R_1^* = \frac{\sum_{k=1}^{K} p_{1k}^* \int_0^\tau S_{1k}^*(t) h_{1k}^*(t)\, dt}{\sum_{k=1}^{K} p_{1k}^* \int_0^\tau S_{1k}^*(t)\, dt}.$$

Assume that F is not affected by E, which implies that $p_{1k}^* = p_{1k}$ for all k. Also assume there is no residual confounding within strata of F, that is, $h_{1k}^*(t) = h_{2k}(t)$ for all k. An identity such as the preceding one is meant to indicate equality of functions, that is, equality for all t. It follows that $S_{1k}^*(t) = S_{2k}(t)$ for all k, and so

$$R_1^* = \frac{\sum_{k=1}^K p_{1k} \int_0^\tau S_{2k}(t)h_{2k}(t)\,dt}{\sum_{k=1}^K p_{1k} \int_0^\tau S_{2k}(t)\,dt}.$$

According to the counterfactual definition, confounding is present when $R_1^* \neq R_2$, in which case ρ is said to be confounded. The condition for no confounding, $R_1^* = R_2$, is

$$\frac{\sum_{k=1}^K p_{1k} \int_0^\tau S_{2k}(t)h_{2k}(t)\,dt}{\sum_{k=1}^K p_{1k} \int_0^\tau S_{2k}(t)\,dt} = \frac{\sum_{k=1}^K p_{2k} \int_0^\tau S_{2k}(t)h_{2k}(t)\,dt}{\sum_{k=1}^K p_{2k} \int_0^\tau S_{2k}(t)\,dt}. \tag{G.4}$$

Each of the following conditions is sufficient to ensure that (G.4) is true:

  (i) $h_{2k}(t) = h_2(t)$ for all k
  (ii) $p_{1k} = p_{2k}$ for all k.

When there is no confounding, ρ is an overall measure of effect for the exposed and unexposed cohorts.

G.1.2 Proportional Hazards Assumption

Suppose that $h_{1k}(t) = \psi h_{2k}(t)$ for all k, for some constant ψ. By definition, ψ is the proportional hazards constant. Then $S_{1k}(t) = S_{2k}(t)^\psi$ where, for brevity, we denote $[S_{2k}(t)]^\psi$ by $S_{2k}(t)^\psi$. It follows that

$$\frac{h_1(t)}{h_2(t)} = \psi \frac{\left[\sum_{k=1}^K p_{1k} S_{2k}(t)^\psi h_{2k}(t)\right] \Big/ \left[\sum_{k=1}^K p_{1k} S_{2k}(t)^\psi\right]}{\left[\sum_{k=1}^K p_{2k} S_{2k}(t) h_{2k}(t)\right] \Big/ \left[\sum_{k=1}^K p_{2k} S_{2k}(t)\right]}$$

and

$$\rho = \psi \frac{\left[\sum_{k=1}^K p_{1k} \int_0^\tau S_{2k}(t)^\psi h_{2k}(t)\,dt\right] \Big/ \left[\sum_{k=1}^K p_{1k} \int_0^\tau S_{2k}(t)^\psi\,dt\right]}{\left[\sum_{k=1}^K p_{2k} \int_0^\tau S_{2k}(t) h_{2k}(t)\,dt\right] \Big/ \left[\sum_{k=1}^K p_{2k} \int_0^\tau S_{2k}(t)\,dt\right]}. \tag{G.5}$$

Suppose that condition (ii) is satisfied, in which case F is not a confounder. Assume that, for each i and k, survival is governed by the exponential distribution. With $h_{2k}(t) = \lambda_{2k}$ and $h_{1k}(t) = \psi \lambda_{2k}$, (G.5) becomes

$$\rho = \psi \frac{\left[\sum_{k=1}^K p_{1k}(1 - e^{-\psi \lambda_{2k} \tau})\right] \Big/ \left[\sum_{k=1}^K p_{1k}(1 - e^{-\psi \lambda_{2k} \tau})/\lambda_{2k}\right]}{\left[\sum_{k=1}^K p_{1k}(1 - e^{-\lambda_{2k} \tau})\right] \Big/ \left[\sum_{k=1}^K p_{1k}(1 - e^{-\lambda_{2k} \tau})/\lambda_{2k}\right]}. \tag{G.6}$$

Let

$$\xi_{1k} = \frac{p_{1k}(1 - e^{-\psi \lambda_{2k} \tau})}{\psi \lambda_{2k}}$$

and

$$\xi_{2k} = \frac{p_{1k}(1 - e^{-\lambda_{2k} \tau})}{\lambda_{2k}}.$$

Then (G.6) can be written as

$$\rho = \psi \frac{\left(\sum_{k=1}^{K} \xi_{1k} \lambda_{2k}\right) \big/ \left(\sum_{k=1}^{K} \xi_{1k}\right)}{\left(\sum_{k=1}^{K} \xi_{2k} \lambda_{2k}\right) \big/ \left(\sum_{k=1}^{K} \xi_{2k}\right)}.$$

Assume that $\psi > 1$. Since $\psi \lambda_{2k} > \lambda_{2k}$, it follows that

$$\frac{1 - e^{-\psi \lambda_{2k} \tau}}{\psi \lambda_{2k}} = \int_0^\tau e^{-\psi \lambda_{2k} t} dt < \int_0^\tau e^{-\lambda_{2k} t} dt = \frac{1 - e^{-\lambda_{2k} \tau}}{\lambda_{2k}} < \left(\frac{1 - e^{-\psi \lambda_{2k} \tau}}{\psi \lambda_{2k}}\right) \psi$$

and hence

$$1 < \frac{\xi_{2k}}{\xi_{1k}} < \psi.$$

Now consider the case where the p_{1j} are all equal. In what follows, λ denotes a continuous variable. For given $\psi > 0$ and $\tau > 0$, define

$$\xi_1(\lambda) = \frac{1 - e^{-\psi \lambda}}{\psi \lambda}$$

and

$$\xi_2(\lambda) = \frac{1 - e^{-\lambda}}{\lambda}$$

for λ in $[0, \tau]$. Although $\xi_1(\lambda)$ and $\xi_2(\lambda)$ are not defined at 0, the limiting values exist and are given by $\xi_1(0) = \xi_2(0) = 1$. The probability functions, $f_1(\lambda)$ and $f_2(\lambda)$, and survival functions, $S_1(\lambda)$ and $S_2(\lambda)$, corresponding to $\xi_1(\lambda)$ and $\xi_1(\lambda)$ involve integrals that do not have closed forms. However, numerical results indicate that the relationships observed in Appendix A for $\xi_1(\omega)$ and $\xi_2(\omega)$ have counterparts in terms of $\xi_1(\lambda)$ and $\xi_2(\lambda)$. Figures G.1(a) and G.1(b) show the probability functions and survival functions for $\psi = 5$ and $\tau = 10$. There is an obvious similarity to Figures A.1(a) and A.1(b). However, the interval $[0, 10]$ is not typical of the range of values of λ that might be observed in practice. For example, when the units are deaths per person-year, τ might be as small as .001, or even smaller. Figures G.2(a)

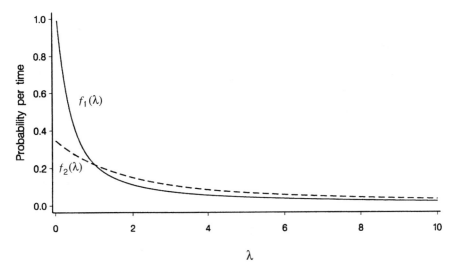

FIGURE G.1(a) Probability functions, with $\psi = 5$ and $\tau = 10$

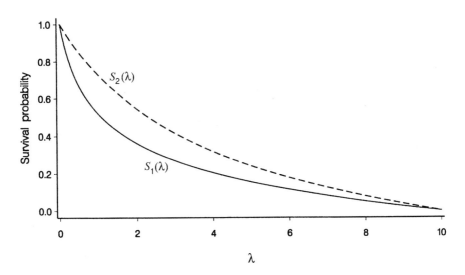

FIGURE G.1(b) Survival functions, with $\psi = 5$ and $\tau = 10$

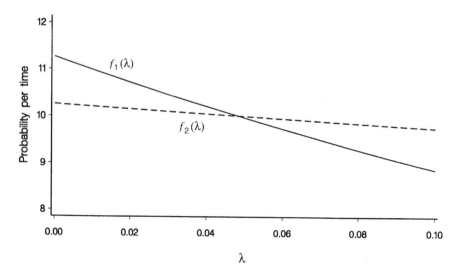

FIGURE G.2(a) Probability functions, with $\psi = 5$ and $\tau = .1$

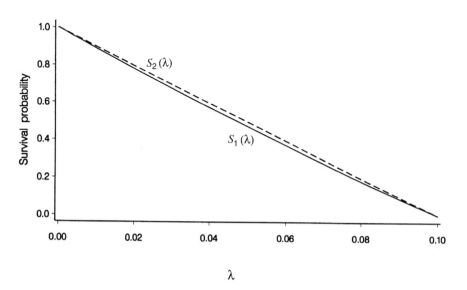

FIGURE G.2(b) Survival functions, with $\psi = 5$ and $\tau = .1$

and G.2(b) show the probability functions and survival functions for $\psi = 5$ and $\tau = .1$. Evidently, when τ is small, the probability functions and survival functions are almost linear.

The preceding observations suggest that inequalities analogous to those derived in Appendix A are likely to hold in the exponential setting—that is, if $\psi > 1$ then $1 < \rho < \psi$, and if $\psi < 1$ then $\psi < \rho < 1$. Inequalities of this type have been demonstrated for the Cox regression model (Gail et al., 1984; Gail, 1986). Recall that the above results are based on the assumption that condition (ii) is true. This means that despite the absence of confounding, $\psi \neq \rho$ unless $\psi = 1$. However, when the λ_{2k} are small, as is usually the case in practice, ρ and ψ will be close in value, as Figures G.2(a) and G.2(b) suggest. We illustrate this with a numerical example. Consider (G.6) with $K = 2$, where we note that $p_{12} = 1 - p_{11}$. Using the original definitions of ξ_{1k} and ξ_{2k}—that is, including the p_{1k}—Table G.1 gives values of ψ/ρ for selected values of p_{11}, λ_{21}, and λ_{22}, with $\psi = 2$. As can be seen, ψ/ρ is very close to 1 unless λ_{21} and λ_{22} are quite large.

Now suppose that condition (i) is satisfied, so that once again F is not a confounder. Then $S_{2k}(t) = S_2(t)$ and $S_{1k}(t) = S_1(t)$ for all k. Therefore

$$R_1 = \frac{\int_0^\tau S_1(t) h_1(t)\, dt}{\int_0^\tau S_1(t)\, dt} = \psi \frac{\int_0^\tau S_2(t)^\psi h_2(t)\, dt}{\int_0^\tau S_2(t)^\psi\, dt}$$

$$R_2 = \frac{\int_0^\tau S_2(t) h_2(t)\, dt}{\int_0^\tau S_2(t)\, dt}$$

and

$$\rho = \psi \frac{\left[\int_0^\tau S_2(t)^\psi h_2(t)\, dt\right] \Big/ \left[\int_0^\tau S_2(t)^\psi\, dt\right]}{\left[\int_0^\tau S_2(t) h_2(t)\, dt\right] \Big/ \left[\int_0^\tau S_2(t)\, dt\right]}. \tag{G.7}$$

TABLE G.1 ψ/ρ for Selected Values of p_{11}, λ_{21}, and λ_{22}, with $\psi = 2$

p_{11}	λ_{21}	λ_{22}	ψ/ρ
.25	.001	.005	1.000
.50	.001	.005	1.001
.75	.001	.005	1.001
.25	.01	.05	1.004
.50	.01	.05	1.007
.75	.01	.05	1.007
.25	.1	.5	1.039
.50	.1	.5	1.063
.75	.1	.5	1.063

As before, despite the absence of confounding, $\psi \neq \rho$ unless $\psi = 1$. In (G.7) it appears that "follow-up time" is behaving like a confounder. The distribution of "follow-up time"—that is, person-time—in the ith cohort is given by

$$P_i(t) = \frac{S_i(t)}{\int_0^\tau S_i(t)\,dt}.$$

Corresponding to (2.19), the observed number of deaths in the exposed cohort can be written as

$$O = \left[\int_0^\tau P_1(t)h_1(t)\,dt\right]\left[\int_0^\tau l_1(t)\,dt\right]$$

$$= \psi \left[\int_0^\tau P_1(t)h_2(t)\,dt\right]\left[\int_0^\tau l_1(t)\,dt\right].$$

Analogous to (2.20), the standardized expected number of deaths is defined to be

$$sE = \left[\int_0^\tau P_1(t)h_2(t)\,dt\right]\left[\int_0^\tau l_1(t)\,dt\right]$$

where standardization is according to the distribution of follow-up time in the exposed cohort. Since $O/sE = \psi$ we see that, even after "adjusting" for follow-up time, the resulting parameter is still not equal to ρ. This is not surprising because, as was pointed out by Greenland (1996b), follow-up time is "affected" by exposure status (unless $\psi = 1$) and thus does not satisfy one of the assumptions underlying the definition of confounding given above.

G.2   CASE-CONTROL STUDIES

Denote the age distribution of the population at (calendar) time t by $n(x, t)$, where x is continuous age. By definition, $n(x, t)$ equals the number of individuals in the population "per unit age" at time t. The amount of person-time experienced by the population in the rectangle $[x_1, x_2] \times [t_1, t_2] = \{(x, t)|x_1 \leq x \leq x_2, t_1 \leq t \leq t_2\}$ is

$$\int_{x_1}^{x_2}\int_{t_1}^{t_2} n(x, t)\,dt\,dx.$$

The hazard function $r(x, t)$ is defined so that

$$\int_{x_1}^{x_2}\int_{t_1}^{t_2} r(x, t)n(x, t)\,dt\,dx$$

equals the number of deaths in the population that take place in the rectangle $[x_1, x_2] \times [t_1, t_2]$, for any choice of rectangle. The hazard rate for the rectangle is defined to be

$$R = \frac{\int_{x_1}^{x_2} \int_{t_1}^{t_2} r(x,t)n(x,t)\,dt\,dx}{\int_{x_1}^{x_2} \int_{t_1}^{t_2} n(x,t)\,dt\,dx}. \tag{G.8}$$

Now consider a cohort study of $l(0)$ individuals followed from birth until the upper limit of the life span τ. Denote the survival function and hazard function by $S(x)$ and $h(x)$, where x is continuous age. Suppose that the cohort consists of exposed ($i = 1$) and unexposed subcohorts ($i = 2$). Denote the crude hazard rates by R_1 and R_2, and let $\rho = R_1/R_2$. We can "create" a stationary population from the cohort by defining $r(x,t) = h(x)$ for all t. With $l(0)$ defined to be the number of births in the stationary population per unit time, it follows that $n(x,t) = l(0)S(x)$ for all t. With $x_1 = 0$ and $x_2 = \tau$, (G.8) simplifies to

$$R = \frac{\int_0^\tau h(x)S(x)dx}{\int_0^\tau S(x)dx}.$$

The correspondence between cohorts and stationary populations also works in the other direction. Starting with a stationary population it is possible to "create" a cohort. The idea is to follow a birth cohort in the stationary population "along the diagonal" as described in Section 13.1.

In Section 11.1.3 it was demonstrated that, for an incidence case-control study that is nested in a stationary population, the odds ratio equals R_1/R_2, which is the ratio of hazard rates for the stationary population. The above correspondence makes it possible to give this result a cohort interpretation. Similarly, the counterfactual definition of confounding for an open cohort study can be translated into the case-control setting.

Odds Ratio Estimate in a Matched Case-Control Study

H.1 ASYMPTOTIC UNCONDITIONAL ESTIMATE OF MATCHED-PAIRS ODDS RATIO

In this section an estimate of the odds ratio for a matched-pairs case-control study is derived using the unconditional maximum likelihood equations (5.2) and (5.3). Due to the matched-pairs design, $m_1 = 1$ in Table 5.1. It follows from (5.3) and (5.4) that

$$1 = \frac{\widehat{OR}_u \hat{\pi}_{2j} r_{1j}}{\widehat{OR}_u \hat{\pi}_{2j} + (1 - \hat{\pi}_{2j})} + \hat{\pi}_{2j} r_{2j}. \tag{H.1}$$

and

$$\hat{a}_{1j} = \frac{\widehat{OR}_u \hat{\pi}_{2j} r_{1j}}{\widehat{OR}_u \hat{\pi}_{2j} + (1 - \hat{\pi}_{2j})}. \tag{H.2}$$

Recall the types of configurations depicted in Table 11.6.

Configurations of Type $(1, 1)$
Since $r_{1j} = 2$ and $r_{2j} = 0$, (H.1) becomes

$$1 = \frac{2\widehat{OR}_u \hat{\pi}_{2j}}{\widehat{OR}_u \hat{\pi}_{2j} + (1 - \hat{\pi}_{2j})}.$$

Solving for $\hat{\pi}_{2j}$ gives $\hat{\pi}_{2j} = 1 / (1 + \widehat{OR}_u)$, and substituting in (H.2) gives $\hat{a}_{1j} = 1$.

Configurations of Types $(1, 0)$ and $(0, 1)$
Since $r_{1j} = r_{2j} = 1$, (H.1) becomes

$$1 = \frac{\widehat{OR}_u \hat{\pi}_{2j}}{\widehat{OR}_u \hat{\pi}_{2j} + (1 - \hat{\pi}_{2j})} + \hat{\pi}_{2j}.$$

Solving for $\hat{\pi}_{2j}$ gives $\hat{\pi}_{2j} = 1 \big/ \left(1 \pm \sqrt{\widehat{OR}_u}\right)$. Only the solution with the positive root lies in [0, 1]. Substituting $\hat{\pi}_{2j} = 1 \big/ \left(1 + \sqrt{\widehat{OR}_u}\right)$ in (H.2) gives $\hat{a}_{1j} = \sqrt{\widehat{OR}_u} \big/ \left(1 + \sqrt{\widehat{OR}_u}\right)$.

Configurations of Type $(0, 0)$
Since $r_{1j} = 0$, it follows from (H.2) that $\hat{a}_{1j} = 0$.

The left-hand side of (5.2) is

$$\left[f_{(1,1)} \times 1\right] + \left[f_{(1,0)} \times 1\right] + \left[f_{(0,1)} \times 0\right] + \left[f_{(0,0)} \times 0\right].$$

With the above values of $\hat{a}_{1j}$, the right-hand side is

$$\left[f_{(1,1)} \times 1\right] + f_{(1,0)} \left[\frac{\sqrt{\widehat{OR}_u}}{1 + \sqrt{\widehat{OR}_u}}\right] + f_{(0,1)} \left[\frac{\sqrt{\widehat{OR}_u}}{1 + \sqrt{\widehat{OR}_u}}\right] + \left[f_{(0,0)} \times 0\right].$$

The unconditional maximum likelihood equation is therefore

$$f_{(1,0)} = \frac{\left[f_{(1,0)} + f_{(0,1)}\right]\sqrt{\widehat{OR}_u}}{1 + \sqrt{\widehat{OR}_u}} \tag{H.3}$$

which has the solution

$$\widehat{OR}_u = \left[\frac{f_{(1,0)}}{f_{(0,1)}}\right]^2 = (\widehat{OR}_c)^2.$$

Observe that (H.3) is the same as (11.5) with $\sqrt{\widehat{OR}_u}$ in place of $\widehat{OR}_c$.

H.2 ASYMPTOTIC CONDITIONAL ANALYSIS OF $(1 : M)$ MATCHED CASE-CONTROL DATA

The notation and arguments below are analogous to those of Section 11.2.1. From (5.21) and (5.22),

$$E_{(1,m)} = \frac{(m+1)OR}{(m+1)OR + M - m} \qquad V_{(1,m)} = \frac{OR(m+1)(M-m)}{[(m+1)OR + M - m]^2}$$

$$E_{(0,m)} = \frac{mOR}{mOR + M + 1 - m} \qquad V_{(0,m)} = \frac{ORm(M+1-m)}{(mOR + M + 1 - m)^2} \tag{H.4}$$

$(m = 0, 1, 2, \ldots, M)$.

Point Estimate

The left-hand side of (5.23) is

$$a_{1\bullet} = \sum_{m=0}^{M} \left[f_{(1,m)} \times 1 \right] + \sum_{m=0}^{M} \left[f_{(0,m)} \times 0 \right]$$

$$= \sum_{m=1}^{M} f_{(1,m-1)} + f_{(1,M)}.$$

From

$$E_{(1,m)} = \begin{cases} E_{(0,m+1)} & \text{if } m < M \\ 1 & \text{if } m = M \end{cases}$$

and $E_{(0,0)} = 0$ it follows that

$$\sum_{m=0}^{M} f_{(1,m)} \hat{E}_{(1,m)} = \sum_{m=0}^{M-1} f_{(1,m)} \hat{E}_{(1,m)} + f_{(1,M)}$$

$$= \sum_{m=0}^{M-1} f_{(1,m)} \hat{E}_{(0,m+1)} + f_{(1,M)}$$

$$= \sum_{m=1}^{M} f_{(1,m-1)} \hat{E}_{(0,m)} + f_{(1,M)}$$

and

$$\sum_{m=0}^{M} f_{(0,m)} \hat{E}_{(0,m)} = \sum_{m=1}^{M} f_{(0,m)} \hat{E}_{(0,m)}.$$

The right-hand side of (5.23) is

$$\sum_{m=0}^{M} f_{(1,m)} \hat{E}_{(1,m)} + \sum_{m=0}^{M} f_{(0,m)} \hat{E}_{(0,m)} = \sum_{m=1}^{M} \left[f_{(1,m-1)} + f_{(0,m)} \right] \hat{E}_{(0,m)} + f_{(1,M)}$$

$$= \widehat{OR}_{c} \sum_{m=1}^{M} \frac{\left[f_{(1,m-1)} + f_{(0,m)} \right] m}{m \widehat{OR}_{c} + M + 1 - m} + f_{(1,M)}.$$

$$(\text{H.5})$$

The conditional maximum likelihood equation (5.23) is

$$\sum_{m=1}^{M} f_{(1,m-1)} = \widehat{OR}_{c} \sum_{m=1}^{M} \frac{\left[f_{(1,m-1)} + f_{(0,m)} \right] m}{m \widehat{OR}_{c} + M + 1 - m}$$

(Miettinen, 1970).

Variance Estimate
From

$$V_{(1,m)} = \begin{cases} V_{(0,m+1)} & \text{if } m < M \\ 0 & \text{if } m = M \end{cases}$$

it follows that

$$\sum_{m=0}^{M} \left[f_{(1,m)} \hat{V}_{(1,m)} \right] = \sum_{m=0}^{M-1} \left[f_{(1,m)} \hat{V}_{(0,m+1)} \right] = \sum_{m=1}^{M} \left[f_{(1,m-1)} \hat{V}_{(0,m)} \right]. \qquad \text{(H.6)}$$

From (H.6) and $V_{(0,0)} = 0$, we have from (5.25) that

$$\begin{aligned} \hat{V}_c &= \sum_{m=0}^{M} \left[f_{(1,m)} \hat{V}_{(1,m)} \right] + \sum_{m=0}^{M} \left[f_{(0,m)} \hat{V}_{(0,m)} \right] \\ &= \sum_{m=1}^{M} \left[f_{(1,m-1)} + f_{(0,m)} \right] \hat{V}_{(0,m)} \\ &= \widehat{OR}_c \sum_{m=1}^{M} \frac{\left[f_{(1,m-1)} + f_{(0,m)} \right] m(M + 1 - m)}{(m\widehat{OR}_c + M + 1 - m)^2}. \end{aligned} \qquad \text{(H.7)}$$

From (5.26), an estimate of $\text{var}(\log \widehat{OR}_c)$ is

$$\widehat{\text{var}}(\log \widehat{OR}_c) = \left[\widehat{OR}_c \sum_{m=1}^{M} \frac{\left[f_{(1,m-1)} + f_{(0,m)} \right] m(M + 1 - m)}{(m\widehat{OR}_c + M + 1 - m)^2} \right]^{-1}$$

(Miettinen, 1970).

Mantel–Haenszel Test of Association
Under $H_0 : OR = 1$, from (H.4) the expected counts and variance estimates are

$$\hat{e}_{(1,m)} = \frac{m+1}{M+1} \qquad \hat{v}_{0(1,m)} = \frac{(m+1)(M-m)}{(M+1)^2}$$

$$\hat{e}_{(0,m)} = \frac{m}{M+1} \qquad \hat{v}_{0(0,m)} = \frac{m(M+1-m)}{(M+1)^2}.$$

With $OR = 1$, it follows from the first equation of (H.5) and the second to last equation of (H.7) that

$$\hat{e}_{1\bullet} = \sum_{m=1}^{M} \frac{\left[f_{(1,m-1)} + f_{(0,m)} \right] m}{M+1} + f_{(1,M)}$$

and

$$\hat{v}_{0\bullet} = \sum_{m=1}^{M} \frac{\left[f_{(1,m-1)} + f_{(0,m)}\right] m(M + 1 - m)}{(M + 1)^2}.$$

The Mantel–Haenszel test of association (5.29) is

$$X_{\text{mh}}^2 = \frac{(a_{1\bullet} - \hat{e}_{1\bullet})^2}{\hat{v}_{0\bullet}}$$

$$= \left(\sum_{m=1}^{M} f_{(1,m-1)} - \sum_{m=1}^{M} \frac{\left[f_{(1,m-1)} + f_{(0,m)}\right] m}{M + 1}\right)^2 \Bigg/$$

$$\sum_{m=1}^{M} \frac{\left[f_{(1,m-1)} + f_{(0,m)}\right] m(M + 1 - m)}{(M + 1)^2}.$$

References

Andersen, E. B. (1970). Asymptotic properties of conditional maximum likelihood estimators. *Journal of the Royal Statistical Society, Series B* **32**, 283–301.

Andersen, E. B. (1973). *Conditional Inference and Models for Measuring*. Copenhagen: Mentalhygienisk Forlag.

Anderson, J. A. (1972). Separate sample logistic regression. *Biometrika* **59**, 19–35.

Anderson, S., Auquier, A., Hauck, W. W., Oakes, D., Vandaele, W., and Weisberg, H. I. (1980). *Statistical Methods for Comparative Studies: Techniques for Bias Reduction*. New York: John Wiley & Sons.

Anscombe, F. J. (1956). On estimating binomial response relations. *Biometrika* **43**, 461–464.

Antunes, C. M. F., Stolley, P. D., Rosenshein, N. B., Davies, J. L., Tonascia, J. A., Brown, C., Burnett, L., Rutledge, A., Pokempner, M., and Garcia, R. (1979). Endometrial cancer and estrogen use: Report of a large case-control study. *New England Journal of Medicine* **300**, 9–13.

Armitage, P. (1955). Tests for linear trends in proportions and frequencies. *Biometrics* **11**, 375–386.

Armitage, P. (1966). The chi-square test for heterogeneity of proportions, after adjustment for stratification. *Journal of the Royal Statistical Society, Series B* **28**, 150–163.

Armitage, P. (1975). The use of the cross–ratio in aetiologic surveys. In *Perspectives in Probability and Statistics*, J. Gani (ed.), pp. 349–355. London: Academic Press.

Armitage, P., and Berry, G. (1994). *Statistical Methods in Medical Research*, 2nd edition. London: Blackwell.

Austin, H., Hill, H. A., Flanders, W. D., and Greenberg, R. S. (1994). Limitations in the application of case-control methodology. *Epidemiologic Reviews* **16**, 65–76.

Beaumont, J. J., and Breslow, N. E. (1981). Power considerations in epidemiologic studies of vinyl chloride workers. *American Journal of Epidemiology* **114**, 725–734.

Bernstein, L., Anderson, J., and Pike, M. C. (1981). Estimation of the proportional hazard in two-treatment-group clinical trials. *Biometrics* **37**, 513–519.

Berry, G. (1980). Dose–response in case-control studies. *Journal of Epidemiology and Community Health* **34**, 217–222.

Berry, G. (1983). The analysis of mortality by the subject-years method. *Biometrics* **39**, 173–184.

Birch, M. W. (1964). The detection of partial association I: The 2 × 2 case. *Journal of the Royal Statistical Society, Series B* **26**, 313–324.

Birch, M. W. (1965). The detection of partial association II: The general case. *Journal of the Royal Statistical Society, Series B* **27**, 417–451.

Birkett, N. J. (1992). Effect of nondifferential classification on estimates of odds ratios with multiple levels of exposure. *American Journal of Epidemiology* **136**, 356–362.

Bishop, Y. M. M., Fienberg, S. E., and Holland, P. W. (1975). *Discrete Multivariate Analysis: Theory and Practice*. Cambridge, MA: MIT Press.

Boivin, J.-F., and Wacholder, S. (1985). Conditions for confounding of the risk ratio and of the odds ratio. *American Journal of Epidemiology* **121**, 152–158.

Brenner, H. (1993). Bias due to non-differential misclassification of polytomous confounders. *Journal of Clinical Epidemiology* **46**, 57–63.

Breslow, N. (1975). Analysis of survival data under the proportional hazards model. *International Statistical Review* **43**, 45–48.

Breslow, N. E. (1976). Regression analysis of the log odds ratio: A method for retrospective studies. *Biometrics* **32**, 409–416.

Breslow, N. (1979). Statistical methods for censored survival data. *Environmental Health Perspectives* **32**, 181–192.

Breslow, N. (1981). Odds ratio estimators when the data are sparse. *Biometrika* **68**, 73–84.

Breslow, N. (1984a). Elementary methods of cohort analysis. *International Journal of Epidemiology* **13**, 112–115.

Breslow, N. (1984b). Comparison of survival curves. In *Cancer and Clinical Trials: Methods and Practice*, M. E. Buyse, M. J. Staquet, and R. J. Sylvester (eds.), pp. 381–406. Oxford: Oxford University Press.

Breslow, N. E. (1996). Statistics in epidemiology: The case-control study. *Journal of the American Statistical Association* **91**, 14–28.

Breslow, N., and Day, N. E. (1980). *Statistical Methods in Cancer Research, Volume I: The Analysis of Case-Control Studies*. Lyon: IARC.

Breslow, N. E., and Day, N. E. (1985). The standardized mortality ratio. In *Biostatistics: Statistics in Biomedical, Public Health and Environmental Science*, P. K. Sen (ed.), pp. 55–74. New York: Elsevier.

Breslow, N., and Day, N. E. (1987). *Statistical Methods in Cancer Research, Volume II: The Design and Analysis of Cohort Studies*. Lyon: IARC.

Breslow, N., and Powers, W. (1978). Are there two logistic regressions for retrospective studies? *Biometrics* **34**, 100–105.

Breslow, N., and Storer B. E. (1985). General relative risk functions for case-control studies. *American Journal of Epidemiology* **122**, 149–162.

Bristol, D. R. (1989). Sample sizes for constructing confidence interval and testing hypotheses. *Statistics in Medicine* **8**, 803–811.

Brookmeyer, R., Liang, K.-Y., and Linet, M. (1986). Matched case-control designs and over-matched analyses. *American Journal of Epidemiology* **124**, 693–701.

Brown, C. C. (1981). The validity of approximation methods for interval estimation of the odds ratio. *American Journal of Epidemiology* **113**, 474–480.

Casella, G., and Berger, R. L. (1990). *Statistical Inference.* Belmont, CA: Duxbury.

Chiang, C. L. (1968). *Introduction to Stochastic Processes in Biostatistics.* New York: John Wiley & Sons.

Chiang, C. L. (1980). *Introduction to Stochastic Process and Their Applications.* New York: Krieger.

Chiang, C. L. (1984). *The Life Table and Its Applications.* Malabar, FL: Kreiger.

Clayton, D. G. (1982). The analysis of prospective studies of disease aetiology. *Communications in Statistics—Theory and Methods* **11**, 2129–2155.

Clayton, D., and Hills, M. (1993). *Statistical Models in Epidemiology.* Oxford: Oxford University Press.

Clayton, D., and Schifflers, E. (1987). Models for temporal variations in cancer rates, II: Age–period–cohort models. *Statistics in Medicine* **6**, 469–481.

Cochran, W. G. (1954). Some methods for strengthening the common chi-square tests. *Biometrics* **10**, 417–451.

Cochran, W. G. (1977). *Sampling Techniques*, 3rd edition. New York: John Wiley & Sons.

Cole, P., and MacMahon, B. (1971). Attributable risk percent in case-control studies. *British Journal of Preventive and Social Medicine* **25**, 242–244.

Collett, D. (1991). *Modelling Binary Data.* London: Chapman and Hall.

Collett, D. (1994). *Modelling Survival Data in Medical Research.* London: Chapman and Hall.

Conover, W. J. (1974). Some reasons for not using the Yates continuity correction on 2×2 contingency tables (with discussion). *Journal of the American Statistical Association* **69**, 374–382.

Cornfield, J. (1956). A statistical problem arising from retrospective studies. In *Proceedings of the Third Berkeley Symposium on Mathematical Statistics and Probability, Volume IV*, J. Neyman (ed.), pp. 135–148. Berkeley, CA: University of California Press.

Cornfield, J., Haenszel, W., Hammond, E., Lilienfeld, A., Shimkin, M., and Wynder, E. (1959). Smoking and lung cancer: Recent evidence and a discussion of some questions. *Journal of the National Cancer Institute* **22**, 173–203.

Copeland, K. T., Checkoway, H., McMichael, A. J., and Holbrook, R. H. (1977). Bias due to misclassification in the estimation of relative risk. *American Journal of Epidemiology* **105**, 488–495.

Cox, D. R. (1972). Regression models and life tables (with discussion). *Journal of the Royal Statistical Society, Series B* **74**, 187–220.

Cox, D. R., and Hinkley, D. V. (1974). *Theoretical Statistics*. London: Chapman and Hall.

Cox, D. R., and Oakes, D. (1984). *Analysis of Survival Data*. London: Chapman and Hall.

Cox, D. R., and Snell, E. J. (1989). *Analysis of Binary Data*, 2nd edition. London: Chapman and Hall.

Crowley, J., and Breslow, N. (1975). Remarks on the conservatism of $\sum(O - E)^2/E$ in survival analysis. *Biometrics* **31**, 957–961.

Crowley, J., Liu, P. Y., and Voelkel, J. G. (1982). Estimation of ratio of hazard functions. In *Survival Analysis. Lecture Notes–Monograph Series*, J. Crowley and R. A. Johnson (eds.), pp. 56–73. Hayward, CA: Institute of Mathematical Statistics.

CSHA Working Group. (1994). Canadian Study of Health and Aging: Study methods and prevalence of dementia. *Canadian Medical Association Journal* **150**, 899–913.

Davis, L. J. (1985). Weighted averages of the observed odds ratios when the number of tables is large. *Biometrika* **72**, 203–205.

Day, N. E., and Byar, D. P. (1979). Testing hypotheses in case-control studies—Equivalence of Mantel–Haenszel statistics and logit score tests. *Biometrics* **35**, 623–630.

Diggle, P. J., Liang, K.-Y., and Zeger, S. L. (1994). *Analysis of Longitudinal Data*. Oxford: Oxford University Press.

Dobson, A. J., Kuulasmaa, K., Eberle, E., and Scherer, J. (1991). Confidence intervals from weighted sums of Poisson parameters. *Statistics in Medicine* **10**, 457–462.

Donner, A. (1984). Approaches to sample size estimation in the design of clinical trials—a review. *Statistics in Medicine* **3**, 199–214.

Donner, A., and Hauck, W. W. (1986). The large–sample relative efficiency of the Mantel–Haenszel estimator in the fixed–strata case. *Biometrics* **42**, 537–545.

Dosemeci, M., Wacholder, S., and Lubin, J. H. (1990). Does nondifferential misclassification of exposure always bias a true effect toward the null value? *American Journal of Epidemiology* **132**, 746–748.

Ducharme, G. R., and LePage, Y. (1986). Testing for collapsibility in contingency tables. *Journal of the Royal Statistical Society, Series B* **48**, 197–205.

Dupont, W. D. (1988). Power calculations for matched case-control studies. *Biometrics* **44**, 1157–1168.

Edwards, A. (1972). *Likelihood*. Cambridge: Cambridge University Press.

EGRET. (1999). *A Software Package for the Analysis of Biomedical and Epidemiological Studies*. Cambridge, MA: Cytel Software Corporation.

EGRET SIZ (1997). *Sample Size and Power for Nonlinear Regression Models*. Cambridge, MA: Cytel Software Corporation.

Elandt-Johnson, R. C., and Johnson, N. L. (1980). *Survival Models and Data Analysis*. New York: John Wiley & Sons.

Fienberg, S. E. (1981). *The Analysis of Cross-Classified Categorical Data*, 2nd edition. Cambridge, MA: MIT Press.

Fisher, L., and Patil, K. (1974). Matching and unrelatedness. *American Journal of Epidemiology* **100**, 347–349.

Fisher, R. A. (1925). *Statistical Methods for Research Workers*. Edinburgh: Oliver and Boyd.

Fisher, R. A. (1936). *The Design of Experiments*, 6th edition. Edinburgh: Oliver and Boyd.

Flanders, W. D. (1984). Approximate variance formulas for standardized rate ratios. *Journal of Chronic Diseases* **37**, 449–453.

Fleiss, J. L. (1979). Confidence intervals for the odds ratio in case-control studies: The state of the art. *Journal of Chronic Diseases* **32**, 69–77.

Fleiss, J. L. (1981). *Statistical Methods for Rates and Proportions*, 2nd edition. New York: John Wiley & Sons.

Fleiss, J. L., and Levin, B. (1988). Sample size determination in studies with matched pairs. *Journal of Clinical Epidemiology* **41**, 727–730.

Fleming, T. R., O'Fallon, J. R., O'Brien, P. C., and Harrington, D. P. (1980). Modified Kolmogorov–Smirnov test procedures with application to arbitrarily right-censored data. *Biometrics* **36**, 607–625.

Freedman, D. (1999). From association to causation: Some remarks on the history of statistics. *Statistical Science* **14**, 243–258.

Frome, E. L. (1983). The analysis of rates using Poisson regression models. *Biometrics* **39**, 665–674.

Frome, E. L., and Checkoway, H. (1985). Use of Poisson regression models in estimating incidence rates and ratios. *American Journal of Epidemiology* **121**, 309–323.

Gail, M. (1973). The determination of sample sizes for trials involving several independent 2×2 tables. *Journal of Chronic Diseases* **26**, 669–673.

Gail, M. (1975). A review and critique of some models used in competing risk analysis. *Biometrics* **31**, 209–222.

Gail, M. H. (1986). Adjusting for covariates that have the same distribution in exposed and unexposed cohorts. In *Modern Statistical Methods in Chronic Disease Epidemiology*, S. H. Moolgavkar and R. L. Prentice (eds.), pp. 3–18. New York: John Wiley & Sons.

Gail, M., and Simon, R. (1985). Testing for qualitative interactions between treatment effects and patient subsets. *Biometrics* **41**, 361–372.

Gail, M. H., Wieand, S., and Piantadosi, S. (1984). Biased estimates of treatment effect in randomized experiments with nonlinear regression and omitted covariates. *Biometrika* **71**, 431–444.

Gardner, M. J., and Altman, D. G. (1986). Confidence intervals rather than P values: Estimation rather than hypothesis testing. *British Medical Journal* **292**, 746–750.

Gardner, M. J., and Munford, A. G. (1980). The combined effect of two factors on disease in case-control study. *Applied Statistics* **29**, 276–281.

Gart, J. J. (1962). On the combination of relative risks. *Biometrics* **18**, 601–610.

Gart, J. J. (1970). Point and interval estimation of the common odds ratio in the combination of 2×2 tables with fixed marginals. *Biometrics* **57**, 471–475.

Gart, J. J. (1971). The comparison of proportions: A review of significance tests, confidence intervals and adjustments for stratification. *International Statistical Review* **39**, 148–169.

Gart, J. J. (1972). Addenda and errata to Gart, J. J. (1971). *International Statistical Review* **40**, 221–222.

Gart, J J. (1985). Approximate tests and interval estimation of the common relative risk in the combination of 2 × 2 tables. *Biometrika* **72**, 673–677.

Gart, J. J., and Nam, J. (1988). Approximate interval estimation of the ratio of binomial parameters: A review and correction for skewness. *Biometrics* **44**, 323–338.

Gart, J. J., and Nam, J. (1990). Approximate interval estimation of the difference in binomial parameters: Correction for skewness and extension to multiple tables. *Biometrics* **46**, 637–643.

Gart, J. J., and Tarone, R. E. (1983). The relation between score tests and asymptotic UMPU tests in exponential models common in biometry. *Biometrics* **39**, 781–786.

Gart, J. J., and Thomas, D. G. (1972). Numerical results on approximate confidence limits for the odds ratio. *Journal of the Royal Statistical Society, Series B* **34**, 441–447.

Gart, J. J., and Thomas, D. G. (1982). The performance of three approximate confidence limit methods for the odds ratio. *American Journal of Epidemiology* **115**, 453–470.

Gart, J. J., and Zweifel, J. R. (1967). On the bias of various estimators of the logit and its variance, with application to quantal bioassay. *Biometrika* **54**, 181–187.

Gastwirth, J. L., Krieger, A. M., and Rosenbaum, P. R. (2000). Cornfield's inequality. In *Encyclopedia of Epidemiologic Methods*, M. H. Gail and J. Benichou (eds.), pp. 262–265. Chichester, England: John Wiley & Sons.

George, S. L., and Desu, M. M. (1974). Planning the size and duration of a clinical trial studying the time to some critical event. *Journal of Chronic Diseases* **27**, 15–24.

Glass, R. I., Svennerholm, A. M., Stoll, B. J, Khan, M. R., Hossain, K. M. B., Huq, M. I., and Holmgrem, J. (1983). Protection against cholera in breast-fed children by antibiotics in breast milk. *New England Journal of Medicine* **308**, 1389–1392.

Goodman, S. N. (1993). *p* values, hypothesis tests, and likelihood: Implications for epidemiology of a neglected historical debate. *American Journal of Epidemiology* **137**, 485–496.

Goodman, S. N., and Berlin, J. A. (1994). The use of predicted confidence intervals when planning experiments and the misuse of power when interpreting results. *Annals of Internal Medicine* **121**, 200–206.

Goodman, S. N., and Royall, R. (1988). Evidence and scientific research. *American Journal of Public Health* **78**, 1568–1574.

Grayson, D. A. (1987). Confounding confounding. *American Journal of Epidemiology* **126**, 546–553.

Greenland, S. (1980). The effect of misclassification in the presence of covariates. *American Journal of Epidemiology* **112**, 564–569.

Greenland, S. (1982). Interpretation and estimation of summary ratios under heterogeneity. *Statistics in Medicine* **1**, 217–227.

Greenland, S. (1983). Tests for interaction in epidemiologic studies: A review and a study of power. *Statistics in Medicine* **2**, 243–251.

Greenland, S. (1987). Interpretation and choice of effect measures in epidemiologic analyses. *American Journal of Epidemiology* **125**, 761–768.

Greenland, S. (1988). On sample-size and power calculations for studies using confidence intervals. *American Journal of Epidemiology* **128**, 231–237.

Greenland, S. (1990). Randomization, statistics, and causal inference. *Epidemiology* **1**, 421–429.

Greenland, S. (1991). On the logical justification of conditional tests for two-by-two contingency tables. *The American Statistician* **45**, 248–251.

Greenland, S. (1996a). Basic methods of sensitivity analysis of biases. *International Journal of Epidemiology* **25**, 1107–1116.

Greenland, S. (1996b). Absence of confounding does not correspond to collapsibility of the rate ratio or rate difference. *Epidemiology* **7**, 498–501.

Greenland, S., and Mickey, R. M. (1988). Closed form and dually consistent methods for inference on strict collapsibility in $2 \times 2 \times K$ and $2 \times J \times K$ tables. *Applied Statistics* **37**, 335–343.

Greenland, S., Morgenstern, H., Poole, C., and Robins, J. M. (1989). RE: Confounding confounding (letter). *American Journal of Epidemiology* **129**, 1086–1089.

Greenland, S., and Neutra, R. (1980). Control of confounding in the assessment of medical technology. *International Journal of Epidemiology* **9**, 361–367.

Greenland, S., Pearl, J., and Robins, J. M. (1999). Causal diagrams for epidemiologic research. *Epidemiology* **10**, 37–48.

Greenland, S., and Robins, J. M. (1985a). Confounding and misclassification. *American Journal of Epidemiology* **122**, 495–506.

Greenland, S., and Robins, J. M. (1985b). Estimation of a common effect parameter from sparse follow-up data. *Biometrics* **41**, 55–68.

Greenland, S., and Robins, J. M. (1986). Identifiability, exchangeability, and epidemiologic confounding. *International Journal of Epidemiology* **15**, 412–418.

Greenland, S., and Robins, J. M. (1988). Conceptual problems in the definition and interpretation of attributable fractions. *American Journal of Epidemiology* **128**, 1185–1196.

Greenland, S., Robins, J. M., and Pearl, J. (1999). Confounding and collapsibility in causal inference. *Statistical Science* **14**, 29–46.

Greenland, S., and Thomas, D. C. (1982). On the need for the rare disease assumption in case-control studies. *American Journal of Epidemiology* **116**, 547–553.

Greenwood, M. (1926). The natural duration of cancer. *Reports on Public Health and Medical Subjects* **33**, 1–26. London: Her Majesty's Stationery Office.

Greville, T. N. E. (1948). Mortality tables analyzed by cause of death. *Record of the American Institute of Actuaries* **37**, 283–294.

Grimmett, G. R., and Stirzaker, D. R. (1982). *Probability and Random Processes*. Oxford: Clarendon Press.

Grizzle, J. E. (1967). Continuity correction in the χ^2-test for 2×2 tables. *The American Statistician* **21**, 28–32.

Grizzle, J. E., Starmer, C. F., and Koch, G. C. (1969). Analysis of categorical data by linear models. *Biometrics* **25**, 489–504.

Guerrero, V. M., and Johnson, R. A. (1982). Use of the Box–Cox transformation with binary response models. *Biometrika* **69**, 309–314.

Hadlock, C. R. (1978). *Field Theory and Its Classical Problems*. The Mathematical Association of America.

Haldane, J. B. S. (1955). The estimation and significance of the logarithm of a ratio of frequencies. *Annals of Human Genetics* **20**, 309–311.

Halperin, M. (1977). RE: "Estimability and estimation in case-referent studies." *American Journal of Epidemiology* **105**, 496–498.

Hauck, W. W. (1979). The large sample variance of the Mantel–Haenszel estimator of a common odds ratio. *Biometrics* **35**, 817–829.

Hauck, W. W. (1984). A comparative study of the conditional maximum likelihood estimation of a common odds ratio. *Biometrics* **40**, 1117–1123.

Hauck, W. W. (1987). Estimation of a common odds ratio. In *Biostatistics. Advances in the Statistical Sciences, Volume V*, I. B. MacNeil and G. J. Umphrey (eds.), pp. 125–149. Boston: D. Reidel.

Hauck, W. W. (1989). Odds ratio inference from stratified samples. *Communications in Statistics—Theory and Methods* **18**, 767–800.

Hauck, W. W., Anderson, S., and Leahy, III, F. J. (1982). Finite-sample properties of some old and some new estimators of a common odds ratio from multiple 2×2 tables. *Journal of the American Statistical Association* **77**, 145–152.

Hauck, W. W., and Donner, A. (1988). The asymptotic relative efficiency of the Mantel–Haenszel estimator in the increasing-number-of-strata case. *Biometrics* **44**, 379–384.

Hirji, K. F., Mehta, C. R., and Patel, N. R. (1987). Computing distributions for exact logistic regression. *Journal of the American Statistical Association* **82**, 1110–1117.

Hogg, R. V., and Craig, A. T. (1994). *Introduction to Mathematical Statistics*, 5th edition. New York: Simon and Schuster.

Holford, T. R. (1980). The analysis of rates and of survivorship using log-linear models. *Biometrics* **36**, 299–305.

Holford, T. R. (1991). Understanding the effects of age, period, and cohort on incidence and mortality rates. *Annual Review of Public Health* **12**, 425–457.

Holford, T. R. (1998). Age–period–cohort analysis. In *Encyclopedia of Biostatistics*, P. Armitage and T. Colton (eds.), pp. 82–99. Chichester, England: John Wiley & Sons.

Holland, P. W. (1986). Statistics and causal inference (with discussion). *Journal of the American Statistical Association* **81**, 945–970.

Holland, P. W. (1989). Reader reactions: Confounding in epidemiologic studies. *Biometrics* **45**, 1310–1316.

Holland, P. W., and Rubin, D. B. (1988). Causal inference in retrospective studies. *Evaluation Review* **12**, 203–231.

Hosmer, D. W., and Lemeshow, S. (1989). *Applied Logistic Regression*. New York: John Wiley & Sons.

Hosmer, D. W., and Lemeshow, S. (1999). *Applied Survival Analysis*. New York: John Wiley & Sons.

Joffe, M. M., and Rosenbaum, P. R. (1999). Invited commentary: Propensity scores. *American Journal of Epidemiology* **150**, 327–333.

Jones, M. P., O'Gorman, T. W., Lemke, J. H., and Woolson, R. F. (1989). A Monte Carlo investigation of homogeneity tests of the odds ratio under various sample size configurations. *Biometrics* **45**, 171–181.

Jovanovic, B. D. (1998). Binomial confidence intervals when no events are observed. In *Encyclopedia of Biostatistics*, P. Armitage and T. Colton (eds.), pp. 358–359. Chichester, England: John Wiley & Sons.

Kalbfleisch, J. D., and Prentice, R. L. (1980). *The Statistical Analysis of Failure Time Data*. New York: John Wiley & Sons.

Kalish, L. A. (1990). Reducing mean squared error in the analysis of pair-matched case-control studies. *Biometrics* **46**, 493–499.

Kaplan, E. L., and Meier, P. (1958). Nonparametric estimation from incomplete observations. *Journal of the American Statistical Association* **53**, 457–481.

Keiding, N. (1991). Age-specific incidence and prevalence: A statistical perspective (with discussion). *Journal of the Royal Statistical Society, Series A* **154**, 371–412.

Keiding, N. (1999). Event history analysis and inference from observational epidemiology. *Statistics in Medicine* **18**, 2353–2363.

Keyfitz, N. (1977). *Introduction to Mathematical Demography, With Revisions*. Reading, MA: Addison–Wesley.

Kish, L. (1965). *Survey Sampling*. New York: John Wiley & Sons.

Klein, J. P., and Moeschberger, M. L. (1997). *Survival Analysis: Techniques for Censored and Truncated Data*. New York: Springer.

Kleinbaum, D. G. (1994). *Logistic Regression: A Self-Learning Text*. New York: Springer.

Kleinbaum, D. G. (1996). *Survival Analysis: A Self-Learning Text*. New York: Springer.

Kleinbaum, D. G., Kupper, L. L., and Morgenstern, H. (1982). *Epidemiologic Research: Principles and Quantitative Methods*. Belmont, CA: Lifetime Learning Publications.

Kraus, A. S. (1960). Comparison of a group with a disease and a control group from the same families, in the search for possible etiologic factors. *American Journal of Public Health* **50**, 303–311.

Kupper, L. L., Karon, J. M., Kleinbaum, D. G., Morgenstern, H., and Lewis, D. K. (1981). Matching in epidemiologic studies: Validity and efficiency considerations. *Biometrics* **37**, 271–291.

Lachin (1981). Introduction to sample size determination and power analysis for clinical trials. *Controlled Clinical Trials* **2**, 93–113.

Lachin, J. M. (1992). Power and sample size evaluations for the McNemar test with applications to matched case-control studies. *Statistics in Medicine* **11**, 1239–1251.

Lachin, J. M. (2000). *Biostatistical Methods: The Assessment of Relative Risks*. New York: John Wiley & Sons.

Langholz, B., and Goldstein, L. (1996). Risk set sampling in epidemiologic cohort studies. *Statistical Science* **11**, 35–53.

Lawless, J. F. (1982). *Statistical Models and Methods for Lifetime Data*. New York: John Wiley & Sons.

Lee, E. T. (1992). *Statistical Methods for Survival Analysis*, 2nd edition. New York: John Wiley & Sons.

Liang, K.-Y. (1987). A locally most powerful test for homogeneity with many strata. *Biometrika* **74**, 259–264.

Liang, K.-Y., and Self, S. G. (1985). Tests for homogeneity of odds ratios when the data are sparse. *Biometrika* **72**, 353–358.

Liang, K.Y., and Zeger, S. (1988). On the use of concordant pairs in matched case-control studies. *Biometrics* **44**, 1145–1156.

Lindsey, J. K. (1993). *Models for Repeated Measurements*. Oxford: Oxford University Press.

Lininger, L., Gail, M H., Green, S. B., and Byar, D. P. (1979). Comparison of four tests for equality of survival curves in the presence of stratification and censoring. *Biometrika* **66**, 419–428.

Little, R. J. A. (1989). Testing the equality of two independent binomial proportions. *The American Statistician* **43**, 283–288.

Liu, G. (2000). Sample size for epidemiologic studies. In *Encyclopedia of Epidemiologic Methods*, M. H. Gail and J. Benichou (eds.), pp. 777–794. Chichester, England: John Wiley & Sons.

LogXact. (1999). *A Software Package for Exact Logistic Regression, Version 4*. Cambridge, MA: Cytel Software Corporation.

Louis, T. A. (1981). Confidence intervals for a binomial parameter after observing no successes. *The American Statistician* **35**, 154.

Lubin, J. H. (1981). An empirical evaluation of the use of conditional and unconditional likelihoods for case-control data. *Biometrika* **68**, 567–571.

Lubin, J. H., and Gail, M. H. (1990). On power and sample size for studying features of the relative odds of disease. *American Journal of Epidemiology* **131**, 552–566.

Mack, T. M., Pike, M. C., Henderson, B. E., Pfeffer, R. I., Gerkins, V. R., Arthur, M., and Brown, S. E. (1976). Estrogens and endometrial cancer in a retirement community. *New England Journal of Medicine* **294**, 1262–1267.

Maclure, M., and Greenland, S. (1992). Tests for trend and dose response: Misinterpretations and alternatives. *American Journal of Epidemiology* **135**, 96–104.

Maldonado, G., and Greenland, S. (1993). Simulation study of confounder-selection strategies. *American Journal of Epidemiology* **138**, 923–936.

Mantel, N. (1963). Chi-square tests with one degree of freedom: Extensions of the Mantel–Haenszel procedure. *Journal of the American Statistical Association* **58**, 690–700.

Mantel, N. (1966). Evaluation of survival data and two new rank order statistics arising from its consideration. *Cancer Chemotherapy Reports* **50**, 163–170.

Mantel, N. (1973). Synthetic retrospective studies and related topics. *Biometrics* **29**, 479–486.

Mantel, N. (1977). Tests and limits for the common odds ratio of several 2×2 contingency tables: Methods in analogy with the Mantel–Haenszel procedure. *Journal of Statistical Planning and Inference* **1**, 179–189.

Mantel, N. (1987). Understanding Wald's test for exponential families. *American Statistician* **41**, 147–148.

Mantel, N., and Fleiss, J. L. (1980). Minimum expected cell size requirements for the Mantel–Haenszel one-degree-of-freedom test and a related rapid procedure. *American Journal of Epidemiology* **112**, 129–134.

Mantel, N., and Greenhouse, S. W. (1968). What is the continuity correction? *The American Statistician* **22**, 27–30.

Mantel, N., and Haenszel, W. (1959). Statistical aspects of the analysis of data from retrospective studies of disease. *Journal of the National Cancer Institute* **22**, 719–748.

Mantel, N., and Hankey, B. F. (1975). The odds ratios of a 2×2 contingency table. *The American Statistician* **29**, 143–145.

Marubini, E., and Valsecchi, M. G. (1995). *Analyzing Survival Data from Clinical Trials and Observational Studies*. Chichester, England: John Wiley & Sons.

McKinlay, S. M. (1974). The expected number of matches and its variance for matched-pairs designs. *Applied Statistics* **23**, 372–383.

McNemar, Q. (1947). Note on the sampling error of the difference between correlated proportions or percentages. *Psychometrika* **12**, 153–157.

Mehta, C. R., and Patel, N. R. (1995). Exact logistic regression: Theory and examples. *Statistics in Medicine* **14**, 2143–2160.

Mickey, R. M., and Greenland, S. (1989). The impact of confounder selection criteria on effect estimation. *American Journal of Epidemiology* **129**, 125–137.

Miettinen, O. S. (1969). Individual matching with multiple controls in the case of all-or-none responses. *Biometrics* **25**, 339–355.

Miettinen, O. S. (1970). Estimation of relative risk from individually matched series. *Biometrics* **26**, 75–86.

Miettinen, O. S. (1972a). Standardization of risk ratios. *American Journal of Epidemiology* **96**, 383–388.

Miettinen, O. S. (1972b). Components of the crude risk ratio. *American Journal of Epidemiology* **96**, 168–172.

Miettinen, O. S. (1976). Estimability and estimation in case–referent studies. *American Journal of Epidemiology* **103**, 226–235.

Miettinen, O. S. (1977). The author replies. *American Journal of Epidemiology* **105**, 498–502.

Miettinen, O. S., and Cook, E. F. (1981). Confounding: Essence and detection. *American Journal of Epidemiology* **114**, 593–603.

Moolgavkar, S. H., and Venzon, D. J. (1987). General relative risk regression models for epidemiologic data. *American Journal of Epidemiology* **126**, 949–961.

Muñoz, A., and Rosner, B. (1984). Power and sample size for a collection of 2×2 tables. *Biometrics* **40**, 995–1004.

Neuhaus, J. M., Kalbfleisch, J. D., and Hauck, W. W. (1991). A comparison of cluster–specific and population–averaged approaches for analyzing correlated binary data. *International Statistical Review* **59**, 25–35.

Newman, S. (1986). A generalization of life expectancy which incorporates the age distribution of the population and its use in the measurement of the impact of mortality reduction. *Demography* **23**, 261–274.

Newman, S. (1988). A Markov process interpretation of Sullivan's index of morbidity and mortality. *Statistics in Medicine* **7**, 787–794.

Newman, S. C., and Bland, R. C. (1991). Mortality in a cohort of patients with schizophrenia: A record linkage study. *Canadian Journal of Psychiatry* **36**, 239–245.

Nurminen, M. (1981). Asymptotic efficiency of general noniterative estimators of common relative risk. *Biometrika* **68**, 525–530.

Oleinick, A., and Mantel, N. (1970). Family studies in systemic lupus erythematosus—II. *Journal of Chronic Diseases* **22**, 617–625.

Parmar, M. K. B., and Machin, D. (1995). *Survival Analysis: A Practical Approach*. New York: John Wiley & Sons.

Paul, S. R., and Donner, A. (1989). A comparison of tests of homogeneity of odds ratios in K 2×2 tables. *Statistics in Medicine* **8**, 1455–1468.

Paul, S. R., and Donner, A. (1992). Small sample performance of tests of homogeneity of odds ratios in K 2×2 tables. *Statistics in Medicine* **11**, 159–165.

Pearl, J. (1993). Comment: Graphical models, causality and intervention. *Statistical Science* **8**, 266–269.

Pearl, J. (1995). Causal diagrams for empirical research (with discussion). *Biometrika* **82**, 669–710.

Pearl, J. (2000). *Causality: Models, Reasoning, and Inference*. Cambridge: Cambridge University Press.

Peto, R. (1972). Rank tests of maximal power against Lehmann-type alternatives. *Biometrika* **59**, 472–475.

Peto, R. (1982). Statistical aspects of cancer trials. In *Treatment of Cancer*, K. E. Halnan (ed.), pp. 868–871. London: Chapman and Hall.

Peto, R., and Peto, J. (1972). Asymptotically efficient rank invariant test procedures (with discussion). *Journal of the Royal Statistical Society, Series A* **135**, 185–206.

Peto, R., and Pike, M. (1973). Conservatism of the approximation $\sum (O - E)^2/E$ in the logrank test for survival data or tumour incidence data. *Biometrics* **29**, 579–584.

Phillips, A., and Holland, P. W. (1987). Estimators of the variance of the Mantel–Haenszel log–odds-ratio estimate. *Biometrics* **43**, 425–431.

Poole, C. (1987). Beyond the confidence interval. *American Journal of Public Health* **77**, 195–199.

Prentice, R. L., and Breslow, N. E. (1978). Retrospective studies and failure time models. *Biometrika* **65**, 153–158.

Prentice, R., and Kalbfleisch, J. (1988). Author's reply. *Biometrics* **44**, 1205.

Prentice, R. L., Kalbfleisch, J. D., Peterson, A. V., Flournoy, N., Farewell, V. T., and Breslow, N. E. (1978). The analysis of failure times in the presence of competing risks. *Biometrics* **34**, 541–554.

Prentice, R. L., and Pyke, R. (1979). Logistic disease incidence models and case-control studies. *Biometrika* **66**, 403–411.

Preston, D. (2000). Excess relative risk. In *Encyclopedia of Epidemiologic Methods*, M. H. Gail and J. Benichou (eds.), p. 393. Chichester, England: John Wiley & Sons.

Rao, C. R. (1973). *Linear Statistical Inference and Its Applications*, 2nd edition. New York: John Wiley & Sons.

Robins, J. (1989). The control of confounding by intermediate variables. *Statistics in Medicine* **8**, 679–701.

Robins, J. M. (1998). Structural nested failure time models. In *Encyclopedia of Biostatistics*, P. Armitage and T. Colton (eds.), pp. 4372–4389. Chichester, England: John Wiley & Sons.

Robins, J. M., Blevins, D., Ritter, G., and Wulfsohn, M. (1992). G-estimation of the effect of prophylaxis therapy for Pneumocystis carinii pneumonia on the survival of AIDS patients. *Epidemiology* **3**, 319–336.

Robins, J., Breslow, N., and Greenland S. (1986). Estimators of the Mantel–Haenszel variance consistent in both sparse data and large-strata limiting models. *Biometrics* **42**, 311–323.

Robins, J., and Greenland S. (1989a). The probability of causation under a stochastic model for individual risk. *Biometrics* **45**, 1125–1138.

Robins, J. M., and Greenland S. (1989b). Estimability and estimation of excess and etiologic fractions. *Statistics in Medicine* **8**, 845–859.

Robins, J. M., and Greenland S. (1991). Estimability and estimation of expected years of life lost due to a hazardous exposure. *Statistics in Medicine* **10**, 79–93.

Robins, J., and Greenland S. (1992). Identifiability and exchangeability for direct and indirect effects. *Epidemiology* **3**, 143–155.

Robins, J., Greenland S., and Breslow, N. E. (1986). A general estimator for the variance of the Mantel–Haenszel odds ratio. *American Journal of Epidemiology* **124**, 719–723.

Robins, J. M., and Morgenstern, H. (1987). The foundations of confounding in epidemiology. *Computers and Mathematics with Applications* **14**, 869–916.

Rosenbaum, P. R. (1984a). From association to causation in observational studies: The role of tests of strongly ignorable treatment assignment. *Journal of the American Statistical Association* **79**, 41–48.

Rosenbaum, P. R. (1984b). The consequences of adjustment for a concomitant variable that has been affected by the treatment. *Journal of the Royal Statistical Society, Series A* **147**, 656–666.

Rosenbaum, P. R. (1995). *Observational Studies*. New York: Springer-Verlag.

Rosenbaum, P. R., and Rubin, D. B. (1983). The central role of the propensity score in observational studies for causal effects. *Biometrika* **70**, 41–55.

Rosenbaum, P. R., and Rubin, D. B. (1984). Reducing bias in observational studies using subclassification on the propensity score. *Journal of the American Statistical Association* **79**, 516–524.

Rosenbaum, P. R., and Rubin, D. B. (1985). Constructing a control group using multivariate matched sampling methods that incorporate the propensity score. *The American Statistician* **39**, 33–38.

Rosner, B. (1995). *Fundamental of Biostatistics*, 4th edition. Belmont, CA: Duxbury.

Rothman, K. J. (1974). Synergy and antagonism in cause–effect relationships. *American Journal of Epidemiology* **99**, 385–388.

Rothman, K. J. (1978). A show of confidence. *New England Journal of Medicine* **299**, 1362–1363.

Rothman, K. J. (1986). *Modern Epidemiology*. Little, Brown: Boston.

Rothman, K. J., and Boice, J. D. (1979). *Epidemiologic Analysis with a Programmable Calculator*. NIH Publication 79-1649. Washington, DC: US Government Printing Office.

Rothman, K. J., and Greenland, S. (1998). *Modern Epidemiology*, 2nd edition. Philadelphia: Lippincott–Raven.

Royall, R. M. (1997). *Statistical Evidence: A Likelihood Paradigm*. Boca Raton, FL: Chapman & Hall/CRC.

Rubin, D. B. (1974). Estimating causal effects of treatment in randomized and nonrandomized studies. *Journal of Educational Psychology* **66**, 688–701.

Sackett, D. L., Haynes, R. B., and Tugwell, P. (1985). *Clinical Epidemiology: A Basic Science for Clinical Medicine*. Boston: Little, Brown.

Sahai, H., and Khurshid A. (1996). *Statistics in Epidemiology: Methods, Techniques, and Applications*. Boca Raton, FL: CRC Press.

SAS. (1987). *SAS/STAT™ Guide for Personal Computers, Version 6 Edition*. Cary, NC: SAS Institute Inc.

Sato, T. (1989). On the variance estimator for the Mantel–Haenszel risk difference (letter). *Biometrics* **45**, 1323–1324.

Sato, T. (1990). Confidence limits for the common odds ratio based on the asymptotic distribution of the Mantel–Haenszel estimator. *Biometrics* **46**, 71–80.

Schervish, M. J. (1996). *P* values: What they are and what they are not. *The American Statistician* **50**, 203–206.

Schlesselman, J. J. (1974). Sample size requirements in cohort and case-control studies of disease. *American Journal of Epidemiology* **99**, 381–384.

Schlesselman, J. J. (1978). Assessing the effects of confounding variables. *American Journal of Epidemiology* **108**, 3–129.

Schlesselman, J. J (1982). *Case-Control Studies: Design, Conduct, Analysis*. New York: Oxford University Press.

Schoenfeld, D. A. (1983). Sample-size formula for the proportional-hazards regression model. *Biometrics* **39**, 499–503.

Seber, G. U. H. (2000). Poisson regression. In *Encyclopedia of Epidemiologic Methods*, M. H. Gail and J. Benichou (eds.), pp. 715–723. Chichester, England: John Wiley & Sons.

Self, S. G., and Mauritsen, R. H. (1988). Power/sample calculations for generalized linear models. *Biometrics* **44**, 79–86.

Self, S. G., Mauritsen, R. H., and Ohara, J. (1992). Power calculations for likelihood ratio tests in generalized linear models. *Biometrics* **48**, 31–39.

Shah, B. V., Barnwell, B. G., and Bieler, G. S. (1996). *SUDAAN User's Manual, Release 7.0*. Research Triangle Park, NC: Research Triangle Institute.

Shapiro, S., Slone, D., Rosenberg, L., Kaufman, D. W., Stolley, P. D., and Miettinen, O. S. (1979). Oral-contraceptive use in relation to myocardial infarction. *The Lancet* **April 7**, 743–746.

Shore, R. E., Pasternack, B. S., and McCrea Curnen, M.G. (1976). Relating influenza epidemics to childhood leukemia in tumor registries without a defined population base: A critique with suggestions for improved methods. *American Journal of Epidemiology* **103**, 527–534.

Siegel, D. G., and Greenhouse, S. W. (1973). Validity in estimating relative risk in case-control studies. *Journal of Chronic Diseases* **26**, 219–225.

Siemiatycki, J., and Thomas, D. C. (1981). Biological models and statistical interactions: An example from multistage carcinogenesis. *International Journal of Epidemiology* **10**, 383–387.

Silvey, S. D. (1975). *Statistical Inference*. London: Chapman and Hall.

Simon, R. (1980). RE: Assessing effects of confounding variables (with response). *American Journal of Epidemiology* **111**, 127–129.

Simpson, E. H. (1951). The interpretation of interaction in contingency tables. *Journal of the Royal Statistical Society, Series B* **13**, 238–241.

Sinclair, J. C., and Bracken, M. B. (1994). Clinically useful measures of effect in binary analyses of randomized trials. *Journal of Clinical Epidemiology* **47**, 881–889.

SPSS. (1993). *SPSS® for Windows™ : Advanced Statistics, Release 6.0*. Chicago: SPSS Inc.

STATA. (1999). *Stata Statistical Software: Release 6.0*. College Station, TX: Stata Corporation.

StatXact. (1998). *A Software Package for Exact Nonparametric Inference, Version 4*. Cambridge, MA: Cytel Software Corporation.

Sullivan, D. F. (1971). A single index of mortality and morbidity. *HSMHA Health Reports* **86**, 347–354.

Tarone, R. E. (1981). On summary estimators of relative risk. *Journal of Chronic Diseases* **34**, 463–468.

Tarone, R. E. (1985). On heterogeneity tests based on efficient scores. *Biometrika* **72**, 91–95.

Tarone, R. E., and Gart, J. J. (1980). On the robustness of combined tests for trends in proportions. *Journal of the American Statistical Association* **75**, 110–116.

Tarone, R. E., Gart, J. J., and Hauck, W. W. (1983). On the asymptotic inefficiency of certain noniterative estimators of a common relative risk or odds ratio. *Biometrika* **70**, 519–522.

Thomas, D. C. (1981). General relative risk models for survival time and matched case-control studies. *Biometrics* **37**, 673–686.

Thomas, D. C. (2000). Relative risk modelling. In *Encyclopedia of Epidemiologic Methods*, M. H. Gail and J. Benichou (eds.), pp. 759–767. Chichester, England: John Wiley & Sons.

Thomas, D. C., and Greenland, S. (1983). The relative efficiencies of matched and independent sample designs for case-control studies. *Journal of Chronic Diseases* **36**, 685–697.

Thomas, D. C., and Greenland, S. (1985). The efficiency of matching in case-control studies of risk-factor interactions. *Journal of Chronic Diseases* **38**, 569–574.

Thompson, W. D. (1991). Effect modification and the limits of biological inference from epidemiologic data. *Journal of Clinical Epidemiology* **44**, 221–232.

Thompson, W. D., Kelsey, J. L., and Walter, S. D. (1982). Cost and efficiency in the choice of matched and unmatched case-control study designs. *American Journal of Epidemiology* **116**, 840–851.

Tsiatis, A. A. (1998). Competing risks. In *Encyclopedia of Biostatistics*, P. Armitage and T. Colton (eds.), pp. 824–834. Chichester, England: John Wiley & Sons.

University Group Diabetes Program. (1970). A study of the effects of hypoglycemic agents on vascular complications in patients with adult onset diabetes. *Diabetes* **19 (suppl. 2)**, 747–830.

Ury, H. K. (1975). Efficiency of case-control studies with multiple controls per case: Continuous or dichotomous data. *Biometrics* **31**, 643–649.

Væth, M. (2000). Expected number of deaths. In *Encyclopedia of Epidemiologic Methods*, M. H. Gail and J. Benichou (eds.), pp. 394–396. Chichester, England: John Wiley & Sons.

Vaupel, J. W., and Yashin, A. (1985). Heterogeneity's ruses: Some surprising effects of selection on population dynamics. *The American Statistician* **39**, 176–185.

Walker, A. M. (1985). Small sample properties of some estimators of a common hazard ratio. *Applied Statistics* **34**, 42–48.

Walker, A. M., and Rothman, K. J. (1982). Models of varying parametric form in case-referent studies. *American Journal of Epidemiology* **115**, 129–137.

Walter, S. D. (1980a). Matched case-control studies with a variable number of controls per case. *Applied Statistics* **29**, 172–179.

Walter, S. D. (1980b). Large sample formulae for the expected number of matches in a category matched design. *Biometrics* **36**, 285–291.

Walter, S. D. (1985). Small–sample estimation of log odds ratios from logistic regression and fourfold tables. *Statistics in Medicine* **4**, 437–444.

Walter, S. D. (1987). Point estimation of the odds ratio in sparse 2×2 contingency tables. In *Biostatistics. Advances in the Statistical Sciences, Volume V*, I. B. MacNeil and G. J. Umphrey (eds.), pp. 71–102. Boston: Reidel.

Walter, S. D. (2000). Choice of effect measure for epidemiologic data. *Journal of Clinical Epidemiology* **53**, 931–939.

Walter, S. D., and Cook, R. J. (1991). A comparison of several point estimators of the odds ratio in a single 2×2 contingency table. *Biometrics* **47**, 795–811.

Walter, S. D., and Holford, T. R. (1978). Additive, multiplicative and other models for disease risks. *American Journal of Epidemiology* **108**, 341–356.

Weinberg, C. R. (1993). Toward a clearer definition of confounding. *American Journal of Epidemiology* **137**, 1–8.

Whittemore, A. S. (1978). Collapsibility of multidimensional contingency tables. *Journal of the Royal Statistical Society, Series B* **40**, 328–340.

Wickramaratne, P., and Holford, T. R. (1987). Confounding in epidemiologic studies: The adequacy of the control group as a measure of confounding. *Biometrics* **43**, 751–765.

Wilson, S. R., and Gordon, I. (1986). Calculating sample sizes in the presence of confounding variables. *Applied Statistics* **35**, 207–213.

Woolf, B. (1955). On estimating the relationship between blood group and disease. *Annals of Human Genetics* **19**, 251–253.

Woolson, R. F., Bean, J. A., and Rojas, P. B. (1986). Sample size for case-control studies using Cochran's statistic. *Biometrics* **42**, 927–932.

Yanagawa, T. (1979). Designing case-control studies. *Environmental Health Perspectives* **32**, 143–156.

Yanagawa, T. (1984). Case-control studies: Assessing the effect of a confounding factor. *Biometrika* **71**, 191–194.

Yates, F. (1984). Tests of significance for 2×2 contingency tables (with discussion). *Journal of the Royal Statistical Society, Series A* **147**, 426–463.

Zdeb, M. S. (1977). The probability of developing disease. *American Journal of Epidemiology* **106**, 6–16.

Index

WILEY SERIES IN PROBABILITY AND STATISTICS
ESTABLISHED BY WALTER A. SHEWHART AND SAMUEL S. WILKS

Editors
*Peter Bloomfield, Noel A. C. Cressie, Nicholas I. Fisher, Iain M. Johnstone,
J. B. Kadane, Louise M. Ryan, David W. Scott, Bernard W. Silverman,
Adrian F. M. Smith, Jozef L. Teugels; Vic Barnett, Emeritus,
Ralph A. Bradley, Emeritus, J. Stuart Hunter, Emeritus,
David G. Kendall, Emeritus*

Probability and Statistics Section

*ANDERSON · The Statistical Analysis of Time Series
ARNOLD, BALAKRISHNAN, and NAGARAJA · A First Course in Order Statistics
ARNOLD, BALAKRISHNAN, and NAGARAJA · Records
BACCELLI, COHEN, OLSDER, and QUADRAT · Synchronization and Linearity:
   An Algebra for Discrete Event Systems
BARNETT · Comparative Statistical Inference, *Third Edition*
BASILEVSKY · Statistical Factor Analysis and Related Methods: Theory and
   Applications
BERNARDO and SMITH · Bayesian Statistical Concepts and Theory
BILLINGSLEY · Convergence of Probability Measures, *Second Edition*
BOROVKOV · Asymptotic Methods in Queuing Theory
BOROVKOV · Ergodicity and Stability of Stochastic Processes
BRANDT, FRANKEN, and LISEK · Stationary Stochastic Models
CAINES · Linear Stochastic Systems
CAIROLI and DALANG · Sequential Stochastic Optimization
CONSTANTINE · Combinatorial Theory and Statistical Design
COOK · Regression Graphics
COVER and THOMAS · Elements of Information Theory
CSÖRGŐ and HORVÁTH · Weighted Approximations in Probability Statistics
CSÖRGŐ and HORVÁTH · Limit Theorems in Change Point Analysis
*DANIEL · Fitting Equations to Data: Computer Analysis of Multifactor Data,
   Second Edition
DETTE and STUDDEN · The Theory of Canonical Moments with Applications in
   Statistics, Probability, and Analysis
DEY and MUKERJEE · Fractional Factorial Plans
*DOOB · Stochastic Processes
DRYDEN and MARDIA · Statistical Shape Analysis
DUPUIS and ELLIS · A Weak Convergence Approach to the Theory of Large Deviations
ETHIER and KURTZ · Markov Processes: Characterization and Convergence
FELLER · An Introduction to Probability Theory and Its Applications, Volume I,
   Third Edition, Revised; Volume II, *Second Edition*
FULLER · Introduction to Statistical Time Series, *Second Edition*
FULLER · Measurement Error Models
GHOSH, MUKHOPADHYAY, and SEN · Sequential Estimation
GIFI · Nonlinear Multivariate Analysis
GUTTORP · Statistical Inference for Branching Processes
HALL · Introduction to the Theory of Coverage Processes
HAMPEL · Robust Statistics: The Approach Based on Influence Functions
HANNAN and DEISTLER · The Statistical Theory of Linear Systems
HUBER · Robust Statistics

*Now available in a lower priced paperback edition in the Wiley Classics Library.

*Now available in a lower priced paperback edition in the Wiley Classics Library.

*Now available in a lower priced paperback edition in the Wiley Classics Library.

*Now available in a lower priced paperback edition in the Wiley Classics Library.

*Now available in a lower priced paperback edition in the Wiley Classics Library.

Biostatistics Section

ARMITAGE and DAVID (editors) · Advances in Biometry
BROWN and HOLLANDER · Statistics: A Biomedical Introduction
CHOW and LIU · Design and Analysis of Clinical Trials: Concepts and Methodologies
DUNN and CLARK · Applied Statistics: Analysis of Variance and Regression, *Second Edition*
*ELANDT-JOHNSON and JOHNSON · Survival Models and Data Analysis
*FLEISS · The Design and Analysis of Clinical Experiments
FLEISS · Statistical Methods for Rates and Proportions, *Second Edition*
FLEMING and HARRINGTON · Counting Processes and Survival Analysis
KADANE · Bayesian Methods and Ethics in a Clinical Trial Design
KALBFLEISCH and PRENTICE · The Statistical Analysis of Failure Time Data
LACHIN · Biostatistical Methods: The Assessment of Relative Risks
LANGE, RYAN, BILLARD, BRILLINGER, CONQUEST, and GREENHOUSE · Case Studies in Biometry
LAWLESS · Statistical Models and Methods for Lifetime Data
LEE · Statistical Methods for Survival Data Analysis, *Second Edition*
MALLER and ZHOU · Survival Analysis with Long Term Survivors
McNEIL · Epidemiological Research Methods
McFADDEN · Management of Data in Clinical Trials
*MILLER · Survival Analysis, *Second Edition*
NEWMAN · Biostatistical Methods in Epidemiology
PIANTADOSI · Clinical Trials: A Methodologic Perspective
WOODING · Planning Pharmaceutical Clinical Trials: Basic Statistical Principles
WOOLSON · Statistical Methods for the Analysis of Biomedical Data

Financial Engineering Section

HUNT and KENNEDY · Financial Derivatives in Theory and Practice
ROLSKI, SCHMIDLI, SCHMIDT, and TEUGELS · Stochastic Processes for Insurance and Finance
SHAFER and VOVK · Probability and Finance: It's Only a Game!

Texts, References, and Pocketbooks Section

AGRESTI · An Introduction to Categorical Data Analysis
ANDĚL · Mathematics of Chance
ANDERSON · An Introduction to Multivariate Statistical Analysis, *Second Edition*
ANDERSON and LOYNES · The Teaching of Practical Statistics
ARMITAGE and COLTON · Encyclopedia of Biostatistics: Volumes 1 to 6 with Index
BARTOSZYNSKI and NIEWIADOMSKA-BUGAJ · Probability and Statistical Inference
BENDAT and PIERSOL · Random Data: Analysis and Measurement Procedures, *Third Edition*
BERRY, CHALONER, and GEWEKE · Bayesian Analysis in Statistics and Econometrics: Essays in Honor of Arnold Zellner
BHATTACHARYA and JOHNSON · Statistical Concepts and Methods
BILLINGSLEY · Probability and Measure, *Second Edition*
BOX · R. A. Fisher, the Life of a Scientist

*Now available in a lower priced paperback edition in the Wiley Classics Library.

*Now available in a lower priced paperback edition in the Wiley Classics Library.

WILEY SERIES IN PROBABILITY AND STATISTICS
ESTABLISHED BY WALTER A. SHEWHART AND SAMUEL S. WILKS

Editors
*Robert M. Groves, Graham Kalton, J. N. K. Rao, Norbert Schwarz,
Christopher Skinner*

Survey Methodology Section

*Now available in a lower priced paperback edition in the Wiley Classics Library.